The **Contract:**
law and administration

Issaka E Ndekugri BSc(Hons)(Civil Eng.), LLB(Hons) MSc, PhD, MCIOB

Principal Lecturer, University of Wolverhampton, UK

and

Michael E Rycroft LLB(Hons), MCIOB, MIMgt, ACIArb

Regional Director, James R Knowles, Contract Consultants, UK

ARNOLD

A member of the Hodder Headline Group

LONDON

First published in Great Britain in 2000 by
Arnold, a member of the Hodder Headline Group,
338 Euston Road, London NW1 3BH

http://www.arnoldpublishers.com

British Library Cataloguing in Publication Data
A catalogue record for this book is available from the British Library

ISBN: 0 340 72008 5

1 2 3 4 5 6 7 8 9 10

Publisher: Eliane Wigzell
Production Editor: Wendy Rooke
Production Controller: Iain McWilliams

Typeset in 10/12pt Times by Phoenix Photosetting, Chatham, Kent
Printed and bound in Great Britain by
MPG Books Ltd, Bodmin, Cornwall

What do you think about this book? Or any other Arnold title?
Please send your comments to feedback.arnold@hodder.co.uk

Contents

Preface

Periodic surveys of standard building contract forms suggest that the Joint Contracts Tribunal stable of forms is in widespread use. In particular its flagship, the Standard Form of Building Contract, in its various versions is popular where works are of more than minimal complexity; and to the extent that the industry can be said to adopt a norm, the administrative principles of the JCT Standard Form appear to be the nucleus of that norm. However, use of the Standard Form has been dogged by a high incidence of disputes. It is an open secret that many of these disputes have been attributed either to lack of understanding of the terms, or to confusion arising from cavalier preparation and alteration of documents. Our personal observations indicate the problem affects all participants in the contracting process including Employers' professional teams and Contractors' site or office management. In recent years there has been a huge increase in the law and procedures content of academic built environment courses; this is matched by a similar increase in continuing professional development courses dealing with contractual issues. Our book results from our considerable involvement in both types of course, together with the realization that difficulties faced by practising construction professionals and individual students are commonplace.

This work is aimed at all construction professionals, whether they be in design or construction organizations, who are involved in the planning, execution or administration of building contracts; it is also aimed at undergraduate and post graduate students studying towards careers in the industry. Of course we must not forget the lawyers; we hope the form and content of this book will assist in particular, those encountering their first construction cases, who may be unfamiliar with the jargon or traditions of the construction world.

We have not dealt with all versions of JCT Standard Form, but have concentrated on the Private Edition With Quantities; comparison of the major differences from other versions are noted in the text. To comply with tradition we have used JCT 98 as the short title. The book was commenced when the 1980 Edition with Amendments 1 to 17 was current, and was necessitated by a dearth at that time of up to date texts. The publication of Amendment 18 in April 1998, followed in November 1998 by the 1998 Edition, seriously delayed drafting; no event could have highlighted more dramatically for us the ongoing problem faced by students and practitioners alike in trying to keep up with change. The inevitable Amendment 1 was issued by the JCT in June 1999. The amendment deals with tax matters which we had previously decided would be outside the scope of the book, and consequently it receives only passing mention in the commentary on preparing documents, and payment obligations. Just before publication, JCT 98 became available on CD with the facility for completing the form including incorporating the standard amendments. This must ease the onerous task described in Chapter 1.

One change in the construction industry which is not reflected in the book, but which deserves special mention, is the welcome increase in the numbers of female practitioners throughout all disciplines. We gave considerable thought to getting around the inadequacy of the English language which thwarts any attempt to neutralise contractual duty holders. Rather than reduce everyone to 'it', or resort to the clumsy 's/he' or 'he/she', we reluctantly decided on the use of 'he' throughout; this was principally because recurring reference needs to be made to the contract terms, and 'he' is the term used exclusively by the JCT.

Change is not confined to the construction industry, and in April 1999 new Court Civil Procedure Rules came into force as a result of Lord Woolf's *Access to Justice* report. The main effect of the Rules on this book is on the terminology. Where reference is made in a general sense to the rights or obligations of a party bringing a case to the Court we have used the term 'claimant' in accordance with the Rules. Where we have used the term 'plaintiff' it refers to the plaintiff in a case preceding the new Rules.

On the matter of style, at the risk of confusing the reader we have adopted the use of capital letters when referring to specific JCT terminology (e.g. the Architect, the Employer, Variation), and lower case where the reference is general (architects, employers, variations).

We owe much to Mr John Wood, a director of James R Knowles, who gave this time freely to act as a sounding board for our ideas and to bench test the user-friendliness of our style. We are also indebted to the editorial and production team who waited so patiently while we struggled to catch up with the Joint Contracts Tribunal.

Finally, we have tried to state the law at September 1999; any errors or omisions are ours.

Issaka Ndekugri
Michael Rycroft   January 2000

Glossary of Latin Terms

ad hoc	'for this special purpose'
caveat emptor	'let the buyer beware'
consensus ad idem	all in agreement as to the same thing
contra proferentem	the doctrine that the least favourable construction of words will be adopted against the person putting them forward
de facto	actual, as opposed to legally recognised
de jure	by right
ex gratia	as a favour
force majeure	acts of God or manmade events beyond people's control
functus officio	someone whose powers have ended
inter alia	among other things
in terrorem	to frighten
nemo dat quod non habet	'you cannot give away what you do not own'
obiter	by the way
obiter dicta	something said by the way; a statement not binding as precedent made in judgment
pari passu	with equal pace; together
pro forma	in legal and business sense a standard format
pro rata	in proportion
quantum meruit	'as much as he has earned' (used in legal sense as a synonym for both 'quantum meruit' and 'quantum valebat')
quantum valebat	'as much as they were worth'
ultra vires	beyond someone's powers

Abstract Abbreviations

Abbreviations

Law Reports

AC	Appeal Cases from 1891 onwards
All ER	All England Law Reports
ALJR	Australian Law Journal Reports
App. Cas.	Appeal Cases up to 1890
BCL	*Building and Construction Law*
BLISS	Building Law Information Subscriber Service
BLR	Building Law Reports
CA	Court of Appeal
CB (NS)	Common Bench New Series
CILL	*Construction Industry Law Letter*
Con.LR	Construction Law Reports
Const.LJ	*Construction Law Journal*
DLR	Dominion Law Reports
ICLR	*International Construction Law Review*
JP	Justice of the Peace
KB	King's Bench
LT	Law Times Reports 1859-1947
LQR	*Law Quarterly Review*
Lloyd's Rep.	Lloyd's Reports, from 1951
NZLR	New Zealand Law Reports
ORB	Official Referee's Business
QBD	Queen's Bench Division
QBENF	Queen's Bench, Court Office internal reference
SJ	*Solicitors Journal*
SARL	South African Law Reports
SLT	*Scots Law Times*
TLR	Times Law Reports
UWA Law Rev.	*University of Western Australia Law Review*
WLR	Weekly Law Reports

General

ACOP	Approved Code of Practice
AI	Architect's Instructions
CDM	Construction (Design and Management) Regulations 1994
CDP	Contractor's Design Portion
CIArb	Chartered Institute of Arbitrators
CIMAR	Construction Industry Model Arbitration Rules
CLCA	Civil Liability (Contribution) Act 1978
COW	Clerk of Works
CPM	Critical path methods
CPR	Civil Procedure Rules
EDI	electronic data interchange
FIDIC	Fédération Internationale des Ingénieurs Conseils (association of national associations of consulting engineers)
HMSO	Her Majesty's Stationery Office
HSE	Health and Safety Executive
ICE	Institution of Civil Engineers
IRS	Information Release Schedule
JCT	The Joint Contracts Tribunal Ltd
LADS	liquidated and ascertained damages
LOSC	labour only sub-contractors
LPA	Law of Property Act 1925
LPCDIA	Late Payments of Commercial Debts (Interest) Act 1998
NEDO	National Economic Development Office
NJCBI	National Joint Council for the Building Industry
PIC	person-in-charge
PSW	Performance Specified Work
RIBA	Royal Institute of British Architects
RICS	Royal Institution of Chartered Surveyors
ROT	Retention of Title
SCC	Scheme for Construction Contracts
SGA	Sale of Goods Act
SMM7	The Standard Method of Measurement of Building Works, 7th Edition
UCTA	Unfair Contract Terms Act 1977

Publications

Hudson's	*Hudson's Building and Engineering Contracts,* 11th edn, ed. I.N. Duncan Wallace, QC (Sweet & Maxwell, 1995)
Keating	*Keating on Building Contracts*, 6th edn, ed. the Hon. Sir Anthony May (Sweet & Maxwell, 1995)

Contracts

DOM/1 and DOM/2	Standard Form for Domestic Sub-contract (the Construction Confederation, 1998)
GC/Works/1	General Conditions of Government Contracts for Building and Civil Engineering Works (HMSO, 1973)
ICE 5th	Institution of Civil Engineers, Association of Consulting Engineers and Association of Civil Engineering Contractors *ICE Conditions of Contract, 5th Edition* (Thomas Telford Services Ltd, 1973)
ICE 6th	Institution of Civil Engineers, Association of Consulting Engineers and Association of Civil Engineering Contractors *ICE Conditions of Contract, 6th Edition* (Thomas Telford Services Ltd, 1991)
JCT *Formula Rules*	*Consolidated Main Contract Formula Rules* (RIBA Publications Ltd)
SFA	RIBA *Standard Form of Agreement for the Appointment of an Architect*

JCT Contracts

IFC 98	*Intermediate Form of Building Contract for works of simple content, 1998 Edition* (RIBA Publications)
JCT 39	RIBA *Standard Form of Building Contract (1939 Edition) 1957 Revision* (the RIBA Form)
JCT 63	*Standard Form of Building Contract 1963 Edition*
JCT 80	*Standard Form of Building Contract 1980 Edition*
JCT 98	*Standard Form of Building Contract 1998 Edition*
MW 98	*Agreement for Minor Building Works 1998 Edition*
NSC/A	*Nominated Sub-contract Agreement*
NSC/C	*Nominated Sub-contract Conditions*
NSC/T	*Nominated Sub-contract Tender*
NSC/W	*Nominated Sub-contract Works*
SCS/98	*Sectional Completion Supplement and Practice Note 1A 1998 Edition*
TC 94	*Amendment TC/94: 1994* (terrorism cover)
WCD 98	*Standard Form of Building Contract With Contractor's Design 1998 Edition*

Style of JCT 98, contract execution and related problems

1.1 Standard form contracts generally

Any transaction between two business concerns will inevitably place a demand on the fixed overhead resources of the parties. Whether the contract is a simple exchange of letters or a lengthy document signed by each party, there is some administrative input; the extent of the input is determined largely by the effort required to negotiate terms and to commit them to writing. The nature of some companies' work may be such that transactions are numerous; other companies' transactions may be few but complicated. In either case it is an advantage to any organization to develop personal terms of trading. A contract which sets out such terms is widely known as a standard form contract. The advantages to the parties issuing such terms are many, e.g. familiarity from use, understanding, saving on drafting time, saving on negotiating time, or saving on senior staff time when transactions can be concluded by clerical staff are all good reasons for the issuing parties to enter into their own standard form contract. This type of standard form is normally issued unilaterally.

However, for the party intending to use the standard form of another (unilateral standard form) there are clearly some disadvantages. Savings on time, both in drafting and negotiating, will certainly remain an advantage, and are perhaps the principal reasons for the use of standard forms, but clearly there is a danger that the terms will also contain onerous provisions for some users. In any event, the main benefit of standard forms is still lost to one party since unless the company enters into a particular contract form regularly it will not develop familiarity or understanding. As a result, unilateral standard forms are often looked upon with suspicion by those with previous bad experiences. The way out of such situations for some is to insist on their own standard terms. There are then two sets of unilateral terms and the scene is set either for a negotiation or a battle[1] with the party in the stronger position dictating the basis of the contract.

---

1   See discussion on 'the battle of forms', Section 1.7.

There is another option; that is to reap the benefits of standard form contracts whilst avoiding the disadvantages. Many industries have developed a different type of standard form contract, a type which is more worthy of the general title 'standard', and which could be called 'industry standard forms'. The terms are settled through negotiation by representative bodies of commercial interests in a particular industry, and are based on mutual experience of the practices of the industry. Such forms, as a class, have been described by the courts as being widely adopted because experience has shown that they facilitate the conduct of trade, and that there is a strong presumption that their terms are fair and reasonable because they are used widely by parties whose bargaining power is evenly matched.[2]

The main virtues of an 'industry standard form contract' can be summarized as:

- being a device for allocating contingent risk whilst saving time and assisting bargaining at arms length;
- being a device for avoiding writing terms for each transaction;
- having the benefit of providing understanding by familiarity and experience in practice;
- being less likely to protect the interests of only one party, having been negotiated by independent bodies representing all interests in an industry;
- producing savings in transaction costs, avoiding the need to negotiate each contract;
- removal of unwanted discretion from individuals, enabling a structured approach to negotiations;
- enabling allocation of risk to be anticipated and provided for in calculations;
- enabling necessary quotations from others such as sub-contractors and suppliers to be obtained with greater accuracy.

Disadvantages of industry standard form contracts as a general class appear to be few, since they go some way towards removing the principal objection to unilateral forms (that is a stronger party imposing its will on a weaker party). However, within an industry the individual industry standard forms in common use may draw criticism, and the building industry is no exception.[3]

1.2 The Joint Contracts Tribunal Limited

One of the most prolific producers of contract forms for the building industry is the Joint Contracts Tribunal Limited (formerly the Joint Contracts Tribunal for the Standard Form of Building Contract), commonly called the JCT.

The constituent members are representative bodies of various commercial interests in the building industry, who settle the terms of the JCT stable of contract forms; at the time of writing they are:

- Association of Consulting Engineers
- British Property Federation
- Construction Confederation
- Local Government Association

---

2   *Per* Lord Diplock, *Schroeder (A) Music Publishing Company Ltd v. Macauley* (1974), 2 All ER 616.
3   See Section 1.3.3.

- National Specialist Contractors Council
- Royal Institute of British Architects (RIBA)
- Royal Institution of Chartered Surveyors (RICS)
- Scottish Building Contract Committee.

In 1931 the JCT was formed comprising the Royal Institute of British Architects and the National Federation of Building Trades Employers (now the Construction Confederation). In 1939 the JCT published a major standard form of contract revising and replacing an earlier form[4] which had been known for many years as 'the RIBA Form'.[5] Although the JCT was self-appointed, its forms were in common usage and in 1963 a further new edition was published which was again revised in 1977. By this time the constituent bodies had grown in number to include the Royal Institution of Chartered Surveyors, Association of County Councils, Association of Metropolitan Authorities, Association of District Councils, the Greater London Council, Committee of Associations of Specialist Engineering Contractors, Federation of Specialists and Sub-contractors, the Association of Consulting Engineers, British Property Federation, and Scottish Building Contracts Committee.

In 1980 a substantial new edition, called the *Standard Form of Building Contract 1980 Edition* (JCT 80), was published in several Private and Local Authority versions – With Quantities, Without Quantities, and With Approximate Quantities.

JCT 80 incorporated changes from the earlier *1963 Edition, 1977 Revision*, notably a decimal numbering system, reference to all paragraphs and sub-paragraphs as clauses, a definitions clause, and separately published fluctuations clauses. Between 1983 and April 1998, JCT 80 was changed by a further approximately 200 items in 18 published amendments and 4 reprints. The most significant of these changes is Amendment 18 issued in April 1998. Amendment 18 was prompted by the need for construction contracts in writing to comply with the Housing Grants Construction and Regeneration Act 1996, Part II.[6] However, the JCT went a step further. The 1996 Act followed some, but not all, of the recommendations contained in a report by Sir Michael Latham called *Constructing the Team*.[7] The JCT took the opportunity to incorporate the spirit of some of the other recommendations, and to introduce provisions in their contracts that reflected modern practice; Amendment 18 was used as the vehicle. Unfortunately Amendment 18 together with its guidance note totalled 60 pages containing 22 items, 2 bond agreements, and 2 misleadingly titled adjudication agreements (perhaps 'adjudicator agreements' would be more accurate). Amendment 18 created a daunting task for those wishing to use JCT 80 as a working document; earlier criticism of JCT 80[8] was reinforced. The JCT immediately realized the size of the problem and quickly published an amalgamated edition, the subject of this book.

In December 1998 the Joint Contracts Tribunal Limited published a new edition of JCT

---

4 An early twentieth-century version of a form used in the late nineteenth century; a copy of the latter is included in *Hudson on Building Contracts*, 3rd edn (Sweet & Maxwell, 1906) vol. 2.

5 This title is still used by some today, e.g. see *Hudson's Building and Engineering Contracts*, 11th edn (Sweet & Maxwell, 1994) at p. cxxvii (thereafter *Hudson's*) and generally throughout; one explanation for common usage may be that the JCT forms are still published by RIBA Publications.

6 The 1998 Act (widely known as the New Construction Act) came into force on 1 May 1998.

7 Published 1994.

8 See criticisms at Section 1.3.3.

80 incorporating all amendments, under the title *Standard Form of Building Contract 1998 Edition*. Following precedent it seems fitting to refer to the 1998 edition as 'JCT 98'.

JCT 98 is published as both Private and Local Authorities versions, 'With Quantities', 'With Approximate Quantities', and 'Without Quantities'.

1.3 JCT 98: use, style and criticism

1.3.1 Use of JCT 98

Factors to be considered when choosing which of the JCT published forms to use are set out by the JCT in Practice Note 20, *Deciding on the appropriate form of JCT Main Contract*. The Practice Note classifies projects by design duty (i.e. design by owner's consultants or by contractor), and by type of work (e.g. major works, maintenance, and minor works). In the cases of minor works and medium size projects, a value range is proposed, although the upper limits of those ranges are not suggested to be a lower limit for the use of the full standard form. JCT 98 falls within the class of 'Employer-Designed Works' together with the Minor Works Agreement, the Intermediate Form, the Management Form and the Form of Prime Cost Contract. This class is then subdivided into lump sum contracts (such as the Standard Form With Quantities, and the Standard Form Without Quantities), and others, such as the Prime Cost Form and the Standard Form With Approximate Quantities.

Practice Note 20 identifies the main difference between the 'With', the 'Without', and the 'With Approximate' Quantities versions.[9] In essence, the 'Without Quantities' form is a pure lump sum contract in which the contractor provides a single price for building what is shown on drawings and described in specifications prepared by the building owner or his consultants. The tenderer is required to bear the cost of his own errors in coming to his price, including underestimation of the amount of work involved; conversely over-estimation of the work will be to his benefit. The 'With Quantities' version is also a lump sum contract, but in order to provide a basis for comparing tenders, the building owner commissions a quantity surveyor to prepare a bill of quantities describing the work required, which the tenderers then price.[10] Where bills are used, the tenderers may rely on the contents at face value in the knowledge that the cost arising out of any errors should be corrected at the building owner's expense. The JCT advise that bills should be used on projects over a certain value,[11] but that bills should still be used under that value if the project is complex. Conversely, bills may not be necessary on high value projects where the work is of a simple or repetitive nature. The 'With Approximate Quantities' form is a 'remeasurement' contract. The Bills are used to provide a market level of pricing and basis for tender comparison; the final price is determined by measuring the work done and applying the Bill rates.

The only guidance as to use printed on the published JCT 98 form itself is a note on the rear cover which explains that the form is only intended for use 'where the Client has engaged a professional consultant to advise on and administer its terms'. The term 'Client'

---

9 See also Section 1.3.4.
10 See Sections 1.4.3 and 1.5.4 for general comment on bills of quantities.
11 Practice Note 20 revised August 1993, para. 15, suggests £120 000 at 1992 prices.

here is used in a generic sense to mean the person who will become the Employer under the Contract; it is not a term used in the form.

In theory JCT 98 can be used for projects of any size and complexity, where the work is designed predominantly by the owner. However, in practice, the criteria of size, value and complexity are not the only factors influencing use of the form. They are all criteria related to the project; but contracts are more to do with relationships than with projects. It has to be emphasized that some construction industry standard forms, including JCT 98, are highly procedural.[12] The one factor which is most likely to affect the successful use of the form on any project, of whatever value or complexity, is the parties' ability to administer the procedural requirements. If the procedures are not understood, or if insufficient administrative resources are allocated, then the parties' relationship is vulnerable to deterioration through misunderstanding. The important criterion then, when deciding suitability, should be the administrative competence both of the contract administrator and the likely contractor. If the nature and complexity of the contract form is such as to impinge on the success of what could otherwise be a project designed and constructed economically and in a time to the satisfaction of all parties, then that contract form is not the correct form for that particular relationship, and the parties should then consider using one of the simpler JCT forms.[13]

1.3.2 Style of JCT 98

JCT 98, in all its forms, retains the style of JCT 80; it is an update, not a complete make-over. The complex and prolific cross-referencing which made JCT 80 difficult to use as a working document remains.

The Fluctuations Clauses 38 to 40, as previously, are published separately in documents called *Fluctuations: Fluctuations clauses for use with the Private versions* and *Fluctuations: Fluctuations clauses for use with the Local Authorities versions*. The strange and (to newcomers to the form) confusing practice of listing the Fluctuations clauses in detail in the Contents section is also retained, albeit with a brief, and inadequate, explanatory note.

Part 4 of the Conditions, 'Dispute Resolution', contains a choice of litigation or arbitration, in addition to the obligatory adjudication option.[14]

Sectional completion is dealt with, as previously in JCT 80, in a separate published supplement, but unlike Fluctuations is not referred to in the Contents or the Conditions. This is surprising since it is common practice to split the whole work into sections to enable beneficial use by the building owner section by section, rather than to rely solely on the concept of partial possession in the unamended form.[15]

A new feature is the introduction of annexes to the Appendix and the Conditions. Annex 1 to the Appendix sets out the terms of two bonds in respect of Advance Payment and Payment for Off-site Materials/Goods. Annex 2 to the Conditions contains su   ʳtal provisions referring to a separate Electronic Data Exchange agreement wh agreement has been executed by the Parties.

---

12 Practice Note 20 does not emphasize complexity, but refers only to some forms being m
    than others; e.g. references to the Intermediate Form and the Minor Works Form, para.
13 E.g. the Intermediate Form (IFC 98), or Minor Works Form (MW 98).
14 Required under the Housing Grants, Construction and Regeneration Act 1996, Part II,
15 See Section 1.4.1 – sub-section 'Appendix', for comment on problems

Amendment TC/94, issued in April 1994, dealing with terrorism cover in insurance clauses is not incorporated in JCT 98, although the loose printed amendment is supplied with the printed JCT 98 form.

Amendment 1, issued by the JCT in June 1999, introduces the Inland Revenue Construction Industry Scheme (CIS) to comply with the Income and Corporation Taxes Act 1988 and The Income Tax (Sub-Contractors in the Construction Industry) (Amendment) Regulations 1998, Statutory Instrument No.2622.

1.3.3 Criticism of JCT 98

JCT 98 is a derivative of JCT 80 and the JCT and RIBA forms which preceded it. The style is the same as its predecessors, and this nullifies some of the principal advantages which the use of any standard form is claimed to bring. Most vulnerable is the general presumption that fairness derives from pre-determined allocation of risk by the parties' representatives, and the idea that familiarity and understanding is gained by regular experience of a particular form.

Dealing first with the question of fairness, it must be remembered that the resources available to the negotiating representative bodies by way of legal and administrative expertise are likely to far outweigh those of their representees. A perfectly balanced form of contract in terms of risk allocation may be negotiated and published for general use with what appears to be little thought given to the practical problems of using it in the workplace. This criticism may be levied at JCT 98. The JCT standard forms, like any other written contract, are a combination of several types of statement. Statements as to obligations, entitlements and liabilities concerning goods and services to be supplied and the price to be paid, are intertwined with statements of procedure, i.e. how the job will be administered.

This last set of statements (procedures) is the code by which the draftsmen of the forms direct the various parties how to go about their business. The intention may be admirable; unfortunately, the execution is not. The standard terms are convoluted and prolix; they are totally unsuitable for the day-to-day requirements of many building sites. The difficulty is that contract forms are required to act, not only as a legal record of the binding agreement, but also as a working tool. This requires clauses to be simple statements wherever possible, with minimal subdivision (but see JCT 98 Clause 13.4.1.2 para. A7.1.1). Unfortunately many forms, including those in the JCT 98 lineage, are written with a mind to dissection in the event of dispute; they are not written in language or in a style capable of being understood, with any justifiable confidence, by those using them.[16] Even the courts have difficulty. In 1967 Lord Justice Sachs said:[17] 'The difficulties arise solely because of the unnecessary amorphous and tortuous provisions of the RIBA contract: those difficulties have for a number of years been known to exist.'

Even more stinging is the observation of Lord Justice Salmon in *Peak Construction Ltd v. McKinney Foundations Ltd*, when criticizing one party's unilateral terms:[18]

---

16 See Sims J., 'Opposing Viewpoints', *Building* magazine, 7 May 1976, p.119.
17 *Bickerton v. N W Metropolitan Hospital Board* (1967) 1 All ER 977 at 979.
(1970) 1 BLR 111 at 114.

The form of this contract has been much criticised during the course of the argument – and not without justification. Indeed, if a prize were to be offered for the form of building contract which contained the most one sided, obscurely and ineptly drafted clauses in the United Kingdom, the claim of this contract could hardly be ignored, even if the RIBA form of contract was amongst the competitors.

The RIBA form referred to is the 1963 Edition (predecessor of JCT 80 and JCT 98). The 1998 Edition standard form is more than twice the length of the 1963 Edition, and its cross-referencing system makes it almost unintelligible.

It is difficult to believe that any document of the length and complexity of JCT 98 plus its amendments and supplements (including Bills of Quantities) is understood by the vast majority of individuals operating them, and it seems in practice that they are not. A survey conducted amongst 70 contractors[19] indicated that 31 per cent of all companies questioned felt the building industry standard form contracts were sufficiently ambiguous about the requirements of the parties to be the cause of claims. There was also strong criticism of professionals, who it was considered did not understand the implications of the contract terms.

In an unpublished 1992 survey[20] from interviews with commercial managers from a wide cross-section of trade contractors, some of whom acted in both main contractor and sub-contractor capacities, the general view of standard form contracts as a class varied considerably:

- JCT forms were perceived to clarify rights and obligations, but in fact they complicate them.
- There are too many procedures and too many sections to complete when tendering.
- Standard forms are fair, but all the amendments make it difficult to keep abreast.
- The procedures (requirement for complicated notices) give fertile ground for excuses (to avoid payment); notices are easily forgotten.
- As a contractor, one can have faith in a standard form.
- Standard forms lead one to believe the contract is being dealt with efficiently.
- Standard forms are sometimes trusted because they are standard forms; as a result they are rarely read in depth.

Turning to the sub-contracting arrangements in JCT 98, there is potentially a greater problem for sub-contractors.

Yet 22 years later the whole of the nominating procedure in the re-drafted form (JCT 80) was described by a court as being 'of inordinate and needless length'.[21]

The nominating procedure itself has drawn criticism,[22] but once he is nominated the sub-contractor faces an even more confusing array of terms than that confronting the contractor. In style, the Nominated Sub-contract documents appear similar to the main contract. In reality they are far more complex, because of the practice of 'stepping down',

---

19 Farrow, J. and Wagland, G. (1990) Occasional Paper No 47, *Effectiveness of Claims Provisions under the JCT Forms of Contract*, Chartered Institute of Building.
20 Survey on use of standard form contracts conducted by Rycroft M., January 1992.
21 *Scobie & McIntosh Ltd v. Clayton Bowmore Ltd* (1990) 49 BLR 119 at 128.
22 In *Mellowes PPG v Snelling Construction Ltd* (1990) 49 BLR 109 at 128, the nominating procedure in JCT 80 was described as being 'of inordinate and needless length'.

that is passing down the obligations of the contractor under the principal contract via the sub-contract, and also in some matters reversing the procedure by making available the rights and benefits of the main contract to a sub-contractor. The effect is to incorporate the principal contract into the sub-contract by reference in so far as the two do not conflict. The resultant consolidated document is far more complex than either of the individual contracts, but the practice introduces another factor to add to the sub-contractor's burden; simply, by introducing a standard document by reference it abrogates the need to attach a copy of it. Thus the main contract terms become remote and the operational staff of the sub-contractor rarely become familiar with them. Just like the operators of the main contract, many sub-contractors then follow some sort of generic system, based on past experience which fits in with their internal company procedures. In the 1992 survey amongst commercial managers referred to above, it emerged that many companies did not (or could not) adjust their standard in-house procedures to match the procedures of particular forms of contract.

Despite these criticisms, the JCT forms invariably work in practice. This may be due largely to the operators' wish to make them work, often by ignoring the strict terms, cutting procedural corners, and by applying some pragmatism and common sense. The JCT has partly addressed this particular issue in JCT 98 by ratifying common practices such as contractors' applications for payment, contractors' evaluation of variations (in the 'With' and 'Without' Quantities versions), and priced schedules to assist in interim valuations. However, the principal criticism remains; JCT 98 is still too long, too complex and too convoluted. It will no doubt remain a lawyers' paradise and a source of dispute like its predecessors. For many of the building industry's practitioners it will be a document to be signed, put in a drawer and forgotten until things go wrong. If that happens JCT 98 will have failed as a set of working rules, not least because the parties will examine the document only to remind themselves of their rights; their duties and obligations will be forgotten until it is too late.

1.3.4 Scope of this book – JCT 98 Private Edition, with Quantities

The commentary in this book is based on JCT 98 Private Edition With Quantities. The Private edition, and the 'With Quantities' version, were chosen solely on the grounds of common usage. The main differences between this form and the other editions and versions are as follows.

Local Authorities Edition: There are no major differences between the Private and Local Authority editions. The Local Authority form contains provision to reflect Local Authority law and practices including public accountability. In use, one of the most significant differences (to contractors, at least) is probably the limitation by a 'non adjustable element' on fluctuations payments, and the introduction of the term 'Contract Administrator' for use where a Local Authority employee performs the role of the Architect. Insurance provisions take account of the probability that Local Authorities will carry their own risks. Unlike the Private editions, the option for advance payment (introduced in Amendment 18 of JCT 80) is not included in the Local Authorities edition; nor is the requirement to place retentions in a separate banking account on request of the Contractor (because a Local Authority cannot become bankrupt).

Without Quantities and With Approximate Quantities Versions: Not surprisingly, from the title, the differences between these forms and the 'With Quantities' form relate to evaluation of the Works and calculation of interim payments. The 'Without Quantities' version contains no reference to Bills or to the Standard Method of Measurement. A significant effect of this is to avoid an implied warranty by the Employer that information provided to the Contractor is accurate and sufficient.[23] Under the 'Without Quantities' version the Contractor is required to price a schedule of work or specification (provided by the Employer), or to give a price backed by an analysis. The principles of valuing variations are similar to those applying to the 'With Quantities' form (JCT Practice Note 14 deals with both versions together).

The 'With Approximate Quantities' version is a remeasurement contract. The 'Tender Price' provides an initial indication of the contract value; the final price, called the 'Ascertained Final Sum', is calculated by applying evaluation principles (similar to the 'With Quantities' version) to measurement of the whole Works. Apart from the concept of total measurement, the most significant features are the absence of provisions for the Contractor to give quotations (as in 'With Quantities' Clause 13A), and Contractor's Price Statements (as in 'With Quantities' Clause 13.4.1.2 Alternative A).

1.4 Documents forming the Contract

1.4.1 JCT 98: the printed form

Contents

JCT 98 contains a contents section at the front of the document listing all the clause headings; regrettably, unlike some of its engineering industry counterparts,[24] it does not contain a subject index.

Articles of Agreement, Recitals, Attestation

The Articles are the most important part of the contract. They are the core statement of what the parties have agreed; without them, and in the absence of some other contractual arrangement, there is no contract. The Articles are also probably the least read section of the contract documents, except by legal advisers and those preparing documents for execution. This is both strange and sad. The Articles not only set out the obligations of the parties, but can also be looked upon as a mission statement.

The Articles of Agreement section begins by recording the names and addresses of the parties, who are introduced immediately as the Employer and the Contractor. The parties' names are not used again except to identify signatories. The parties may be companies or individuals. There then follow the recitals.

- The recitals put the agreement into context. There are seven recitals describing what is required and what events have taken place:

---

23 See Section 1.4.5 dealing with the *Bryant* case.
24 E.g. ICE, *Conditions of Contract 6th Edition* (Thomas Telford Ltd, 1991).

- *The first recital* describes the work required by the Employer, identifies the site and indicates that a design and Bills of Quantities have been prepared on his behalf.
- *The second recital* confirms that the Contractor has supplied the Employer with a priced copy of the said Bills of Quantities, and also where applicable with a priced Activity Schedule (see Chapter 15). The Bills are identified as the Contract Bills.
- *The third recital* identifies the Contract Drawings by number and states that both the Bills and the Drawings have been signed.
- *The fourth recital* identifies whether the Employer is considered to be a contractor for the purposes of tax deduction.
- *The fifth recital* states that the extent of the CDM Regulations is identified in the Appendix. A footnote warns the parties to refer to JCT 80 Practice Note 27 on the actions to be taken by duty-holders under the regulations before the Contractor can start.
- *The sixth recital* is optional and refers to the Information Release Schedule given by the Employer to the Contractor as provided in Clause 5.4.1 (see Chapter 11).
- *The seventh recital* confirms that the Employer has provided copies of relevant bonds if the Employer requires the Contractor to enter into a bond on terms other than those set out in Annex 1 to the Appendix.[25]

The recitals are followed by the articles:

- *Article 1* states the overriding obligation of the Contractor; subject to the Contract Documents, he will carry out and complete the Works. It is this obligation which prevents the Contractor walking off site unless the terms of the Contract provide.[26]
- *Article 2* states the principal obligation of the Employer – to pay the Contract Sum or such other sum as becomes payable under the Contract. JCT 98 is a lump sum contract, subject only to adjustment as provided in the terms. The main adjustment is likely to be in Variations (see Chapter 6) and in loss and expense (see Chapter 13). This still seems to come as a surprise to those professionals who rely on copious hidden approximate quantities in the Contract Bills and to contractors who see wholesale remeasurement of the work as an easy option.
- *Article 3* identifies the person to carry out the duties of the Architect[27] (or the Contract Administrator[28]) under the contract, and gives the Employer the power and the obligation to maintain someone in the role of the so named duty-holder. Thus the Employer is obliged to replace the person named as Architect within a reasonable time and no later than 21 days of his ceasing to be the Architect for whatever reason. Under the Local Authorities versions there is no right of objection for the Contractor, but under the Private versions the Contractor can object within 7 days. The Contractor's objection does not have to be reasonable; the test is that the Contractor's reasons are 'considered to be sufficient' by a person appointed under the disputes resolution procedures in the Contract. It seems Adjudication, Arbitration or Litigation is a pre-requisite to the Contractor's right to influence the replacement name.

---

25 Terms agreed between the JCT and the British Bankers' Association.
26 Subject also to statutory rights to suspend performance under the Housing Grants Construction and Regeneration Act 1996, Part II.
27 See also Section 1.5.2, Executing the Contract: Article 3.
28 See Local Authorities versions.

It is sometimes tempting for an Employer (particularly lay clients who are not familiar with the building industry) towards the end of a project to discharge the Architect to save on fees, and with that in mind strike out the express obligation to re-appoint. Such temptations should be resisted since it has been held that failure to re-appoint, even in the absence of an express obligation, is still a breach of contract.[29] Likewise it is tempting to discharge the Architect for making decisions with which the Employer does not agree, and to re-appoint with a replacement who concurs with the Employer.[30] Again the temptation should be resisted, unless the Employer is prepared to refer the matter under the disputes resolution procedures, since Article 3 expressly provides that the replacement Architect cannot disregard or overrule any certificate, opinion, decision, approval or instruction of the replaced Architect.

- *Article 4* identifies the person to carry out the role of Quantity Surveyor under the contract. The obligation to replace, and the right of objection is the same as for replacement of the Architect under Article 3. However, there is no corresponding express obligation to maintain the replaced Quantity Surveyor's decisions. This is not surprising since the Quantity Surveyor has little power under the Contract to make decisions. What little authority he has is to be found in the Variations procedures Alternative A, where there is express right in any event for either party to refer a matter to adjudication under Clause 41A.

- *Article 5* reminds the parties that they have the right to refer any dispute or difference arising under the Contract to adjudication under Clause 41A. This complies with the Housing Grants, Construction and Regeneration Act 1996, Part II, s.108, and identifies the rules to be applied to such adjudication in order to avoid application of the statutory default rules in the Scheme for Construction Contracts (England and Wales) Regulations 1998 (SCC).[31]

- *Article 6.1* identifies the name of the Planning Supervisor under the CDM Regulations, if it is not the Architect.[32]

- *Article 6.2* identifies the Contractor as being the Principal Contractor[33] for the purposes of the CDM Regulations. This is a strange provision and one can only wonder why the JCT did not adopt wording similar to Article 6.1. Since the Contractor will default to being the duty-holder of Principal Contractor on signing the Contract, the Employer and the professional team are in a vulnerable position if the Contractor does not have the competence or sufficient resources. In practice it is likely that the Contractor will usually be suitable, but there are situations where the Contractor is chosen because of his trade specialism, rather than competence as a CDM duty-holder. A typical and common situation would be where a shopfitting firm, or a mechanical and electrical services contractor, is the main contractor because of the high specialist content. Article 6.2 provides a mechanism for replacing the Contractor if the Contractor ceases to be the Principal Contractor, but not for an alternative from the outset.

---

29 *Croudace Ltd v. London Borough of Lambeth* (1986) 33 BLR 20 CA..
30 The Employer will be in breach of the Contract if interfering with the Architect in the exercise of duties that require the Architect to form an opinion, or that act fairly; see Chapter 2 on the duties and obligations of the Employer and the Architect.
31 Statutory Instruments, 1998 No. 649.
32 See comment on appointment of Planning Supervisor in Section 1.5.2, Article 6.1.
33 See comment on appointment of Principal Contractor in Section 1.5.2, Article 6.2.

- *Article 7* is an additional Article required only if Amendment TC/94 (ie. terrorism cover) is to apply. The words of Article 7 are contained in the user note at the front of Amendment TC/94, and state the Conditions are modified by the amendment attached to the Contract.
- *Article 7A* applies if the Appendix entry against Articles 7A and 7B has not been deleted. It constitutes an Arbitration Agreement within the provisions of the Arbitration Act 1996. With the exception of matters referred to in the Article, the parties must refer any dispute or difference arising under or in connection with the Contract to arbitration. Whilst the parties are given a choice of forum for dispute resolution in addition to adjudication, the default position is the Arbitration Agreement. Article 7A is examined further in Chapter 17.
- *Article 7B* applies if the Appendix entry against Articles 7A and 7B has been deleted, and confirms that the agreed forum for any dispute or difference arising under or in connection with the Contract is litigation. For the right to refer matters to litigation the parties must take the positive step to delete the arbitration option in the Appendix.

The Attestation concludes the articles. Space is left for the parties either to sign the Contract under hand as a 'simple' contract, or to sign as a deed (also known as a 'specialty' contract or a 'contract under seal')[34] (see also section 1.5.2, Executing the Contract: Attestation).

Main Differences between 'Simple'[35] and 'Specialty' contracts (deeds): There are three main differences between simple contracts and deeds. First, a gratuitous promise is binding under a deed, whereas under a simple contract a promise must be supported by consideration. Thus in a simple contract a promise to do work is usually supported by a promise to pay money. In JCT 98 the promise can be seen in Article 1: 'For the consideration hereinafter mentioned the Contractor will ...'. Whilst consideration is not strictly necessary in a deed, a nominal consideration is often included, e.g. 'for consideration of £1 ...'. This is sometimes incorporated to ensure that a simple contract will exist in any event if the document is flawed as a deed by, say, failure to complete the signing formalities properly.[36]

The second difference is that the facts stated in a deed contract cannot be denied. Thus, if JCT 98 is signed as a deed, statements such as those in the recitals, e.g. Sixth Recital: 'the Employer has provided the Contractor with a schedule' cannot be denied later by the Contractor. Clearly it is essential that the parties check factual statements, particularly if the contract is to be executed as a deed.

The third difference is the most significant and affects the parties' exposure to legal action. Under the Limitation Act 1980 s. 5 an action for breach of contract under a simple contract can be commenced at any time within 6 years of the breach occurring; in the case of a deed, under s. 8 the period is 12 years. These periods start when "the cause of action accrues', i.e. the date of the breach of contract. In the case of defective work under a building contract where the obligation is to complete the Works, the point at which the

---

34 The term 'under seal' is still used in some quarters to describe a deed, although since 1989 it is no longer necessary to apply a company seal. See also Section 1.5.2, Attestation.
35 Sometimes referred to as contracts 'under hand'.
36 See Section 1.5.2, Attestation, and Section 1.6.4, Failure to complete formalities.

Contractor's breach occurs is thought to be at practical completion, i.e. the date of the last opportunity to correct the defect before handing it over as being complete.[37] It follows that an action in respect of defective work will normally have to be commenced within 6 years or 12 years of practical completion.

The limitation period for legal action should not be confused with the defects liability period. The 6-year and 12-year periods do not increase the parties' liability; they merely set the time during which the parties are exposed to action for a breach. Thus in the absence of any breach the parties would be free from liability indefinitely. However, the longer the limitation period, the more time is allowed for breaches committed during the course of the contract to emerge. As a result the prudent contractor when signing a deed will ensure that major sub-contractors are also engaged under deeds. In this way he may avoid being in a position where he is pursued after 6 years for a sub-contractor's defect, but is unable to pursue the sub-contractor through being 'statute barred'.

Conditions

The Conditions are split into Parts 1 to 5 and set out the terms, or qualifications, attached to the Agreement:

Part 1: General
Part 2: Nominated Sub-contractors and Nominated Suppliers
Part 3: Fluctuations
Part 4: Settlement of Disputes
Part 5: Performance Specified Work.

Commentary on the principal provisions of the Conditions is dealt with by topic elsewhere in this book.

Code of Practice referred to in Clause 8.4.4 (for identification of defects)

Clause 8.4.4 deals with the opening up and testing of work that the Architect considers may not comply with the Contract. The difficulty for the Architect once he has identified a defective piece of work is to discover whether the defect is present elsewhere without demanding wholesale demolition of good work. The Code of Practice sets out the criteria to be considered by the Architect, the aim being to help in the fair and reasonable operation of Clause 8.4.4.

Appendix

JCT 98 is a standard form which is required to suit different situations, requirements and circumstances. The Appendix is a schedule for the variables in the Agreement and Conditions and contains two main types of entry. Some of the information in the Appendix is data (e.g. the length of Defects Liability Period and Date of Possession), whereas some entries trigger which of alternative clauses in the Articles of Agreement or the Conditions are to apply (e.g. whether the parties have chosen the Court: Clause 41C, or arbitration: Clause 41B), as the proper tribunal to hear disputes.

---

37 See *William Tomkinson & Sons Ltd v. Parochial Church Council of St Michael-in-the-Hamlet* (1990) 6 Const. LJ 319.

The concept of the Appendix entries being no more than a set of variables is important, and it is essential that the variables are compatible with the relevant operating machinery. In practice disputes may arise when either party or the professional team applies appendix information without referring properly to the operating machinery. The case of *Bramall & Ogden Ltd v. Sheffield City Council*[38] provides a salutary example. The contract, like JCT 98, contained a partial possession clause[39] under which the Contractor's total liability for liquidated damages based on the sum in the Appendix would be reduced *pro rata* (Lat. in proportion) the value of work taken over by the Employer in advance of the remainder. The Appendix expressed liquidated damages at a rate of £20 per week for each uncompleted dwelling. Whilst it was possible to make a calculation of damages from the Appendix data, it was not possible to insert that data into the operating clause, since there was no sum to be reduced *pro rata*. The court decided the provisions should be construed strictly, and the parties should not strain to make the calculation work. Since the Appendix entry was not consistent with the operating clause, the clause was held void for uncertainty and the Employer lost his right to any damages.

Annex 1 to Appendix: terms of bonds

The JCT have followed the example of some of the engineering standard form contracts by inserting model associated agreements in the printed form. They are the terms of two bonds for use where the Employer intends to make an advance payment, and where the Employer is prepared to pay for materials before they are brought to site. Both forms of bond are agreed between the JCT and the British Bankers Association, and both are intended to be executed as a deed[40] between the Surety and the Employer; the Contractor is not a party to the bond.

If the Employer requires the Contractor to provide a bond in any other form he must give copies of the required form to the Contractor before entering into contract.[41]

Supplemental Provisions (the VAT Agreement)

The Contract Sum and all adjustments under the Contract are exclusive of Value Added Tax: Clause 15.2. When Value Added Tax was introduced in the Finance Act 1972, the JCT decided to exclude VAT from the contract price calculations and to deal with the new tax under separate provisions. This had the advantage of avoiding complications where the value of work would otherwise appear to be artificially and temporarily increased until the Employer, if he were able, recovered the tax from HM Customs and Excise. The supplement was originally published as a separate document called 'Supplemental VAT Agreement' but is now bound into the printed standard form. Although still termed 'VAT Agreement' there is no need to execute the supplement separately, since it is incorporated adequately by reference in Clause 15.1. The location of the Supplement in the printed form is confusing, coming between 'Annex 1 to Appendix' and 'Annex 2 to the Conditions'; hopefully the JCT will rationalize the position at the first opportunity.

---

38 (1983) 29 BLR 73.
39 JCT 98 Clause 18.
40 See comment on Attestation in Section 1.4.1, Simple and specialty contracts, and Section 1.5.2, Executing the Contract.
41 See Seventh Recital.

Incorporation of the Supplemental Provisions is not optional under the Contract, so there seems no logic in keeping the provisions as a supplement, save that many in the industry tend to deal and calculate in VAT-exclusive figures, leaving VAT to be dealt with like an unwelcome gatecrasher. There is an optional element in Clause 1A, triggered by an Appendix entry, which provides a system for use when the Contractor is aware that all work under the contract will be standard or zero rated. If optional Clause 1A is not used, under Clause 1.1 the Contractor is required to break down the work into categories to identify the VAT zero rated separately from any other rated works.

Annex 2 to the Conditions: supplemental provisions for EDI

The JCT have recognized the trend in the building industry to adopt electronic data interchange as a means of communication.[42] If the parties choose to send drawings, letters, notices, applications for payment, certificates, payment or any other communication under the contract electronically, it is treated as being in writing for the purposes of the Contract, providing the parties have already entered into a collateral contract under one of two standard named EDI (electronic data interchange) agreements. The choice of EDI agreements is between the EDI Association Model Form of Agreement, and the European Model EDI Agreement; they govern the manner in which information is passed between the parties.

However, the JCT Supplemental Provisions do not allow the collateral contract to affect the Main Contract on certain important issues. Express provisions include a statement that the EDI Agreement will not override or modify interpretation of the Contract, and that essential documents must still be in writing; this includes the Final Certificate, notices of determination, suspension and disputes, and any agreement between the parties to amend the JCT conditions.

The disputes procedures of the Main Contract are stated to apply to the EDI Agreement so far as the Main Contract is concerned, and any dispute procedure in the EDI Agreement will not apply. (See Chapter 17 for comment on this point.)

Amendment 1: Construction Industry Scheme (CIS), issued June 1999

Amendment 1 is a necessary requirement to prevent JCT 98 conflicting with the provisions of The Income Tax (Sub-Contractors in the Construction Industry) (Amendment) Regulations 1998. The statutory tax deduction scheme referred to in JCT 98, became invalid after 31 July 1999 and is replaced by the Construction Industry Scheme (CIS), which started on 1 August 1999.

The provisions of CIS are outside the scope of this book, but a useful outline of the scheme is to be found in *Series 2* Practice Note 1.

1.4.2 Contract Drawings

Clause 1.3 'Contract Documents' identifies the drawings referred to in the First Recital, and listed as 'Contract Drawings' in the Third Recital as forming part of the Contract

---

42 For full commentary on EDI see JCT, *Electronic Data Exchange in the Construction Industry* (RIBA Publications,1998).

Documents.[43] The Contract Drawings, prepared on behalf of the Employer, show the extent of work to be carried out by the Contractor under the Contract, but subject to further description of work in the Contract Bills[44] and also subject to the issue of further details.[45]

The latitude allowed by the Contract for the Architect to issue further information as the work progresses has historically been interpreted by some architects and other designers to mean the design need be in no great detail at the contract signing stage. Often the result is *ad hoc* (Lat. special purpose) design as site work proceeds, leading to disruption of the Contractor's progress, and dispute.

A feature of JCT 98 is the introduction of an Information Release Schedule to inform the Contractor when information will be issued by the Architect. The Sixth Recital stating the Employer has provided such a schedule is optional, and the schedule is not required to be annexed to the Contract.

1.4.3 Bills of Quantities

Bills of quantities are quantitative lists of descriptive items making up the entire work to be done by the Contractor. The bills are usually prepared by a Quantity Surveyor from drawings and specifications and are measured and described in accordance with a recognized code or 'method of measurement'. In the case of JCT contracts, where bills of quantities are the basis of the contract, the current edition of the Standard Method of Measurement of Building Works, published by the Royal Institution of Chartered Surveyors and the Building Employers Confederation (from 1997 called the Construction Confederation) is specified in the Contract at Clause 2.2.2.1. Each tenderer is presented with a copy of the bills, and is required to insert unit rates or lump sums. These are then extended by multiplying such rates by the relevant quantities in the bills. The aggregate of all the extensions becomes the amount of the tender and, in the case of the successful contractor, ultimately is transcribed to Article 2 to become the Contract Sum; the bills become the Contract Bills. Strangely, the important connection between the Contract Bills and the Contract Sum is not to be found in the Articles, the Recitals, or even the definitions. Instead it is implied, tucked away in Clause 14.1: 'The quality and quantity of the work included in the Contract Sum shall be deemed to be that which is set out in the Contract Bills'. Strictly, if the total amount in the Contract Bills and the amount inserted as the Contract Sum in Article 2 were not the same, there is no machinery in the contract for making adjustment; under the contract rules the Contract Sum could still be adjusted for Variations and other matters without any more complication than exists if the two amounts are the same. However, if there are any overt pricing errors in the tender bills, it is recommended practice[46] for the Employer's team to notify the contractor before accepting the tender to give him the chance to correct or stand by the error.

Bills of quantities are a common feature throughout all contractual levels of the United Kingdom Building Industry. Such is the place of bills of quantities in the education of

---

43 Not all contracts contain drawings as contract documents in their own right. For example a contract under the JCT form 'With Contractor's Design' may incorporate the Employer's drawings but they must be contained with other information in a Contract Document called 'Employer's Requirements'.

44 See Clause 2.1; for commentary on discrepancies in or divergences between documents see also Chapter 6, Section 6.14.

45 See Clause 5.4.2.

46 See Code of Procedure for Single Stage Selective Tendering, Clauses 6.1 to 6.5.

building industry professionals that even under principal contracts not containing Contract Bills, the Contractor will invariably produce his own bills of quantities, either for his own cost control purposes, or for use in sub-contracts. In the early 1960s the National Federation of Building Trades Employers agreed with its members that they would not tender for work over £8000 in value unless the contract incorporated bills.[47] However, the wide range of present-day standard form contracts suggests that the use of bills as the basis of the contract is on the wane. This is probably due partly to changing techniques in project procurement led by the development of design build contracts[48], partly by recognition of practical shortcomings of bills,[49] and partly because Employers are not prepared to pay for full Quantity Surveyors' services. The two main advantages of using bills of quantities are (1) a means of comparing tenders like for like, and (2) a pricing structure for valuing variations and calculating the amounts for interim payments. In practice there is a problem in the use of bills for calculating the value of variations, particularly when the bills have been produced before the design is complete.

As long ago as 1944, the 'Simon Report'[50] described the purpose of bills as 'to put into words every obligation or service which will be required in carrying out the building project'. There is no need, therefore, for a separate 'specification' document under JCT 98 'With Quantities'; indeed there is no place for it, since the specification will be contained in the Contract Bills. Except for adding and omitting quantities for variations the bills are supposed to describe the work to be done, but often, due to time constraints and incomplete design the bills are little more than an informed guess as to the content of the work. Since the bills are prepared by the Employer's team the risk in their accuracy is expressly placed on the Employer; thus errors are required to be corrected as though they were a variation.[51] This provides a buffer which enables the tender enquiry documents to be prepared more quickly to meet demanding time targets. Unfortunately when such tender bills become the Contract Bills the strict and limited rules for their adjustment can become unmanageable, and particularly so when there are numerous variations to be adjusted in addition to the task of correcting the original quantities to match the developing design.

Other benefits of bills are not so affected by their content. Even an inaccurate bill will at least give a uniform basis for tendering, enabling the Quantity Surveyor to make adjustments in his cost advice to his client; and provided the descriptions contained in the bills are representative of the type of work required, the unit rates are likely to be suitable for valuing variations. Of greater effect is the practice adopted by some contractors of inserting unit rates either to suit anticipated cashflow (i.e. 'front loading' to increase the value of interim payments for early trades), or to 'gamble' on the items likely to be increased or omitted,[52] in the hope of increasing the quantity of high profit items.

---

47 In *re Birmingham Association of Building Trade Employers' Agreement* [1963] 1 WLR 484, this agreement was held to be against the public interest under the Restrictive Trade Practices Act 1956.

48 Following suggestions in the 'Banwell Report' that design and construction processes should be less segregated: *The Placing and Management of Contracts for Building and Civil Engineering Work* (HMSO, 1964).

49 For detailed criticism of bills of quantities see Duncan Wallace, I. N., *Construction Contracts: Principles and Policies in Tort and Contract*, (Sweet & Maxwell, 1986), Chap 26.

50 *The Placing and Management of Building Contracts*, HMSO.

51 Clause 2.2.2.2.

52 See Chapter 6, Section 6.12.2.2 for further comment, and *dicta* of Lord Justice Pearce in *The Mayor Aldermen and Burgesses of the Borough of Dudley v. Parsons and Morrin Ltd.*

One feature of bills which has attracted debate is the tendency of Quantity Surveyors to incorporate 'special conditions', sometimes conflicting with the JCT Conditions. A typical example is the incorporation of design duties for specified work in the absence of special provisions in the Conditions. The effectiveness of such provisions is considered later in this chapter in Section 1.5.5, when dealing with amendments and potential conflict between the Bills and Clause 2.2.1.

1.4.4 Fluctuations Clauses

Fluctuations in price due to the effect of inflation are dealt with in Part 3 Clauses 37 to 40 of the Conditions, but only Clause 37 referring to 38, 39 and 40 appears in the published JCT 98 form. Unlike other supplemental clauses (e.g. VAT) which are printed at the back of the printed JCT 98 form, fluctuations Clauses 38, 39 and 40 are published separately.

Clauses 38, 39 and 40 are choices:

> *Clause 38:* fluctuations in Employer's contribution, levies and taxes calculated by reference to the Contractor's records and actual cost incurred;
> *Clause 39:* fluctuations in the cost of labour and materials and taxes calculated by reference to the Contractor's records and actual cost incurred;
> *Clause 40:* a price adjustment formula calculation using monthly published indices.

For commentary on the application of the fluctuations clauses see Chapter 14.

1.4.5 Other documents: Standard Method of Measurement

Clause 2.2.2.1 states 'the Contract Bills ... are to have been prepared in accordance the *Standard Method of Measurement of Building Works, 7th Edition*, published by the Royal Institution of Chartered Surveyors and the Building Employers Confederation (now Construction Confederation)'. This document is known throughout the industry as 'SMM7'.

One purpose of SMM7 is to standardize the way in which work is measured and described, so that contractors and Employers' professional teams understand what is included, and what is not. In this way, in theory, disputes over issues such as whether a unit rate in the Contract Bills includes all associated items of labour can be avoided.

Another purpose of SMM7 is to ensure that the tenderer whose task it is to price the bills knows the extent of the work for which he is tendering. This squares with the description of bills of quantities in the Simon Report:[53] 'to put into words every obligation or service which will be required in carrying out the building project'. As a result probably the most important section of SMM7 is Rule 1.1: 'Bills of Quantities shall fully describe and accurately represent the quantity and quality of the works to be carried out. More detailed information than is required by these rules shall be given where necessary in order to define the precise nature and extent of the required work.'[54]

---

53 See commentary on bills of quantities in Section 1.4.3.
54 See *C. Bryant & Son Ltd v. Birmingham Hospital Saturday Fund* [1938] 1 All ER 503. In that case the equivalent clause using similar words in the Standard Method of Measurement of the time, when read with the Contract which stated the Standard Method had been used, was construed by the Court to be a warranty by the Employer that information provided to tenderers was both accurate and sufficient for their needs. In other words such information could be relied on at face value.

SMM7 also influences the Contractor's entitlement to extension of time and to recovery of loss and expense. Whilst SMM7 is mainly concerned with accurate measurement, it also provides for inclusion of approximate quantities[55] and provisional sums[56] in circumstances where work is not designed, or is designed in whole or in part but cannot be properly measured for any reason. General Rules 10.1 to 10.6 of SMM7 are sufficiently important to justify repetition verbatim in the Contract at Footnote [n] to Clause 1.3. The Rules provide that:

- Where work is capable of description but quantities cannot be calculated accurately, the quantities are to be estimated and described as approximate.
- Where work cannot be described and given in items it must be given as a provisional sum and identified as being for defined or undefined work.
- Where a provisional sum is identified as being for defined work the Contractor will be deemed to have made allowance for the work in his programme and in pricing Preliminaries.

1.4.6 Other documents: Adjudication Agreements

Clause 41B refers to the adjudication agreement to be executed by the parties and the adjudicator appointed to decide any dispute referred to him under the Contract. There are two forms of agreement; one for use where the Adjudicator is named in the contract and one for use either where a name is agreed by the parties or where a name is nominated by the nominating body in the Contract.

The Adjudication Agreements are published in Amendment 18 to JCT 80 together with an amendment to the Appendix for use when the Adjudicator is named in the Contract. Under JCT 98, the adjudication agreements are published as separate documents and are not incorporated in the published JCT 98 form. The Appendix amendment for naming the Adjudicator in the Contract has not found its way into JCT 98, although Footnote [ww] to Clause 41A advises that the appropriate version of the Adjudication Agreement is available from the retailers of JCT forms.

See Chapter 17 for further comment.

1.4.7 Other documents: Model Arbitration Rules

Clause 41B refers to the JCT 1998 of the Construction Industry Model Arbitration Rules (CIMAR). See Chapter 17 for comment.

1.4.8 Optional documents: Sectional Completion Supplement[57]

JCT 98 contains one completion date for the whole of the Works, and it is this date which is the reference point for release of retention, transfer of risk for the Works, the start of the

---

55 See Clauses 25.4.14 and 26.2.8 (extension of time and loss and expense in respect of inaccurate forecast of approximate quantities).
56 See Clauses 25.4.5.1 and 26.2.7 (extension of time and loss and expense not allowed in respect of provisional sum for defined work).
57 See also Chapter 3: Sections 3.10, Partial possession, and 3.11, Sectional completion.

defects liability period, and the Employer's entitlement to liquidated damages etc. (see Chapter 3). The only provision for the Employer to take over part of the Works before completion of the whole is Clause 18 under which the Employer is entitled, with the Contractor's consent, to take over some of the Works 'early', when they have reached a suitable stage.

If the Employer wishes to place an obligation on the Contractor to release parts of the Works before others, the use of the Sectional Completion Supplement is essential.

The supplement provides for sections of the Works to be identified in a modified Appendix, and sets out the start and completion dates of each section, together with the relevant liquidated damages.

The essence of the supplement is that each section identified in the Appendix has to be treated rather like an individual contract in respect of both time and loss and expense. This includes requirements for separate notices of delay or loss, separate awards of extensions of time, and separate practical completion certificates.

Unlike the design and build form WCD 98, the Sectional Completion Supplement for use with JCT 98 is not published in the printed contract, but is a separate publication incorporating its own Practice Note 1 (SCS/(W/WA) 98). The supplement contains additions and amendments to be made in the standard Recitals and Articles, and an entire replacement Appendix. An additional Article 6 to be inserted into the Agreement lists amendments to be made in the Conditions.[58]

1.4.9 Optional documents: Contractor's Designed Portion Supplement

Where the Contractor is to design or complete the design of a substantial element of the Works, the Designed Portion Supplement should be used. The general effect of the Supplement is to treat an identified portion of the Works like a design and build contract along similar lines to those found in the full design and build form, WCD 98. Unlike the provisions for Performance Specified Work (PSW) (see Chapter 4), the Supplement is not triggered by an Appendix entry, but requires incorporation into the JCT 98 form. A Practice Note CD/2 is published separately.

The Supplement introduces three new documents into the contract: the Employer's Requirements, the Contractor's Proposals, and the CDP Analysis (i.e. breakdown of price). In respect of the contractor-designed work, these documents replace the traditional drawings and bills, although drawings and specifications could form part of either the Requirements or the Proposals.

The Contractor is liable for his own design to the same extent that an independent designer would be liable, ie. he is required to use reasonable care and skill. Whilst this is not as high an obligation as preparing a design which is fit for its purpose, it is arguable that in applying reasonable care and skill to the interpretation of the Employer's Requirements in order to complete substantive design, the overall obligation comes very close to fitness for purpose.[59] Whilst the Contractor is responsible for his own design, the Architect is responsible for incorporating the Contractor's design into the whole Works.

---

58 Note that the Agreement already contains Articles numbers 6.1 and 6.2, dealing with the Planning Supervisor and Principal Contractor.
59 See also Chapter 4: Sections 4.9.1 and 4.9.3, *Performance Specified Work*.

The Supplement contains replacement Recitals and Article 1, which latter refers to a Schedule to the Articles and a Supplementary Appendix. The Schedule in turn lists amendments to be made in the Conditions. The Supplementary Appendix identifies the Employer's Requirements, the Contractor's Proposals and the CDP Analysis.

1.5 Preparing the Contract Documents

1.5.1 Generally

JCT 98 is drafted for execution as a formal contract using the printed form. There is no standard alternative such as that found in some forms[60] where provision is sometimes made for a letter of acceptance together with the Tender to create the contract. Nevertheless, and unfortunately, there is a common trait, particularly amongst industrial and occasional Employers, to forego using the printed form. Instead an Order form is issued, making reference to the standard JCT form. The result is often confusion when an optional clause has not been chosen. Some of the problems resulting from such misuse of the standard form are dealt with later in this chapter.

The principal documents forming the contract are described earlier in this chapter, and it is essential that they all are constantly kept in mind as a group when preparing the JCT form for execution. Indeed it is well worth having the various documents available as a complete set so that they can be checked against one another to reduce risk of discrepancy and misunderstanding.

The main sections of the form requiring completion are the Articles of Agreement and the Appendix, but there may also be loose printed amendments[61] to be inserted unless they are a matter of choice.[62] In addition there are various parts of the form at which a selection is offered sometimes without a default position being stated; failure to deal with these options can lead to misunderstanding or dispute.

In preparing documents for execution the following actions are required.

(*Note: reference in the following sections 1.5.2 to 1.5.4 to 'Footnote [ ]' is reference to the relevant footnote printed in the JCT 98 printed form*).

1.5.2 Agreement, Conditions, Appendix, Annex 2

Agreement

Articles of Agreement: The names and addresses of the parties are entered in the appropriate spaces. The date is normally inserted after the second party has completed the attestation at the end of the Articles of Agreement.

---

60 E.g. ICE *Conditions 6th Edition* which incorporates the Tender as a contract document.
61 JCT 1998 edn incorporates all amendments to JCT 80 up to Amendment 18. Prior to the publication of JCT 98 the current printed version of JCT 80 (printed version *P With 10/95*) incorporated amendments 1–2, 4–13, and 15; Amendments 3, 14, 16–18 were loose documents. JCT 98, at the time of first publication, contains loose 'Amendment TC/94 Issued April 1994' consisting of six items related to insurance in respect of terrorism.
62 Some amendments may be printed as optional, e.g. 'Amendment 18', 'Item 20', 'clause 41A' where the parties wish to name an adjudicator in the contract.

First Recital: A brief description of the intended works and address of the site is entered.

Second Recital: Ensure a fully priced copy of the Contract Bills has been supplied to the Employer and copies are available.

If a priced Activity Schedule has not been provided by the Contractor, reference must be struck out, as Footnote [e].

Third Recital: The numbers of Contract Drawings are to be inserted. It is important to ensure the Drawings numbers and revisions referred to are the same as those referred to in the Contract Bills, and that the copies for incorporation in the bundle of Contract Documents are copies of the specified Drawings. It is not unknown for the latest edition of a Drawing to be appended in error to the contract in place of the earlier revision specified.

The Contract Drawings and the Contract Bills are required to be signed by the parties for identification at the same time as they sign the Attestation.

Fourth Recital: This recital should be amended in accordance with standard Amendment 1 (CIS), unless the amendment is to be incorporated by attachment to the Form of Contract and referred to in an additional article (see 'Article next number' below).

Fifth Recital: See Footnote [f] referring to JCT Practice Note 27.

Sixth Recital: If an Information Release Schedule has not been provided by the Employer, reference must be struck out, as Footnote [e].

Seventh Recital: If the Employer wishes to use terms of bonds other than those printed in the JCT 98 published form, copies of the required terms must be given to the Contractor before the Contract is signed. It is not necessary to incorporate the terms in the Contract.

Article 2: The Contract Sum must be inserted from the Contract Bills. Since it is common for last minute amendments to be made to the Bills, it is wise to transcribe the figure from the Contract Bills which have been prepared for signing.

Article 3: The name and address of the Architect should be inserted, but care should be exercised in ensuring the named person is entitled to be called an architect in accordance with the Architects Registration Acts 1931 to 1969 amended by the Housing Grants, Construction and Regeneration Act 1996, Part III. If the named person does not qualify to be called an architect the term 'Architect' should be amended to 'Contract Administrator'. The Local Authorities versions contain alternative Articles 3A and 3B to deal with this.

Article 4: The name and address of the Quantity Surveyor should be inserted. There are no restrictions on the use of the title for the purposes of the Contract.

Article 6.1: The name and address of the Planning Supervisor should be inserted here only if the Architect is not going to fill the role. Care should be taken before the Architect is appointed automatically, to ensure that the Client has satisfied himself that the Architect is not only competent but has sufficient resources to be a duty-holder under the CDM Regulations.

This article may be deleted only when the Works are not subject to the CDM Regulations apart from Regulations 7 and 13.

Article 6.2: This article does not require any action providing the Contractor is the Principal Contractor for the purposes of the CDM Regulations. There is no provision for appointing a Principal Contractor other than the Contractor. It is essential, therefore, that the Contractor is assessed for his competence and availability of resources to fulfil the role of Principal Contractor. If necessary the words of Article 6.2 should be amended to provide for the appointment by the Employer, from the outset, of a Principal Contractor other than the Contractor.

Article 7: This is an additional Article to be written into the Articles if Amendment TC/94 applies. The words of Article 7 are contained in the user note at the front of Amendment TC/94.

Article 8: If the JCT 98 Form itself is not amended, and amendments are attached to the Form of Contract, an additional article should be inserted here to incorporate the relevant amendments.

Attestation: Space is provided to execute the Agreement as a simple contract or as a deed (see commentary on Attestation in Section 1.4.1).

If the Contract is to be signed as a simple contract the parties and witness(es) are required to sign at 'AS WITNESS THE HANDS OF THE PARTIES HERETO'.

If the Contract is to be signed as a deed the parties are required to sign in the respective spaces below 'EXECUTED AS A DEED BY THE (EMPLOYER/CONTRACTOR)'. There are three options, i.e. where a party is a company using its common seal, where a party is a company not using its common seal, and where a party is an individual. The Notes [2] to [6] printed in the Attestation against the relevant spaces provide detailed guidance on application of the alternatives.

The reason for the alternative spaces for companies which may or may not use their common seal is that under the Companies Act 1989 a company need not use its seal. A document without a seal but stated to be intended as a deed, and signed by two directors or one director and the company secretary, will have effect as a deed on delivery; it is presumed to be delivered when it is executed, provided there is no evidence of intention that delivery should take place later or upon some condition being met.

Under the Law of Property (Miscellaneous Provisions) Act 1989, if a party is an individual it is not necessary for him to affix a seal; provided the document makes clear that it is intended as a deed and it is signed by the individual in the presence of, and signed by, a witness, the document will have effect as a deed.

Care needs to be taken to ensure both parties complete all the relevant formalities. This includes signing any Contract Documents which require signing (e.g. the Contract Drawings, Contract Bills, certain supplemental agreements), and initialling all amendments. If the relevant formalities are not completed the contract may not have effect as a deed. However, that is not to say that no contract exists; it is likely that a simple contract will still be concluded.[63]

One situation to be avoided is where one party has signed the Contract, and the other party has either sealed the Contract or signed as a deed. The Contract is still binding, but the party who has signed will be bound under a simple contract, whereas the party signing under seal or as a deed will be bound under a deed (specialty). The result is that the party under the simple contract will be vulnerable to pursuit for a period of 6 years, and the party bound under the deed will be exposed for 12 years.[64] For this reason, a prudent practical measure when documents are being prepared for execution is to delete the alternative not being used before a signature is entered in error.

Conditions

Clause 1.3 Public Holiday: The meaning of Public Holiday should be amended if appropriate.

Clause 1.10: This clause should be amended if the law applicable to the Contract is not the law of England.

Clause 5.3.1.2: This clause should be deleted if the Contractor is not required to provide a master programme.

Clause 5.3.2: The words in parentheses should be deleted if the Contractor is not required to provide a master programme.

Clauses 22A, 22B, 22C: See Chapter 7 for commentary on use of alternative clauses. Under Clause 22.1 the application of the alternatives is identified in the Appendix. Strictly, there is no need to delete the redundant alternatives, although in practice it is often done in an attempt to avoid confusion. Unfortunately it sometimes has the reverse effect. It is not unknown for the wrong alternative to be deleted leaving a discrepancy in the Conditions. Clearly, if Clause 22.1 remains intact the Appendix entry will apply; but that will only leave uncertainty if the relevant alternative in Clause 22 is deleted. It is unlikely that the intention of the parties will be relevant since the Contract as signed will operate satisfactorily, and the operative clause will be the one not deleted.

Footnotes [ff] and [gg] advise the parties that it may not be possible to obtain insurance as specified. The parties should arrange between them what the cover should be, and the definition of, and all clauses referring to, 'All Risks Insurance' and/or 'Specified Perils' should be amended accordingly.

Footnote [hh] advises that some Employers such as tenants may not be able to comply with the obligations regarding insurance of existing structures; Clause 22C.1 may require amendment accordingly.

Amendment TC/94 Issued April 1994 is published as a separate document and contains amendments to insurance clauses (Clauses 22.2, 22A, 22B, 22C.1, 22C.2 in all versions;

---

63 See commentary in section 1.4.1, Attestation, regarding consideration in simple contracts and specialties.
64 See commentary in section 1.4.1, Attestation, regarding statutory limitation periods.

additionally Clauses 22B.2, 22C.3 in Local Authorities versions) to provide for terrorism cover. If *TC/94* is to apply, an additional Article must be inserted[65] and the Amendment should be attached to the Conditions. Alternatively the individual amendments may be made to the published form and initialled by the parties.

Clause 30.4.1.1: The Contract incorporates a Retention Percentage of 5 per cent, but Footnote [nn] advises no more than 3 per cent if at the tender stage the Employer estimates the Contract Sum to be £500 000 or more. To incorporate the lower rate an entry has to be made in the Appendix. If a higher rate (e.g. 7 per cent) is inserted in the Appendix it will be of no effect, since Clause 30.4.1.1 expressly states the rate to be 5 per cent unless a lower rate has been agreed and inserted. If the Employer wishes to apply a higher rate, Clause 30.4.1.1 must be amended.

Clause 30A: In accordance with standard Amendment 1, Clause 30A should be inserted, unless Amendment 1 is to be incorporated by attachment to the Form of Contract.

Clause 31: In accordance with standard Amendment 1, the existing printed clause should be deleted and new Clause 31 inserted, unless Amendment 1 is to be incorporated by attachment to the Form of Contract.

Clause 35.13.5.3.4: Footnote [qq] advises that this clause relates to winding up of the Contractor and may require amendment if the Contractor is a person subject to bankruptcy laws. The events relating to bankruptcy commence with an act of bankruptcy followed by a Petition presented to the Court for an order for the protection of the estate.

Clauses 38, 39, 40: The provisions of the relevant Fluctuations clause are incorporated in the Contract by Clause 37.1 as identified by the choice identified in the Appendix, and strictly do not need to be attached to the Contract. However, if the Contract Documents are to be a working tool, for completeness and the avoidance of doubt, it is better to insert a copy of the required clause. Footnotes [rr], [ss] and [tt] advise which clause to be used. (For commentary see Chapter 14).

Clause 41B.5: Footnote [yy] reminds the user that the Arbitration Act 1996 does not extend to Scotland. If the site of the Works is in Scotland, the parties are advised to use the forms issued by the Scottish Building Contract Committee. However, if the parties wish the Arbitration Act 1996 to regulate the proceedings of dispute resolution irrespective of the address of the site, Clause 41B.5 need not be amended.

Appendix

The Appendix entries are variables, inserted to enable the relevant operating clause in the Articles or the Conditions to have effect. The Appendix entries are not substitutes for the operating clause; nor do they operate in isolation as terms of the Contract. It is essential therefore, that each entry is compatible with its operating clause.[66]

---

65 See user note in *TC/94* at p. 2.
66 See example in *Bramall & Ogden v. Sheffield City Council* given in Section 1.4.1, Appendix.

Appendix, Fourth Recital: The text requires to be amended in accordance with standard Amendment 1, to incorporate reference to the Inland Revenue Construction Industry Scheme.

The alternative standard entries should be deleted as appropriate to identify whether the Employer is or is not a 'contractor' for the purposes of the tax regulations.

Appendix, Fifth Recital: The alternative standard entries should be deleted as appropriate to identify either whether all CDM Regulations apply, or only Regulations 7 (notification of project) and 13 (requirements on designer) apply.

Appendix, Articles 7A and 7B and Clauses 41B, 41C: The standard entry should be deleted if disputes are to be decided by legal proceedings. If the standard entry is not deleted, Clause 41B (Arbitration) will apply. The parties are referred to the Guidance Note to JCT 80 Amendment 18 for factors to be taken into account when deciding between litigation and arbitration. Note, the parties rights to refer a dispute to adjudication are not affected by this decision.

Appendix, Clause 1.3 Base Date: The Base Date is to be inserted here being the date of the tender or a date agreed between the parties.

Appendix, Clause 1.3 Date for Completion: The date to be inserted here is the date fixed by the parties at the time the contract is executed. It is not a fixed date in the sense that it cannot be adjusted. Time is not 'of the essence'[67] since express provision is made for adjustment in Clause 25.[68]

Appendix, Clause 1.11: The alternative standard entry should be deleted as appropriate to identify whether or not the Supplemental Provisions for EDI apply.

If the Supplement does apply, the type of Agreement executed by the parties must be identified by deleting the relevant alternative. If the form of agreement is other than one of the two standard forms indicated, a brief description of the applicable form should be entered here.

Note that Clause 1 of the Supplement requires the EDI agreement to be entered into no later than the date of this Agreement.

Appendix, Clause 15.2: The alternative standard entry should be deleted as appropriate to identify whether or not Clause 1A of the VAT Agreement applies. It can only apply when the Contractor, at the date of the Contract, is satisfied that all work supplied under the Contract will be at either a zero or a positive rate. Footnote [x] explains the position on work for a charity.

Appendix, Clause 17.2: The length of the Defects Liability Period must be agreed between the parties and entered here. A period of 6, 9 or 12 months is common. Many building projects have a high mechanical and electrical services or process plant content, in which

---

67 'An essential condition or stipulation in a contract without which the contract would not have been entered into.': *Osborn's Concise Law Dictionary*, 7th edn (Sweet & Maxwell, 1983).
68 See commentary in Chapter 11.

defects do not become apparent until many months after Practical Completion; it is common, therefore, for the services elements of the Works to have a Defects Liability Period of at least 12 months. If no period is inserted the Appendix states that the period is 6 months.

Appendix, Clause 19.1.2: The alternative standard entry should be deleted as appropriate to identify whether or not an assignee to whom the Employer has assigned his benefits under the Contract after Practical Completion may commence proceedings in the Employer's name to enforce such benefits.

Appendix, Clause 21.1.1: The amount inserted is the minimum cover required in respect of 'third party' claims.

Appendix, Clause 21.2.1: The alternative standard entry should be deleted as appropriate to identify whether or not a Joint Names Policy insurance *may* be required. The amount inserted is the amount of insurance cover to be provided if the Employer requires the Contractor to insure in joint names against injury or damage to property which is the liability of the Employer.

Appendix, Clause 22.1: The alternative standard entry should be deleted as appropriate to identify which of Clauses 22A, 22B and 22C applies. See also comment under 'Conditions, Clauses 22A, 22B, 22C' above.

Appendix, Clauses 22A, 22B.1, 22C.2: The percentage entry is the percentage to be added to insurance cover to allow for professional fees required to reconstruct the work.

Appendix, Clause 22A.3.1: The annual date supplied by the Contractor for renewal of insurance.

Appendix, Clause 22D: The alternative standard entry should be deleted as appropriate to identify whether or not the Employer requires insurance to cover the loss of liquidated damages in the event that the Contractor is awarded an extension of time as a result of damage caused by one of the Specified Perils.

Appendix, Clause 22FC.1: The alternative standard entry should be deleted as appropriate to identify whether or not the Joint Fire Code applies.

If the Joint Fire Code applies, the alternative standard 'yes/no' entry should be deleted as appropriate to identify whether or not the insurer under Clause 22A, 22B or 22C.2 has specified that the Works are a 'Large Project'.

The note to the Appendix reminds the parties that where Clause 22A applies, these entries are made on information supplied by the Contractor.

Appendix, Clause 23.1.1: The date to be inserted here is the date of possession fixed by the parties at the time the contract is executed.

Appendix, Clauses 23.1.2, 25.4.13, 26.1: The alternative standard entry should be deleted as appropriate to identify whether Clause 23.1.2 applies (entitlement of Employer to defer possession of the site for up to 6 weeks).

If Clause 23.1.2 does apply, a maximum period other than 6 weeks may be entered provided it is less than 6 weeks.

***Appendix, Clause 24.2*:** The entry for liquidated damages must be a rate (£) per unit of time to be compatible with Clause 24.2 and Clause 18 (partial possession by Employer); see Section 1.4.1, Appendix, regarding confusion between the concepts of partial possession and sectional completion in so far as they affect the Appendix entry for liquidated damages.[69]

If the Employer does not wish to apply liquidated damages at all, but instead wishes to rely on his common law entitlement to recover unliquidated damages (i.e. such damages as he can prove) he must amend the Contract; it is not sufficient for the Employer to insert 'Nil' in the Appendix. In *Temloc v. Errill Properties*[70] the Court of Appeal decided the liquidated damages provision was exhaustive of the employer's rights to damages, but in any event, a '£Nil' entry indicated to the contractor at the time of entering the contract that the employer would suffer no damage. The Court held the employer lost his right to any damages.

***Appendix, Clause 28.2.2*:** The period for which the Employer is entitled to suspend the Works as a result of his own culpable events without giving cause for the Contractor to give notice of determination should be entered here. If no entry is made, a standard period of 1 month applies.

***Appendix, Clauses 28A.1.1.1 to 28A.1.1.3*:** The period for which the Employer is entitled to suspend the Works as a result of *force majeure* (Fr. act of God) Specified Perils and civil commotion without giving cause for the Contractor to give notice of determination should be entered here. If no entry is made, a standard period of 3 months applies.

***Appendix, Clauses 28A.1.1.4 to 28A.1.1.6*:** The period for which the Employer is entitled to suspend the Works as a result of default by statutory undertakers, hostilities or terrorist activity without giving cause for the Contractor to give notice of determination should be entered here. If no entry is made, a standard period of 1 month applies.

***Appendix, Clause 30.1.1.6*:** The alternative standard entry should be deleted as appropriate to identify whether the Employer wishes to make advance payment. If he does then he must enter details of the amount (£) or, if applicable, the percentage of the Contract Sum to be paid. He must also enter the date or the time when payment will be made and the time(s) at which the advance payment will be reimbursed by the Contractor.

If an advance payment bond is required the appropriate alternative entry must be deleted.

***Appendix, Clause 30.1.3*:** The period between interim certificates may be agreed between the parties and stated here but if there is no entry made, the standard period is stated to be 1 month.

---

69 Referring to *Bramall & Ogden v. Sheffield City Council* (1983).
70 (1988) 39 BLR 30.

Appendix, Clause 30.2.1.1: The alternative standard entry should be deleted as appropriate to identify whether a priced Activity Schedule is attached. The schedule is provided by the Contractor. See also the Second Recital.

Appendix, Clause 30.3.1: If no off-site materials payment bond is required for 'uniquely identified' listed items, the standard entry should be deleted.
   If a bond is required the amount of the bond should be entered.

Appendix, Clause 30.3.2: If provision for payment for 'not uniquely identified' listed items off site does not apply, the standard entry should be deleted.
   If provision for payment does apply, a bond is required and the amount of the bond should be entered.

Appendix, Clause 30.4.1.1: An entry is only required if the Retention Percentage is less than 5 per cent. If a percentage higher than 5 per cent is required Clause 30.4.1.1 must be amended. See also comment above under 'Conditions': Clause 30.4.1.1.

Appendix, Clause 35.2: If the Contractor wishes to tender for work reserved for Nominated Contractors, a list of such work must be entered here. If necessary a separate list may be annexed but it must be identified at this entry.

Appendix, Clause 37: The alternative standard entries should be deleted as appropriate to identify which fluctuations clause applies. See comment above under 'Conditions Clauses 38, 39, 40'.

Appendix, Clauses 38.7 or 39.8: A percentage addition to amounts payable under Fluctuations Clauses 38 and 39 (i.e. recovery of actual costs) should be entered.

Appendix, Clause 40.1.1.1: The Base Month for the purposes of Rule 3 of the Formula Rules should be entered. See commentary on Clause 40 in Chapter 14.
   The alternative standard entry should be deleted as appropriate to identify whether formula adjustment is to be made under the Work Category Method (i.e. Part I) or the Work Group Method (Part II) of Section 2 of the Formula Rules. Note the method used should be the method stated in the tender documents.

Appendix, Clause 41A.2: The alternative standard entries should be deleted as appropriate to identify which Adjudicator nominating body is selected. If no selection is made here by the parties, the standard selection is stated to be the RIBA.
   If the parties wish to agree on another nominating body they are free to do so, but care should be taken to ensure the Adjudicator will comply with the provisions stated in Clause 41A.2.2 (i.e. to enter into the JCT Adjudication[71] Agreement).

Appendix, Clause 41B.1: The alternative standard entries should be deleted as appropriate to identify which Arbitrator nominating body is selected. If no selection is made here by the parties the standard selection is stated to be the RIBA.

---

71 Note the 'JCT Adjudication Agreement' is the agreement between the parties and the Adjudicator, not the rules by which the adjudication process is run.

Appendix, Clause 42.1.1: Space is provided to insert each item of Performance Specified Work to be carried out by the Contractor. The list should be checked against the Contract Bills to ensure the requirements of Clause 42.1.4 are met.

Footnote [zz] refers the parties to Practice Note 25 paras 2.6 to 2.8 for a description of work which is not to be treated as Performance Specified Work. Whilst the footnote and the Practice Note describe good practice, neither are terms of the Contract. It is, therefore, open to the parties to agree on any part of the Works to be treated as Performance Specified Work.

Annex 2 to the Conditions, Supplemental Provisions for EDI

The provisions refer to a separate Electronic Data Interchange Agreement which the parties must have entered into no later than the conclusion of this contract.

Clause 1.3 provides alternative references to such separate agreement to suit either the 'EDI Association Standard EDI Agreement' or the 'European Model EDI Agreement'. The alternative which is not applicable should be deleted.

1.5.3 Drawings

The Contract Drawings must be collected together and checked to ensure the drawing numbers correspond exactly with the numbers in the Third Recital. If, for example, drawing No. 04B is listed, then drawing No. 04B should be annexed, not No. 04A or 04C. This point is so obvious that it should not need to be made, but the error is common, often as a result of late changes. Another necessary check, which is a result of growing flexibility of computer-aided design and electronic communication, is to ensure that amendments to drawings are registered and 'frozen' on a drawing revision. The danger is that 'advance copy' drawings may be communicated freely between design disciplines, or designer and Quantity Surveyor, without control. It is not uncommon to find two or more copies of a drawing in circulation, bearing the same reference, but containing different information; which version is the basis of the contract can easily be forgotten.[72]

The correct drawings must all be copied and signed by the parties[73] to comply with the statement in the Third Recital.

1.5.4 Bills of Quantities

The Contract Bills should be checked to ensure:

- the Bills are arithmetically correct;
- the total price is the price transcribed into Article 2 (the Contract Sum);
- any drawing numbers listed as being the drawings on which the Bills have been prepared are the same numbers as those listed in the Third Recital;
- any work to be treated as Performance Specified Work is identified and either a provisional sum or information is included to comply with the provisions of Clause 42.1.4;

---

72 See Section 1.6.2, When things go wrong.
73 See Section 1.5.2, Attestation above, signing formalities.

- Appendix data given in the Bills do not conflict with the Contract Appendix (whilst the Contract Appendix will override the Bills, it should be remembered that in day-to-day use, the project team are more likely to refer to the Bills than the Contract; it is wise to avoid unnecessary potential dispute by removing the likely causes);
- if the Bills provide that certain work must be carried out by persons listed in the Bills, such a list must contain not less than three names to comply with Clause 19.3.2.1 of the Conditions;
- if the Employer wishes to pay for materials off site, that the list supplied to the Contractor showing uniquely identified and not uniquely identified materials in pursuance of Clause 30.3 of the Conditions is annexed to the Contract Bills.

The Contract Bills must be signed by the parties[74] to comply with the Third Recital.

1.5.5 Amending the Standard Contract Form

Generally

Whilst JCT 98 is published by an organization including representatives of various construction industry client bodies, individual building owners do not necessarily see the entire standard form as their own. Many will introduce amendments or additions to the form to bring it into line with their own commercial practices. The normal received wisdom is that the parties should avoid tinkering with a standard form. This is partly for fear of destroying the delicate balance of risk allocation; it is also to avoid legal uncertainty created by changing individual clauses which form part of a complex web of interacting cross-referred terms. However, reallocation of risk is a matter for the parties if that is what they are prepared to agree upon, and one can only advise caution on the manner of its reallocation. Amendment can sometimes have unexpected results, including prohibition by statute.

Amendment by introduction in the Bills and other documents

Employers (or their professional teams) often introduce changes to obligations through description in the Contract Bills or other descriptive schedules. A typical example is the tendency sometimes to incorporate design duty and liability in the absence of express provisions in the Conditions. It is argued by many contractors that the purported imposition of design duty and liability in the bills is an attempt to modify the Conditions, and that Clause 2.2.1 prevents it.[75] Clause 2.2.1 states 'Nothing contained in the Contract Bills shall override or modify the application of that which is contained in the Articles ... the Conditions, or the Appendix.' However, in the recent case of *Haulfryn Estate Co. Ltd v. Leonard J. Multon & Ptnrs and Frontwide Ltd*,[76] it was held that such a provision to place design obligations on the Contractor did not override or modify the interpretation of the Conditions; rather they 'added to but were consistent with the obligations imposed by the conditions.' In other words, design and build is an extension of build only, not a contradiction; the concepts are not mutually exclusive. The contract was the JCT Minor

---

74 See Section 1.5.2, Attestation above, signing formalities.
75 The authority used for this view is invariably *John Mowlem & Co. Ltd v. British Insulated Callenders Pension Trust Ltd* (1977) 3 Con. LR 63.
76 ORB; 4 April 1990, Case No. 87-H-2794.

Works form in which Clause 4.1 contains words very similar to Clause 2.2.1 of JCT 98. This decision will please those who view the Bills as a specific description of the work required on a particular project, and that should in any event have priority over the standard terms which are not project specific.[77]

Another example of an attempt to increase the Contractor's obligation can be found in *M. J. Gleeson (Contractors) Ltd v. London Borough of Hillingdon*.[78] In this case the Bills, by a hand-written amendment, contained obligations to complete various parts of the Works by specified dates; the Conditions (JCT 63) contained only a single Date for Completion, and in Clause 12(1) provided similar words to those in JCT 98 Clause 2.2.1. It was held that, by operation of Clause 12(1), the higher obligation as to time in the Bills did not override that in the conditions. Here the concept of one completion date did contradict the concept of several dates, so Clause 12(1) could be operated.

A cavalier approach to inserting special terms in the Bills should be avoided if success is to be certain; it is better to amend the Conditions, if amendments are necessary at all.

Amendments to the Conditions

It is beyond the scope of this book to consider the possible problems which may arise when the standard Conditions are amended by the parties, save to suggest that if amendments are necessary at all, they should always be done with professional guidance. There are three principal areas of risk. First is the danger of affecting multiple clauses where they interact through cross-references; only careful diagnosis and tracking will avoid ambiguity or confusion.[79] Next is interface with the common law. In *Peak Construction v. McKinney Foundations*[80] it was held that the contractor was entitled to payment for inflation up to practical completion, even though he may not be entitled to extension of time, where the printed text of the extensions clause had been amended.[81] It was also held that if the Employer obstructs the Contractor by his act or omission, and there is no corresponding ground in the contract to grant extension of time, the Contractor's obligation is to do no more than complete in a reasonable time in all the circumstances; the Employer then loses his right to recover liquidated damages. Clearly, deleting standard clauses, particularly those relating to extension of time, can be costly for the Employer.

The third area of risk is conflict with statute, particularly the Housing Grants Construction and Regeneration Act 1996, Part II, and the Unfair Contract Terms Act 1977, which are dealt with in turn below.

Amendments to the Conditions in breach of the 'Construction Act' 1996[82]

Generally. Amendment of standard form contracts may fall foul of the Housing Grants Construction and Regeneration Act 1996, Part II if provisions protected by the Act are

77 In the Scottish case of *Barry D. Trentham v. McNeil* (1995) GWD 26-1366, it was held that written amendments prevailed over Clause 2.2.1 of JCT 80.
78 23 April 1970; (1970) 215 EG 165.
79 For example of difficulty see Section 1.4.1 dealing with the Appendix and referring to *Bramall & Ogden v. Sheffield City Council*; the partial possession clause in the conditions conflicted with the appendix entry for liquidated damages, rendering the clause void and unenforceable and depriving the Employer of his rights to damages.
80 (1970) 1 BLR 114.
81 JCT 98 Clauses 38.4.8, 39.5.8 and 40.7.2, incorporate this principle (see Chapter 14: Section 14.5)
82 Housing Grants Construction and Regeneration Act 1996, Part II.

removed. The main purpose of the 'Construction Act' is to influence the culture of the construction industry, to ensure contracts incorporate minimum provisions with regard to payment and dispute resolution. It was enforcement of the Act from 1 May 1998 which prompted the JCT to produce Amendment 18 of JCT 80 in April 1998, although the JCT took the opportunity to introduce other changes at the same time.[83] Both JCT 80 amended by Amendment 18, and JCT 98 are fully compliant. Further amendment, albeit agreed between the parties, may contravene the requirements of the Act, triggering the application of statutory 'default provisions' contained in a statutory instrument generally known as the 'Scheme for Construction Contracts'.[84] The Scheme contains a Schedule in two parts. Part I provides rules for the selection of an adjudicator with powers to make enforceable decisions; Part I will apply where the contract contains no adjudication provisions, or where the provisions do not comply with the Act. Part II provides payment rules which will apply either where the contract fails to make provision at all, or where the parties have failed to agree on details such as timing of payments.

The Act requires that, unless one of the parties is a residential occupier, a contract in writing for a 'construction operation' must contain rules sufficient to enable the party receiving payment to know in advance when payments will become due, how each payment is calculated, and the latest date by which payment must be received. In addition, the remedy of suspending performance is provided for late or under payment, and the contract must expressly give either party the right to refer disputes to an adjudicator.[85] Further, a clause making payment conditional on payment from a third party is unenforceable unless the third party is insolvent. These limitations on the parties' agreements are dealt with in more detail in the following paragraphs.

Contracts in which the parties' amendments may not be effective. The scope of the Act is wide, applying to contracts for 'construction operations' which are defined in s.105. A lengthy list of operations is set out in s.105(1), including amongst other things, construction, maintenance and demolition of permanent and temporary buildings and structures, works ranging from power lines to railways, from heating installations to communications systems, from site clearance to dismantling scaffolding. In short, the Act applies to contracts for most of the activities for which JCT 98 would be used (except exclusions listed in s.105(2) – see following paragraph). If the contract is one to which the Act applies, any amendments which have the effect of avoiding compliance with the Act will cause the relevant default provisions of the Scheme for Construction Contracts to bind the parties. This may, in some cases, result in 'cherry picking' individual details from the Scheme (e.g. the length of the payment period where the contract does not provide one), or it may mean the complete replacement of a non-compliant adjudication clause with the adjudication provisions in the Scheme.

Contracts in which the parties' amendments may be effective. S.105(2) of the Act lists exclusions from the definition of 'construction operations'. Contracts for supply and delivery only are not affected automatically by the Act, and nor are contracts for

83 In Amendment 18, changes to comply with the 'Construction Act' are identified by annotation ' ♦ '
84 Scheme for Construction Contracts (England and Wales) Regulations 1998; Statutory Instrument 1998 No. 649.
85 See Chapter 17.

- drilling for, or extraction of, oil or natural gas;
- extraction of minerals;
- assembly, installation or demolition of plant or machinery, and its support or access steelwork, where the main activity of the site is nuclear processing, power generation, water/effluent treatment, manufacture or storage (other than warehousing) of chemicals, pharmaceuticals, oil, gas, steel, food or drink;
- wholly artistic works.

If a JCT 98 contract for one of the excluded operations is amended or added to, even though it is done in such a way as to contravene the substantive provisions of the Construction Act (i.e. s. 108 to s. 116), the amendment will be not be prohibited or replaced by the Act. This is because the contract is not a 'construction contract', by definition, for the purposes of the Act.

Mandatory contractual provisions where the Act applies: Payment. Unless the duration of the work is less than 45 days, there is an entitlement to interim stage or periodic payments (s.109). In any event the contract must provide an adequate mechanism for calculating the amount of a payment, the date when payment first becomes due, and the date by which payment is to be made (s.110(1)). Further, the contract must contain an obligation on the paying party to give a notice to the party receiving payment specifying the amount due, together with the basis on which it is calculated (s.110(2)); the notice must be given no later than 5 days after the date when the payment first becomes due.

If the party making payment wishes to withhold any part of the amount due he is required to give another notice no later than a time provided in the contract, specifying the amount or amounts to be withheld, and the ground for withholding each amount (s.111).

Amendment of Clause 30 of JCT 98 (dealing with payment) which removes any of these provisions from the Contract, will be replaced by the relevant corresponding provision in the Scheme for Construction Contracts, Schedule Part II, paras 1 to 10.

Mandatory contractual provisions where the Act applies: Conditional payment. 'Pay when paid' clauses are often associated with sub-contracts, but are not unknown in main contracts. Where an Employer relies on a third party for funding the project, he may wish to insert a clause making payment to the Contractor conditional on his own receipt of the necessary funds. Such a clause is unenforceable under the Act, unless the third party on whom the Employer relies is insolvent.

Mandatory contractual provisions where the Act applies: Suspension. The Act does not expressly require a construction contract to provide for suspension in the event of late payment, although JCT 98 contains such a provision at Clause 30.1. However, s.112 of the Act does give statutory rights to suspend performance of obligations where a payment due is not made in full by the final date for payment. Entitlement to suspend is conditional upon giving 7 days advance written notice of intention; entitlement ceases at the end of the day in which payment is made in full.

Amendment of Clause 30 of JCT 98 (dealing with payment), which removes any of these express provisions from the Contract, will still leave the Contractor with a statutory right.

Under the Act s.112(4), a party who suspends is also entitled to an extension of time for the period of the actual suspension, so deletion of Clause 25.4.18 (the relevant event of suspension) will not deprive the Contractor totally of his right to an extension. However, the statutory right to extension is not so wide as Clause 25.4.18 of the Contract; the Act provides only for the net period of the suspension, whereas the right in Clause 25.4.18 relates to delay arising from a suspension, which may incorporate time lost bringing back resources.

Under the Act there is no statutory right to recover loss and expense. It follows that if Clause 26.2.10 (i.e. entitlement to loss and expense as a result of suspension) were to be deleted, a Contractor delayed by his own rightful statutory suspension would need to demonstrate a breach of contract by the Employer, and claim damages at common law.

Mandatory contractual provisions where the Act applies: Adjudication. Adjudication is discussed in detail in Chapter 17. The Act s.108(1) gives express right to the parties to refer any dispute arising under the contract for adjudication, and sets out in s.108(2) to (4) the matters to be covered in an adjudication clause in the contract. These matters are covered in Chapter 17. However, the essential point here is that both s.108(5) of the Act and para. 2 of the Scheme state that adjudication provisions in Part I of the Schedule to the Scheme will apply where the requirements of s.108(1) to (4) are not met.

Amendments and additions in breach of the Unfair Contract Terms Act 1977

The imposition of liability on the other party may sometimes seem unfair, but the general position is that if two parties wish to make a bad bargain, the courts will not seek to make it good. The Unfair Contract Terms Act 1977 (UCTA) does not always help; the title is misleading. UCTA does not govern general fairness in contracts between two businesses, but it does control some exemption and limitation clauses which attempt to reduce liability for negligence and breach of contract. The Act applies to contracts where one party is a consumer, and to some contracts where both parties are businesses.

A clause attempting to exclude or limit liability for death or personal injury caused by negligence is unenforceable under s. 2(1). S. 2(2) makes any clause limiting any loss or damage caused by negligence unenforceable, except where such a clause can pass a test of reasonableness. Under s. 3, where a party is either a consumer or deals on the other's standard terms, the other cannot rely on a term which attempts to exclude or limit liability for his breach, unless the term 'satisfies the requirements of reasonableness'. The guidelines in UCTA for application of the 'reasonableness test' include strength of bargaining position,[86] knowledge of the relevant term, and the practicalities of complying with any conditions on which liability would rely. Under UCTA s.11(5) it is for the party alleging a term is reasonable to demonstrate that it passes the test.[87]

Putting UCTA into context, amendment of JCT 98 to exclude or limit liability for death or personal injury will always be unenforceable. Other amendments may be subject to the reasonableness test, if they are considered to be one party's own terms. JCT 98 is generally considered to be a consensus contract, since it is published by an organization made up from bodies representing both parties' interests. However, when amendments are

---

86 In *Chester Grosvenor Hotel Co. Ltd v. Alfred McAlpine Management Ltd* (1992) 56 BLR 115, McAlpine's own form limited liability, but it was held the parties had equal bargaining power so the term was not unreasonable.
87 In *AEG (UK) Ltd v. Logic Resource Ltd*; CA 20 October 1995; terms were held unreasonable since the plaintiff failed to plead its case in such a way as to adduce evidence showing that its terms were reasonable.

introduced, they may have the effect of converting the contract into the unilateral terms of one of the parties. Sometimes amendment will be by genuine agreement after negotiation, but it is not unusual for Employers unilaterally to impose new or changed terms. These may appear in a tender enquiry in which the Employer also states that qualification or amendment by the tenderer will disqualify the entire tender. The terms may then be the Employer's own, and their effectiveness will depend, amongst other things, on whether the parties have equal bargaining power. However, for the purposes of UCTA, the size of a company is not relevant to equality of bargaining power, since legal advice is available to both. Indeed it can be very difficult to establish inequality, since there is no obligation to enter into the contract at all,[88] unless there is already a legal commitment known to the other party and on which the other party then relies to impose its terms.[89]

UCTA provides some protection to the parties, but in practice the protection is limited to a very narrow band. In the context of JCT 98, the tendering procedure under which the contract is normally brought about means that the relevance of UCTA is more likely to be protection for the Contractor than for the Employer. However, a contractor who knowingly enters into a contract containing terms which he considers to be onerous does so at his peril.

1.5.6 Custody of Contract Documents

Clause 5.1 requires the Contract Drawings and the Contract Bills (but not the Agreement) to remain in the custody of the Architect or the Quantity Surveyor to be available for inspection by the Employer or the Contractor. This rarely happens in practice, and it is more likely that the Employer will retain the original Contract Documents intact, a copy of the full set being given to the Contractor if the Contractor requests it. Such practice has the advantage of keeping together all the original documents, whereas the procedure described in Clause 5.1 unwisely requires the Contract Documents to be split. Some Employers, usually when represented by lawyers, prepare two sets of Contract Documents for exchange, particularly when the Contract is executed as a deed.

Under Clause 5.2 the Contractor is provided with a copy of the Contract Documents at no charge, together with two further copies of both the Contract Drawings and the unpriced Contract Bills.

1.6 When things go wrong

1.6.1 Two versions of a drawing revision, the wrong version of which is signed

If this occurs the Contractor is both obliged and entitled to carry out the work on the signed, or annexed, drawing. The general position is that there is an error of intended scope

---

88 In *Denholm Fishselling Ltd v. Anderson*, 2 November 1990, 1991 SLT 24, it was held that the buying power of the parties was equal even though the buyer could not purchase on any other terms locally, since he was not obliged to buy from any individual seller.

89 For example, in *Northern Construction Ltd v. Gloge Heating and Plumbing* (1984) 6 DLR (4th) 450, a sub-contract tenderer was held to be in breach of a collateral contract when it revised its tender knowing the contractor was committed under a main contract won on the basis of the sub-contractor's original figures.

of work which both parties have accepted in signing the Contract, and which neither is entitled to have corrected unless the other is in agreement. The position might be different if one of the parties were aware of the error and took advantage of what he saw as a bargain. In that instance the remedy of rectification may be available.[90]

1.6.2 Contract Drawings listed do not correspond with Drawings signed

An example would be the listing of Drawing No. 123–27B, and the incorporation in the signed drawings of Drawing No. 123–27C (or even No. 123–27A). The difficulty is in identifying which drawing revision number represents the contractual obligation. Reference to the Third Recital seems to indicate the Contract Drawings are the drawings signed by the parties: 'the said drawings numbered ... (hereinafter referred to as 'the Contract Drawings') ... have been signed by'. Thus the drawing to be followed is the signed drawing rather than the number listed, which may be a drawing in existence and in the possession of the Contractor but which may have been discarded temporarily. This situation can occur easily when last minute changes are made, particularly when they have been made for budgetary purposes and are later reintroduced.

The difficulty in practice is that the Contract Documents are often separated.[91] Indeed the Contract at Clause 5.1 requires the Architect to keep the Contract Drawings available for inspection. Contractors often seem to forget their right, or are reluctant, to insist on viewing the original, particularly when they have already been provided with a copy of the drawings listed in the Contract Bills or the Recitals. Under Clause 5.2.1 the Contractor is provided with a certified copy of the Contract Documents; if the Contractor's copy is prepared and handed over at the time of signing, transcription errors are likely to be discovered at that point. The prudent contractor should have checked the Contract Documents thoroughly before signing in any event; but unfortunately the euphoric atmosphere on signing, together with the strain on resources in getting the job under way at the start of a new project, can easily push final checks to the back of the mind.[92]

1.6.3 Failure to complete the Contract Documents

The documents for a contract under JCT 98 require considerable effort to bring to a state ready for signing, leading to documents not being completed in time or errors in and dispute over the parity of the documents. Many building owners seek to buy time, perhaps while finalizing a design or completing contract details, or even while concluding financing arrangements; their method is by use of a device called a 'letter of intent'. The popularity and notoriety of letters of intent justify more than passing comment, so a section devoted to the topic is included at Section 1.8.

Sometimes the parties do not sign the contract at all, but a contract may still exist. JCT 98 contains a form of agreement and, unlike some standard forms,[93] does not refer to a

---

90 For full discussion on equitable remedy of Rectification see *Keating on Building Contracts*, 6th edn (Sweet & Maxwell, 1995), pp. 288–91 (hereafter *Keating*).
91 See Section 1.5.6, Custody of Contract Documents.
92 See Section 1.8.1, referring to cavalier approach to letters of intent.
93 E.g. Mechanical and Electrical Engineering Form MF/1, ICE *Conditions 6th Edition*.

letter of acceptance; but nevertheless an unequivocal acceptance by the building owner of a contractor's valid tender will create a binding agreement, i.e. a contract. The terms will be the terms of the tender which in most cases are likely to be the terms of the Employer's enquiry, but if the contractor has qualified his tender in any way, even if the bills, as they often do, state the tender must be unqualified, then the qualification will be binding.

An example can be seen in the engineering case of *Yorkshire Water Authority v. Sir Alfred McAlpine & Son (Northern) Ltd.*[94] The tender was based on the ICE *Conditions 5th edn* 1973, reprinted 1986, which provides for a programme to be submitted by the Contractor after entering into contract. A programme and method statement were attached to the tender. The tender was accepted by the Employer by letter, but no formal contract was signed. A dispute arose over the method of carrying out the works when site circumstances prevented the Contractor from following the method statement. The court held the method statement was incorporated in the contract and was the specified method of construction; the Contractor was entitled to have the change of working method treated as a variation.

There are occasions when work commences while the final details of the contract are still being negotiated, but agreement is reached at a later date. In those circumstances it has been held the agreement acts retrospectively and the terms bind both parties in their performance of the contract prior to their agreement. In *Trollope and Colls Ltd v. Atomic Power Construction Ltd*[95] work started in June 1959 but agreement was not reached until April 1960. It was suggested in the judgment in that case that the parties would have said as a matter of course, if they had been asked: 'This contract is to be treated as applying, not only to our future relations, but also to what has been done by us in the past since the date of the tender in the anticipation of the making of this contract.'

1.6.4 Failure to complete the formalities

There are three common situations in which one or both of the parties regularly fail to complete the formalities of signing a contract:

Orders or agreements 'subject to contract'. The words 'subject to contract' are usually used in relation to the sale of land, but occasionally the term is used in construction agreements to signify that a formal contract is to follow. It is a term normally used to give comfort, without committing to being contractually bound. However, there are a few instances where all the terms are agreed and the parties commence performance. In those circumstances there can be a binding contract based on the agreement as it stands, irrespective of whether the formal document has been completed. In *Lewis v. Brass*[96] an architect wrote to a tenderer stating that he was instructed to accept the tender, and that the contract would be drawn up by his client's solicitors. The contractor tried to withdraw his tender when he realized it contained a mistake, but the Court of Appeal held his tender and the architect's acceptance had concluded a contract.

---

94 (1986) 32 BLR 114.
95 (1962) All ER 1035.
96 CA; (1877) 3 QBD 667.

Failure to sign the proposed contract agreement after starting work. When work starts before the Contract is signed, the incentive to sign may wane, and this occurs regularly in association with letters of intent.[97] This is particularly so when problems in performing proposed obligations provide one of the parties with hindsight as to the risks under the contract. However, failure to sign does not necessarily mean that no contract exists. If the essential matters are agreed, or if negotiations cease without clear dissent on some point, then a contract may be concluded if work is started or continued. An example can be seen in *Birse Construction Ltd v. St David Ltd*[98] in which the contractor received documents referring to the standard form JCT 80 after work had started. Negotiation on final matters including programme came to an end at a meeting a week later, but the documents were never signed and returned. It was held that a contract had been concluded both by offer and acceptance, and by conduct:[99]

> The offer was accepted by the plaintiff's confirmation that the dates were feasible ... at the meeting ... Even if there had been no such agreement ... then alternatively I conclude that there was acceptance of (various matters) by conduct by one party or the other in either permitting or continuing the execution of work.

It is often a mixture of fact and law that determines whether the parties have reached agreement, and the judge in the *Birse* case saw the legal analysis as artificial; he recognized that there was no distinct offer and acceptance, but that the parties were building their contract like the pieces of a jigsaw, the last piece of which had been put in place. The continuance of work without dissent was also crucial: 'given that work was continuing on a basis apparently acceptable to both parties, it would require clear evidence to displace the inference that it was not being carried out pursuant to some contract.'

If work is being done (or allowed to be done) while negotiations continue, an express statement that it is in the absence of a contract is necessary if either party wishes to be certain of avoiding agreement.[100]

Failure to complete the formalities of a deed. Sometimes contracts signed as a deed will be completed by the parties together, but it is not unusual for one party to sign, then return the document to the other who does not get around to signing. In those circumstances a confusing situation can develop. The party signing as a deed will be bound to his promises under the deed, whereas the other party who did not sign will only be bound if a simple contract has been created in place of the deed. The main effect of a deed is to bar actions after 12 years, whereas the limitation period on legal actions on simple contracts is 6 years (see Section 1.4.1, Attestation, on simple contracts and 'specialties'). Where only one party has completed a deed, only that party will be exposed to action for its breaches for the longer 12-year period. The other party will be exposed under its simple contract for 6 years.

1.6.5 Discrepancies

A fruitful area for misunderstanding and dispute is in ambiguity or conflicting requirements built into and between the various parts of the contract documentation.

---

97 See Section 1.8.1.
98 [1999] BLR 194.
99 *Per* Lloyd HH J, [1999] 194 at 202.
100 See also Section 1.7, with reference to *Con Kallergis Pty v. Calshonie Pty Ltd* for another example on this point.

Simple examples would be, say, a specification requiring carpets throughout whilst a drawing showed floor tiles in toilet areas; or drawings indicating use of engineering bricks where the bills describe facings. The issue is seeded during the preparation of the contract documents, and it normally emerges in the context of Variations, triggered by the Contractor asking for extra payment or resisting a credit Variation. Accordingly, treatment of discrepancies and inconsistency is dealt with earlier in this Chapter at Section 1.5.5 dealing with Amendment by introduction in the Bills, and Chapter 6, Variations and provisional sums.[101]

1.7 Forming a contract and the 'Battle of Forms'

In English law the first requisite of a contract is that the parties should have reached agreement;[102] an agreement is made either (a) when a statement of agreement is signed, as in the case of signing the JCT 98 Articles of Agreement, or (b) when one party makes an unambiguous offer capable of being accepted, and the other accepts it unequivocally. In the context of JCT 98, the contractor's tender is usually an offer even if it does not comply with the enquiry, just as long as it fulfils the legal requirements of a valid offer.[103] If the building owner were to say to the tendering contractor 'I accept your tender', a contract would be formed on the basis of the tender as it stood. However if, instead, the owner were to respond 'I accept your tender on the basis that the completion date as stated in the enquiry document applies – not as amended in your tender', that would not be an unequivocal acceptance and no contract would be formed. That is not to suggest the owner's response can be ignored – it may signal the use of a common practice deriving from the desire of businessmen to contract on their own terms.

A qualified acceptance of an offer in law is not acceptance at all, and no contract is concluded; but provided the response fulfils the criteria for an offer, it may be construed as a counter-offer which would in turn kill the previous offer and be open for acceptance by the other party. The procedure has become known as the 'battle of forms'. Whether or not a contract is formed will depend on the facts and the courts may apply the rules of contract construction in order to decide whether one party has accepted the other's terms by words or conduct. The difficulty which arises is that the counter-offer may itself be countered by a further counter-offer reverting to the terms of the original offer or introducing new terms. Such a situation was examined by the Court of Appeal in the case of *Butler Machine Tool Co Ltd v. Ex-Cell-O Corporation (England) Ltd*.[104] The dispute involved the sale of machine tools quoted on the seller's terms, which included a price variation clause. The buyer purported to accept the quotation by a standard purchase order incorporating its own standard conditions which stated that the price was on a fixed basis. The buyer's order contained a tear-off acknowledgement slip stating 'we accept your order on the terms and conditions stated thereon', which the seller duly signed and returned with a covering letter. The seller's covering letter said that the official order was being entered

---

101 Section 6.14.
102 See *Courtney & Fairburn Ltd v. Tolaini Bros (Hotels) Ltd*: CA; (1975) 2 BLR 97.
103 Described in *Hudson's* (11th edn, 1994), as 'something which invites, and is intended by the offerer to invite, acceptance, and must be sufficiently definite to be capable of resulting in a contract if accepted.'
104 (1979) 1 All ER 965.

in accordance with the seller's quotation. The machine was then delivered and a price increase was claimed in accordance with the variation clause. A dispute arose over whose terms formed the basis of the contract. In the Court of Appeal it was held that the seller's covering letter, although referring to the initial quotation, did so merely to identify the subject matter. Lord Denning said:

> In some cases the battle (of forms) is won by the man who fires the last shot. He is the man who puts forward the latest terms and conditions: and, if they are not objected to by the other party, he may be taken to have agreed to them.

Clearly, 'shots' can only include those responses capable of being construed as offers; in this case the seller's covering letter was decided by the court to be something short of a counter-offer, thus leaving the acknowledgement of the buyer's counter-offer as acceptance of the buyer's terms.

The *Butler Machine Tool* case involved relatively simple facts compared with situations arising on building projects. The exchange of communication was short and the obligation was fulfilled by one delivery. In building contract negotiations the communication can extend to many letters and meetings over a period of many months leading to numerous changes to documents which become annexed to the proposed contract conditions.

The battle of forms is used by many to get a foot in the door, backed by a general philosophy that the exchange of standard terms or tender qualifications prevents acceptance of the other party's terms, but without real expectation that either's terms will be agreed. The contractor is then relying on receiving some sort of fair payment,[105] in the belief that it will be better than signing a contract with all its obligations. In short, the 'last shot' doctrine becomes a powerful tactic in the risky game of contract avoidance. Unfortunately for those employing the tactic there is a rule of contract construction which may prove fatal if ignored; that is that a party can impliedly accept the other's terms by its conduct. Application of the rule is not without its difficulties.

If, having responded to an offer by issuing a counter-offer, a contractor immediately commences work, then clearly such conduct cannot be construed as acceptance since a contractor cannot accept his own offer. He starts at risk. Nevertheless the conduct of the person receiving the offer, in allowing the work to continue without dissent, may be construed as acceptance, but in the short intervening period more correspondence may have passed changing the terms yet again. In the building industry such situations occur frequently, exacerbated by the desire to commence preliminary work to meet building or fiscal deadlines, and cushioned by the use of letters of intent (some of which may be construed as further counter-offers).

The risks of acting hastily are clear; so too are the risks of delaying. Adoption of the 'last shot' doctrine requires fine timing. In the event of a dispute the relationships in contract will depend entirely on the sequence of 'shots' and how it relates to the point at which one of the parties conducts himself in a manner signifying his acceptance of the other's terms. One example can be seen in the case of *Chichester Joinery Ltd v. John Mowlem and Co plc*.[106] Chichester's quotation for joinery was submitted under cover of a letter which referred to standard conditions printed on the reverse. Some months later Mowlem sent out to Chichester and others an invitation to tender referring to printed

---

105   See Section 1.8.2 dealing with *quantum meruit* in the absence of a contract.
106   (1987) 23 Con. LR 30.

conditions. Following a telephone conversation during which further terms were agreed, Chichester confirmed they would do the work in accordance with their original quotation and the oral agreement. Mowlem then sent an order referring to its own terms (different from those in their own invitation). It was held that Mowlem's order and Chichester's subsequent acknowledgement containing their earlier terms were each counter-offers killing the other side's previous offers. The judge went on to reject the suggestion that Chichester accepted Mowlem's terms by their conduct when they started preparing for manufacture, deciding the evidence fell far short of that necessary to establish acceptance by conduct. On considering the later conduct of Mowlem in accepting delivery of the joinery on site, he found it constituted acceptance of the terms in Chichester's counter-offer (the acknowledgement of order). The alternative would have been that no contract existed at all.

The significance of this case is cited purely as an example. It is not new law, but the application of established principles in the analysis of situations, with the possibility of unexpected results. The risk for the parties was recognized in the judgment:

> This case highlights the risk facing parties who seek to impose their own respective conditions rather than some well established form of contract commonly used. ... Unless such a form of contract is used, the courts may be faced (as here) with what is essentially an artificial state of affairs.

The artificial state of affairs referred to is the urging by legal advisers of their clients to respond to each document from the other party with an endorsement or annexation of their own terms. In other words, the judge was openly criticizing the promotion of the 'battle of forms'.

For those engaged in the battle, however, it is not so simple. The insertion of such matters as completion dates, which a tenderer knows during negotiations cannot be met, can be resisted only with counter-offers or express rejection. The former is seen by many as an opportunity to secure work, albeit without agreeing terms, but in the hope that the other side will eventually agree either expressly or by conduct. Express rejection carries with it the risk of the other party taking its custom elsewhere, so rejection is frequently withheld until sufficient work is done to make it too late to involve another contractor without the owner incurring significant extra cost. An example can be found in *Kitsons Insulation Contractors Ltd v. Balfour Beatty Buildings Ltd*.[107] In that case the contractor sent a Standard Form Contract DOM/2[108] to the sub-contractor, inviting him to sign the document and return it. Certain amendments which the sub-contractor then saw for the first time were not acceptable to him. Work had already started but acceptance of the document would have retrospective effect. The sub-contractor wrote to the contractor refusing to sign the contract since it did not contain a number of terms from their offer concerning payment and further contained a scope of work different from that in the offer. The main issue in the case concerned an early letter of intent and whether or not it gave rise to a contract. That letter was potentially one of the 'shots' in a series of offers and counter-offers which culminated in the issue of formal documents. It was held there was

107 (1989) Unreported; ORB 1461; 17 January 1991.
108 The standard form of domestic sub-contract published by the Construction Confederation (formerly Building Employer's Confederation) for use where the main contract is the JCT Standard Form 'With Contractor's Design'.

no contract; the parties had never reached agreement. Where there is no contract as a result of a battle of forms ending in stalemate, the contractor's entitlement to payment is governed by the same principles as those applying where letters of intent are used.[109]

A similar result can be seen in *Midland Veneers Ltd v. Unilock HCP Ltd.*[110] in which negotiations, a letter of intent, an order and an acknowledgement of order all came to nought. Negotiations between the two companies for the supply of veneered heater fittings resulted in a letter of intent. It was a pure letter of intent stating no more than an intention and confirming costs incurred would be met if no order was placed. Some goods were then supplied. A month after the letter Unilock placed an order referring to the terms of the main contract, Unilock's own terms on the back, and other matters including liquidated damages and a guarantee. Midland Veneers responded immediately stating the order had been passed to the production unit for processing and they would sent an Order Acceptance shortly. Further letters were exchanged ending with a letter from Unilock confirming an agreement on the damages and guarantee, and also that the rest of their order was acceptable to Midland Veneers. On receipt of this letter Midland Veneers made the remaining deliveries which were accepted. Disputes developed over delivery dates and the court was asked to rule on the basis of the contract, if any. It was held no contract was concluded, and Unilock appealed. The Court of Appeal could find no evidence of agreement; it held that the acknowledgement letter from Midland Veneers was no more than that, and it was clear that an acceptance was intended to follow from another department in the future. Thus the letter itself could not be an acceptance of the order. It was also held that the conduct of the parties in delivering and accepting the goods did not signify agreement; indeed the only thing agreed was a price. In short, there was no contract. The amount to be paid was not at issue in this case because Unilock had always maintained there was a contract based on an agreed price. The dispute was over the liquidated damages, which in the absence of a contract could not be deducted. The court was only asked to decide whether or not a contract existed.

In the Unilock case neither express nor implied agreement could be found, but nor was there any express rejection of the other's terms. The possible effect of the absence of express rejection can be seen in the Australian case of *Con Kallergis Pty Ltd v. Calshonie Pty Ltd.*[111] A supplier quoted to an electrical contractor for luminaires, referring to the contractor's drawing. The contractor sent an order attaching a different drawing and requesting samples. The supplier responded with a document describing itself as 'cost variation subsequent to our original offer' and sent samples. The contractor then issued a Variation, stating it was to be valued by negotiation based on information to be provided by the supplier. The supplier claimed its 'cost variation' was a counter-offer which had been accepted when the contractor accepted the samples. The court looked at the conduct of the parties, and held it showed they had reached a concluded agreement in the terms of the contractor's order. A letter from the supplier signifying the contractor's order was not accepted would probably have protected the supplier against a contract being formed on the other's terms.[112]

---

109 See Section 1.8.2 dealing with *quantum meruit*.
110 CA; 12 March 1998, QBENF 96/1652 CMSI; Unreported; [1998] 15 BLISS 7.
111 25 March 1997; (1998) 14 BCL 201–214; cited [1998] 28 BLISS 6.
112 See p. 42 dealing with the *Kitsons* case.

Again, the issue of an order was the subject of dispute in the American case of *ICC Protective Coatings v. A.E. Staley Manufacturing Co.*[113] In response to an invitation from Staley to quote for protective coatings to plant, ICC submitted a proposal. The proposal contained a price and terms dealing with correction of defects. Staley issued a purchase order referring to their own terms on the back, and stating that any additional or different terms proposed by ICC were rejected unless expressly accepted in writing. A dispute arose when Staley did not give ICC the opportunity to put right their own work. The court held that ICC had accepted Staley's terms without question, and they were the basis of the contract, to the exclusion of ICC's terms.

Timing is critical in the battle of forms. For example, in the *Kitsons* case the letter rejecting the sub-contract documents is central to the conclusion that there was no contract. If the timing of commencing substantive work had been different, the result of that case could have been reversed. If the sub-contractor had not started work so early, but instead had waited until after receiving the sub-contract document, and then had started before commenting on the amendments, the court may have construed a contract, as it did in the *Con Kallergis* case. Whether or not that is the desired result will depend on the reason for engaging in the battle; for both Kitsons and Midland Veneers it could have been fatal, since one of their main objectives in bringing the action was to avoid contractual obligations as to time.

1.8 Letters of intent and the right to payment

1.8.1 Letters of intent: meaning and effect

Strictly, letters of intent do not feature in the formation of contracts using the JCT forms, but their use is so common that they deserve a section to themselves.

Parties who are negotiating or are about to negotiate a contract will often start work towards fulfilling the obligations of that contract before all the terms of contract are resolved or a contract is signed. They do so usually in the knowledge and with the intention that the contract, when it is signed, will act retrospectively. In other words all their actions when performing within the scope of the envisaged contract will become actions carried out under and subject to the terms of the actual contract.[114] In these circumstances one of the parties will commit resources and expenditure before the other; since it is unlikely that payment will be made before any reciprocal performance the first to commit himself will be usually the supplier. As there is no contract the purchaser gains a distinct advantage – receipt of work that he wants to be done, with no clear express contractual obligations as to either the amount or timing of payment for it, whilst at the same time being able to avoid the pressure of negotiating unwanted terms in a tight timescale. In short, the purchaser gains time. Such time enables him to prepare lengthy documentation without delaying commencement of the work and to hold out in negotiations for his most favourable terms. The supplier, too, is given time to negotiate, but has the distinct disadvantage of having committed himself to cost; for him the

---

113 695 N.E.2d 1030 (Ind. App. 1998); *Construction Claims Monthly* 20 (9) September 1998, 2.
114 *Trollope & Colls Ltd v. Atomic Power Constructions Ltd* (1963) 1 WLR 33.

attraction of the situation is the start of work against an anticipated order which he hopes will become more certain with the passage of time.

A device often used to create such situations is the 'letter of intent'. A principal may send a letter stating that he intends to enter into a contract with the recipient. In its purest form that is all a letter of intent comprises. When the recipient knows that he is one of a number of tenderers in competition, the letter is taken often as a sign that the uncertainty is over, the contract is secured and that it is simply a question of completing formalities; the choice of contractor has been made and the work can get under way. If a contract is subsequently concluded, then the common intention has been fulfilled and clearly any dispute which arises will be a dispute over rights and obligations under the contract. The difficulty arises where the formalities are not concluded, but the parties have proceeded as though they were and a dispute then arises. Often the dispute will concern the rights or obligations of one of the parties under the terms of the contract that was envisaged. Whether or not that contract ever came into existence becomes a critical issue and represents a significant cause for concern when letters of intent form the basis of a relationship.

The traditional view of the English courts is to treat letters of intent as a means of postponing legal liability.[115] It is a fundamental requirement in English law that for a contract to be concluded there must be *consensus ad idem* (Lat. agreement to the same thing); the parties must have reached agreement and negotiations on all but minor terms must be finished.[116] A statement from one party to another that there is an intention to enter into a contract is a statement that the contract is not yet concluded; it is a thing for the future.[117]

Even more speculative is the letter which states 'it is our intention to place an order'. Since an order is often no more than an offer[118] open for acceptance by the recipient, there is no relationship whatsoever created by a mere statement of intention to make an offer. Accordingly there is no legally binding obligation on either party in so far as such obligations are seen as fulfilling terms of the envisaged contract.

Whilst this is the position when a letter of intent is in its simple and pure form, complications arise in those instances where the letter contains more information or instructions. There are comparatively few reported cases concerning letters of intent, and those which find their way into the law reports generally do so because of the uncertainty created by statements other than of simple intention to enter into a contract or place an order. The issuing party will often refer to what the terms will be with the possible effect of turning what purports to be a letter of intent into an offer. Whether this effect is deliberate or not is immaterial. If the letter contains the constituent elements of an offer capable of unequivocal acceptance by the recipient, then it may be construed as an offer.[119] The statement of intention does not cut across this concept. Indeed, a statement of

---

115 See Ball S.N., Work Carried Out in Pursuance of Letters of Intent – Contract or Restitution? (1983) 99 LQR, 572.

116 *Per* Lord Denning MR, *Courtney & Fairburn Ltd v. Tolaini Bros (Hotels) Ltd* (1975), 2 BLR 97 at 101–102.

117 *Peter Lind v. Mersey Docks and Harbour Brd* (1972) 2 Lloyd's Rep. 234.

118 *Butler Machine Tool Co v. Ex-Cell-O Corporation (England) Ltd* (1979) 1 All ER 965.

119 *Per* Bigham J, *Crowshaw v. Pritchard and Renwick* (1899), 16 TLR 45; cited by Powell-Smith, V. and Furmston, M.P., *A Building Contract Casebook*, 2nd edn, (BSP Professional Books, Blackwell Scientific Publications Ltd, 1990) p.17.

intention to enter into a contract contained in an offer, merely serves to reinforce the presumed intention to create legal relations.[120]

The significance of the interpretation of letters of intent can be seen in two recent cases. In the *Kitsons Case*[121] the court was required to decide whether a letter of intent formed the basis of a contract. The letter sent by Balfour Beatty to Kitsons stated an intention to enter into a contract using a standard form sub-contract DOM/2 with amendments to be forwarded in due course; the letter also requested Kitsons to accept the letter as authority to commence the sub-contract works. The formal contract was drawn up some months later, but it was never signed and certain terms were rejected by Kitsons. Meanwhile work continued and Balfour Beatty made payment consistent with the payment rules of the standard DOM/2 terms. Notwithstanding the actions of the parties, when a dispute arose the court had to decide a preliminary point as to whether the letter of intent created a contract. Regarding intention, it was said:[122] 'I am satisfied that Balfour Beatty were not prepared to enter into a concluded sub-contract with Kitsons on 23 March 1988, although they were anxious to get the work started'.

On the question of a contract being concluded later the evidence showed that the documentation differed from that referred to in the letter and was rejected by Kitsons. The judge consequently found that the rejection precluded any possibility of a contract: 'it remained one of the matters upon which the parties never agreed'. In this case the letter of intent gave rise to no contractual relationship whatsoever, and moreover had by its terms fixed Kitson's expectations which resulted in their late rejection of the formal documents.

In the comparable case of *C. J. Sims Ltd v. Shaftesbury plc*,[123] Sims had made an offer based, as requested, on standard JCT 80 terms. In response Sims were sent a letter of intent and were asked to start work immediately. The letter provided for payment of reasonable costs in the event that the contract did not proceed, but also that such costs were to be 'substantiated in full to the reasonable satisfaction of [the defendants] quantity surveyor.' Work proceeded while precise terms were negotiated. No agreement was reached and a dispute arose over entitlement to payment. It was agreed by the parties that a contract existed although they also agreed that the terms were to be found strictly in the letter of intent. The court found that the letter created a contract and that the requirement for costs to be substantiated to the satisfaction of the quantity surveyor was a condition precedent to the entitlement to any payment.

In contrast with the *Kitsons* case, the court in the *Sims* case found a relationship in contract based on the wording of the letter of intent together with the subsequent actions of the parties in reliance on that wording. Clearly, the words of a letter of intent are critical if the nature of a relationship is to be certain before commitments are made.

Letters of intent and standard form contracts

In both the *Kitsons* and the *Sims* cases reference had been made to standard form contracts. In both cases the result turned on the content of the letter of intent. It must therefore be questioned what would have been the result of either case if the terms contained in the

---

120 In the case of a business there is a presumption that there is an intention to create a legal relationship when it makes an offer; when a private individual makes an offer there is a presumption that there is no such intention. In either case the presumption is reversed by an express intention.
121 Unreported, Judgment 17 January 1991.
122 *Per* his Honour Judge Thayne Forbes, Judgment 17 January 1991, at 9, para. 3.
123 (1991) 25 Con. LR 72.

letter had been precise, clear and complete. The mere fact of the letter being a pre-contractual document does not of itself prevent a contract existing. In *Charles Church Developments Ltd v. Pacific Western Oil Corporation*,[124] Lord Templeman stated: 'A preliminary arrangement which contemplates a formal contract may itself constitute a binding agreement.'

Likewise if a written communication, albeit headed as a letter of intent or containing words to that effect, contains all the necessary elements required to constitute a valid offer, then it may be construed as an offer irrespective of its title. The consequent risk for issuer and recipient alike is that a letter of intent is seldom clear but is nevertheless capable in some instances of being an offer, acceptance of which, including implied acceptance by conduct,[125] will be binding.

This point raises yet another aspect of risk. Where standard forms are contemplated an architect, engineer, or project manager with apparent authority will be involved in the pre-contractual stages. It is possible therefore that a letter of intent, if sent to a contractor will be issued by the architect or other professional. A salutary example can be seen in the recent unreported case of *A. Monk Building and Civil Engineering Ltd v. Norwich Union Life Insurance Society*.[126] The judge was required to consider various preliminary issues including the status of a letter of intent and the authority of the issuer to make a binding contract. He decided on the facts that although the project managers had been given express authority to send the letter of intent, notwithstanding its content they had no authority to commit their client, Norwich Union, to a binding agreement. A small unpublished survey[127] of features occurring in letters of intent identified that in six out of sixteen letters examined, the letter had been issued by a person other than one of the parties. So even if the letter gives every appearance of being a valid offer capable of acceptance, the question of authority to bind a building owner may reverse an otherwise clear situation long after the parties have committed themselves to commercial courses of action.

The letter of intent is then a useful instrument to get work under way while documents are prepared, albeit there are risks attached; this is one of the classic purposes of such letters. Reciprocally, letters of intent can be made more clear and their issue may be facilitated by referring to what the terms will be. This creates greater confidence in what might otherwise be no more than a bare promise to negotiate. The well publicized nature of standard forms such as JCT 98 assists incorporation into any contract by mere reference to them; indeed in *Killby & Gayford Ltd v. Selincourt Ltd*[128] the Court of Appeal held the words in a letter 'subject to the normal standard form of RIBA contract' to be sufficient to incorporate the then current JCT terms into the contract. Consequently, a recipient of a letter of intent referring to industry standard terms is given illusory comfort first by the stated intention, and second by the passing reference to a standard form contract; this may be particularly so when such reference is made by a respected professional, albeit he has no authority. Not wishing to appear mistrustful of a potential customer the recipient is consequently induced into commencing preparatory work, design or even installation.

---

124 (1980) Unreported, Case No. 914 (Transcript: Association), 8 July 1980.
125 See Section 1.7, Forming a contract and the 'Battle of Forms'.
126 62 BLR 107.
127 Survey of letters of intent conducted by Rycroft, M. January 1992.
128 (1973), 3 BLR 104.

Thus the letter enables matters to progress without the recipient having any sight at all of the actual contract terms. This factor is significant for those not familiar with the standard terms, for unless the recipient of the letter has a copy to hand, a set of both standard main contract and sub-contract must be obtained. This takes time and creates a remoteness from the issue; accordingly there is a temptation to wait until the actual documents arrive from the principal. Such action may not be legally prudent, but legal issues are not to the fore in the minds of many who receive letters of intent. It is comfort that is sought and it is comfort that is gained. In this, letters of intent appear to succeed in their primary function to prevent delays in commercial transactions.

From the principal's point of view some of those delays are unavoidable, but the letter of intent is a facility which can by itself promote delay. It eases the pressure, and there seems little doubt that once a letter of intent (particularly one referring to standard terms) has been acted upon, both the incentive and enthusiasm to pour resources into preparing and concluding a formal contract must surely wane. The result is a period of uncertainty; but often the effects of the uncertainty are appreciated only after it is too late to remedy the situation and a dispute has arisen.

Much depends on the circumstances of any ensuing dispute and the entitlement to be paid. An example can be seen in the High Court decision in *British Steel Corporation v. Cleveland Bridge and Engineering Co Ltd*.[129] The defendant Cleveland Bridge sent British Steel a letter stating an intention to enter into a sub-contract for the supply and delivery of castings. The price was stated and authority was given to commence work pending preparation of sub-contract documents. Before the completion of documents a schedule of delivery dates was given to British Steel and all but one of the deliveries were made. Disputes arose over quality and delivery dates, Cleveland claiming that deliveries had been made late compared with the schedule. The court decided that no contract existed but that the content of the letter gave rise to an entitlement to be paid a *quantum meruit* (see below). The decision was significant for British Steel, for had it been decided that a contract was concluded, then the terms may well have included the delivery schedule, and a counterclaim from the defendant for faulty performance might have succeeded.

Similarly in the earlier unreported case of *Hammond and Champness Ltd v. Hawkins Construction (Southern) Ltd*,[130] it was argued by the defendant contractor that if the court found there was no contract and therefore no completion date, then at least the sub-contractor's work should be carried out in a reasonable time; there was an implied obligation to that effect. The court disagreed:

> if a party does work not pursuant to an agreement but at the request of another party he is free to stop work ... when he wishes and that is a consideration which would militate against there being any obligation to do a specified body of work or to do it within a reasonable time, or any other time.

It is clearly an attractive proposition for any contractor to be in a position where the difficult requirements of a contract can be avoided, but payment can still be claimed. The attraction is greater still if the requirement which can be avoided is related to progress and completion. After all, that is the area where contractors are most at risk of having contractual counterclaims or damages levied against them, and where they have the

---

129 (1981) 24 BLR 94; (1984) 1 All ER 504.
130 Unreported, Judgment 24 April 1972; *Construction Law Yearbook* (Sweet & Maxwell, 1995), pp. 181–88.

greatest difficulty in complying with strict procedural requirements of standard form contracts. It follows that if the potential confusion and extreme administrative requirements of forms like JCT 98 can be avoided and payment can be received for work done simply by working in reliance on a letter of intent, it may be wondered why a contractor should ever wish to enter into a contract, standard or otherwise. One reason is the question of payment.

It can be seen from the *British Steel* and *Hammond and Champness* cases that the recipient of a letter of intent may be entitled to payment in the absence of a contract. However, even though there may be valid authority to carry out work, there is some uncertainty as to the amount of payment that may be due. It is the absence of an agreed price and the lack of any rules to determine one that many recipients of letters of intent do not recognize or simply choose to disregard. Yet many contractors accept letters of intent almost without question. Perhaps the rationale behind such an apparently cavalier approach is summed up in the suggestion:[131] 'Businessmen compare a letter of intent with the alternative, which is to provide no guarantee at all and therefore see it as better than nothing and significant in forming a limited relationship.'

1.8.2 Letters of intent: right to payment

Whenever a contractor, in order to keep to a tight programme, starts work before a contract is concluded, he does so in anticipation of being paid eventually under that contract. A problem arises when the contract has not come into existence but the contractor having carried out work wishes to realize his expectations; often he will claim the price he would have recovered under the contract. Likewise the building owner will often act as though the contract existed and make payments in accordance with the proposed terms. The difficulties start when one of the parties considers the other is not fulfilling all its obligations. A typical situation would be where the programme of work is not being maintained by a contractor and as a consequence the building owner suffers losses which he then attempts to set off against payments being made. Since the set-off is in respect of what would be an alleged breach of contract, it is a convenient defence for the contractor to revert to the true position and argue that since there is no contract there can be no breach; consequently there can be no entitlement to set-off damages. The question which then arises, if there is no contract, is whether there is an entitlement to be paid anything for the work done, and if so, how much.

The position in these circumstances was considered in *William Lacey (Hounslow) v. Davis*:[132]

> In neither case was the work to be done gratuitously, and in both cases the party from whom payment was sought requested the work and obtained the benefit of it. In neither case did the parties actually intend to pay for the work otherwise than under the supposed contract. .... When the beliefs of the parties were falsified, the law implied an obligation – to pay a reasonable price for the services which had been obtained.

Similarly in *British Steel v. Cleveland Bridge*[133] it was said:

---

131 See n. 115.
132 (1957) 1 WLR 932; *per* Barry J.
133 (1981) 25 BLR 94.

Both parties confidently expected a formal contract to eventuate ... one then requested the other to expedite the contract work, and the other complied with the request ... if, contrary to their expectation, no contract was entered into, then the performance of the work is not referable to any contract ... and the law simply imposes an obligation on the party who made the request ... such an obligation sounding in quasi-contract or, as we now say, in restitution.

It should be noted at this point that whilst most contractors probably think subjectively in terms of an 'entitlement to receive' in the absence of a contract, the courts clearly place emphasis on an 'obligation to give' by the recipient of the service. The two concepts do not necessarily produce the same results, but it seems something becomes due and the rationale is to be found in what is variously referred to as *quasi-contract* or 'restitution'.

Quasi-contract or restitution?

Historically the basis of liability and entitlement in arrangements of a contractual nature when no actual contract exists has been the subject of two main theories; i.e. that there is some sort of implied contract, alternatively that a person should not be unjustly enriched at the expense of another. The former theory, endorsed by the idea that a contract was necessary to enable a claim to be classified,[134] gives rise to what is known as a 'quasi-contractual' claim and covers various circumstances such as work done under a void or illegal contract, or when the contract had been terminated for breach, but excludes work done voluntarily. The particular area relevant to work on a letter of intent, however, may be seen as an exception to the general rule that work voluntarily done does not confer a benefit for which payment must be made. The exception occurs where there is an express or implied request for services to be rendered. This will include the situation where the performance of services is requested in anticipation of a contract which does not materialize, such as a request or authority in a letter of intent. The second theory falls into the area known as 'restitution'. Restitutionary remedies may be applied throughout all areas of law and the nature of restitution has been defined as[135] 'the law relating to all claims, quasi- contractual or otherwise, which are founded on the principle of unjust enrichment'. The idea that a person ought not to gain at another's expense is broadly in accord with the general views of many in the building industry who find themselves in the position of working without a contract. It is seen by many as being 'fair'. This is not so far from the rationale behind the concept. In *Moses v. Macferlane*,[136] Lord Mansfield said:

The gist of this kind of action is, that the defendant under the circumstances of the case, is obliged by the ties of natural justice and equity to refund the money.

Unfortunately for many contractors, they are prone to apply their concept of natural justice and equity to their own actions of providing a service and a consequent obligation to pay, omitting to take into account the circumstances of the case. The omission is crucial. Restitution is about refunding a gain (or in many circumstances paying for it) but is arguably an exercise for the recipient of the service and taken from his standpoint. What

---

134 Classification in civil actions in tort or in contract; see *Halsbury's Laws of England*, 4th edn, (Butterworth), vol. 9, para. 636.
135 Goff, R., and Jones, G., 'The Law of Restitution', 3rd edn, p.3.
136 (1760) 2 Burr, 1005 at 1012; cited by Owen-Conway, S., *Restitution and Quantum Meruit (1985–86)* (Sweet & Maxell, 1896); 16 UWA Law Rev., 155.

then is a fair value of a contractor's work? There are a number of possible answers. The choice is between the actual cost to the contractor, the cost plus a reasonable profit for the contractor, the usual contract price between the parties, a typical price in the local market place, the benefit or gain by the owner, or a mixture of all these choices. Each may be a *quantum meruit* in different circumstances, but which type applies in any particular case will probably depend on the reason for claiming a *quantum meruit*.

What is *quantum meruit?*

The term *quantum meruit* is a synonym meaning 'as much as he has earned' and 'as much as they were worth';[137] but in the context of restitution the latter should provide the more accurate guide. Unfortunately the courts appear reluctant to assist, for rarely do they provide any rules or general guidance as to the calculation of a *quantum meruit* in the absence of contract. The furthest they appear prepared to go is to identify the various parts of a service provided and to attach obligation in principle by looking at the nature of the enrichment. An example can be seen in *Marston Construction Co. Ltd v. Kigass Ltd*,[138] from which several principles emerge. The court considered the provision of various estimates and drawings by a contractor. The work was done in expectation of being awarded a design and build contract for a factory. The judge found that all the work was carried out in reasonable contemplation of payment (to be made through the future contract)[139] and much of the work was requested either expressly or impliedly by the owner. He determined that the owner had to pay for the drawings prepared for planning approval which had realizable value in that they were capable of being used, even though in fact the project was unlikely ever to proceed. On the other hand, the owner was not obliged to pay for the work carried out in the preparation of revised tenders which was of value only to the contractor for the purpose of gaining further work.

It is clear from the *Marston* judgment that in order to succeed in a *quantum meruit* claim in the absence of any contract, there must be an express or an implied instruction to do the work; there must be actual expectation of payment in circumstances where it is reasonable to expect payment; and the principal must be left with something of value to them which is realizable. Unfortunately the Court was only asked to decide certain preliminary issues, so the question of how a reasonable sum was to be calculated was not addressed. The approach in determining liability is nevertheless indicative of value-based rather than cost-based calculations. This idea must sound warning bells to those who rely with blind faith on recovering a *quantum meruit* when all else fails, e.g. all those who work in reliance on letters of intent.

Most contractors assume that quantification of their work on a *quantum meruit* should be done on a time and materials basis (cost plus), in order to compensate them for the resources provided. However, there are three main situations which give rise to different types of *quantum meruit*, the 'time and material' basis being only one. First, there is a situation where a contract has been broken and remuneration is being sought for work done. Second, there is the position where a contract exists but there are no precise

---

137 *Osborn's Concise Law Dictionary*, 7th edn: *Quantum meruit* is a synonym for *quantum meruit et quantum valebat*.

138 (1989) 46 BLR 109.

139 This aspect of the decision is open to the criticism that it was not reasonable for the contractor to expect payment, since in this case the owner had said the development relied on a successful insurance claim.

provisions as to price or payment. The third situation is where work has been done on request, in the absence of a contract but in the knowledge that the work is not being done gratuitously. The first situation, i.e. breach of contract, is probably the only one which always involves a 'time and materials' claim.[140]

In the first case the broken agreement requires an assessment of damages. Time and materials, or cost recovery claims, are essentially a form of damages recoverable by the injured party from the party in breach of a contract. Thus time and materials claims could logically be associated with a 'quasi-contractual' relationship (but not a claim based on restitution).

The second situation also gives rise to a type of contractual claim, this time based on an implied promise to pay 'a reasonable sum'. It has been said that a reasonable sum in these circumstances is 'the value between a willing seller and a willing buyer'[141] where the court is required to take up a bargaining stance on behalf of both parties and to mentally enact the negotiation that never took place. Thus in *British Bank for Foreign Trade Ltd v. Novinex Ltd*[142] it was considered that a commission basis was a proper mode of reasonable remuneration in line with the normal practice of the particular trade in question. Clearly, in the evaluation of *quantum meruit* in this second analysis the calculation must take into account the circumstances not only of both parties, but also external influences.

The third situation, i.e. works known not to be done gratuitously (relevant to letters of intent and to the battle of forms if no contract is formed), may give rise to a restitutionary claim. The aim is 'to strip the defendant of a benefit which he has unjustly gained at the plaintiff's expense'.[143] This time the value of the *quantum meruit* should be determined by viewing the position from the 'outcome' of the arrangement, i.e. the value of the benefit conferred on the recipient. The method of calculation, however, is a problem, since there is little unequivocal guidance from the courts.

This problem was addressed by the Court of Appeal in *Crown House Engineering Ltd v. Amec Projects Ltd*.[144] The issue concerned circumstances which are common in the building industry. Crown House had carried out work on instruction in the absence of a contract. Amec had made payments on certificates as though a contract existed, but on the last certificate had set off sums in respect of counterclaims which included a charge for delays, allegedly resulting from Crown's tardy performance. Amec in defending, argued that if there was no contract and a *quantum meruit* claim could be made against them, then at least regard should be had to any necessary cost incurred by them as a result of receiving the service later than they had required. This, they said, had the effect of reducing the value of the benefit. At trial the judge said the value should be the 'objective value of the job that has been done'; he looked for genuine reduction which could be made in the objective valuation of the job, but found none. He considered tardiness could only be related to a contractual obligation; consequently, in the absence of a contract, Amec's counterclaim had no basis. An appeal from Amec was rejected, but in his speech Lord Justice Bingham said:

---

140 See Hugh Tomlinson on this point, 'Assessing a Quantum Meruit' (1986) 130 SJ 171.
141 *Per* Lord Denning MR in *Seager v. Copydex (No 2)* (1969) 1 WLR 809 at 813.
142 (1949) 1 All ER 155.
143 See Goff & Jones, *Law of Restitution*, 3rd edn, pp. 19, 148.
144 (1989) 48 BLR 32.

Crown has argued ... that these matters [tardiness] are wholly irrelevant to the assessment of what reasonable remuneration Crown should recover. It may well be that they are right ... But the answer does not seem to me to be obvious ... The doctrine of unjust enrichment ... does no doubt require [that] ... a customer should not take the benefit of a contractor's services rendered at his request without making fair recompense ... but it does not so obviously require that assessment should be made without regard to the acts or omissions of the contractor when rendering those services which have served to depreciate or even eliminate their value to the customer.

The message appears to be clear; fair value should take into account the worth of the service provided to the recipient. This seems particularly relevant when the service has little or no value, albeit that the contractor may have incurred substantial costs. The principle may be applied both ways. A service which is inexpensive to provide may be of great value. An example of this can be seen in the next case, where the challenge of defining a *quantum meruit* is met head on.

In *Costain Civil Engineering Ltd and Tarmac Construction Ltd v. Zanen Dredging and Contracting Co. Ltd*[145] there was a sub-contract for a tunnel under a river which required a casting basin as temporary works. It was decided to convert the casting basin into a marina. It was held that the work to the marina was not part of the original contract and it was not a variation. No contract existed, but the sub-contractor was entitled to a *quantum meruit*. In deciding what a *quantum meruit* included the court had to consider two unusual circumstances. Since the sub-contractor was on site it did not have to incur substantial setting-up costs, and since the conversion to a marina was a beneficial afterthought by the client, the Joint Venture contractor received additional profit. If the value was to be based on costs, the sub-contractor would not be out of pocket, but it would be a bargain for the contractor since the costs were much lower than any other sub-contractor would have incurred; if the value was to be based on benefit, the inclusion of the profit gained by the Joint Venture could put a value on the sub-contractor's work higher than if done under a sub-contract. The court held that a *quantum meruit* in the absence of a contract should be restitution-based and therefore valued as a benefit. The difficulty always lies in valuing, and it is convenient to relate to costs, even when the calculation is not theoretically cost-based. Here Zanen was awarded a 'top up', to make the costs comparable with what another sub-contractor would have had to charge as setting-up costs. Zanen was also awarded a split of the Joint Venture's profit. It is important to remember that what is being paid here is the value from the recipient's point of view, not the value to the provider.

In the Zanen case the *quantum meruit* could be higher than a corresponding contractual price would have been; but it could in other circumstances be lower, and the relevant circumstances are often outside the control or even the knowledge of the provider. The only certainty about working in expectation of a *quantum meruit* is the uncertainty. Contractors who deliberately work without a contract are always at risk. In many of the cases cited it was clear that the court looked upon the failure to achieve agreement as an unfortunate and unexpected event. The parties had gone about their business in anticipation of a contract being concluded, albeit that that stage had not yet been reached. However, if a contractor on receipt of a letter of intent or an unacceptable counter-offer deliberately decides to 'hook' the building owner with no intention of ever concluding a

---

145 (1996) 85 BLR 77.

contract,[146] it must be questioned whether restitution as an equitable principle is available to him at all; the contractor may be entitled to nothing. An example of this principle can be seen in the Australian case of *Sabemo v. North Sydney Municipal Council*[147] in which judgment was given for the plaintiff contractor who was awarded restitutionary *quantum meruit* where the failure to conclude a contract was caused by the defendant's action.

In summary, it appears that there is no clear principle on which to calculate a reasonable sum to be paid for work carried out in the absence of a contract, and despite indications that restitutionary principles are gaining ascendance over 'damages' principles, neither is likely to oust the other completely. The resultant calculation emerges as an amalgam. The 'time and material' (cost plus) basis is a convenient starting point, as in the Zanen case, but costs should be modified by any diminution or enhancement of value to the recipient of the service. Costs should then be moderated further, if applicable, by external factors such as the general market rate or the type of costs which would be included in a general market price. One thing is clear, however; that is the amount to be received cannot be certain, and the principles behind the calculation provide an extremely unstable bedrock from which to build the foundations of any relationship.

---

146 In *Barclays Bank plc and Another v. Hammersmith and Fulham LBC, The Times*, 27 November 1991; [1991] 12 CL 64, Neill LJ suggested one of the factors to be taken into account might be the state of knowledge of the parties.
147 [1977] 2 NSWLR 880; cited in *Building Law Monthly*, June 1989, 4.

Participants in the project and their roles under the Contract

Execution of a project let on the JCT 98 requires performance of distinct roles by a variety of participants. In most cases, their roles and duties are determined by a contract with the Employer and/or statutory provisions. The contractual and administrative relationships among them are shown in Fig. 2.1. It is important to note that the labels for the participants refer to holders of certain duties rather than types of professional or business organization and that it is possible for one organization or professional to discharge the role of more than one participant. For example it is possible, although not advisable, for the Architect also to be the Employer's Representative, a designer and the Planning Supervisor. The aim in this chapter is to examine these roles.

The Architect is referred to in the Conditions mainly in his capacity as the administrator of the Contract. His role as designer is not dealt with expressly. Similarly, the role of specialist design consultants is not explicitly referred to in the Conditions although they play a major role during construction. From the Contractor's standpoint, the generic role of designer is treated as part of the Architect's. Whilst this arrangement has the advantage of administrative convenience, it is artificial in some respects because decisions often have to be made in the same proceedings as to whether a mishap or other undesirable event is the responsibility of the Architect, the Contractor or a specialist design consultant. For this reason, the role of the Architect in only his administrator's hat is covered in Section 2.4 whilst the generic role of designer, which covers both the specialist design consultant and an Architect who also contributed to the design of the Works, is dealt with separately in Section 2.8. This means that where the Architect also provided or provides some design input, the reader needs to read both sections to develop a full picture of his role.

2.1 Employer

The Employer plays easily the most dominant role on the whole project because it is he who initiated it and secured the necessary funding. This section considers only his role and duties during the construction phase. Readers are directed to textbooks on project management for examination of his wider role in the total procurement process.

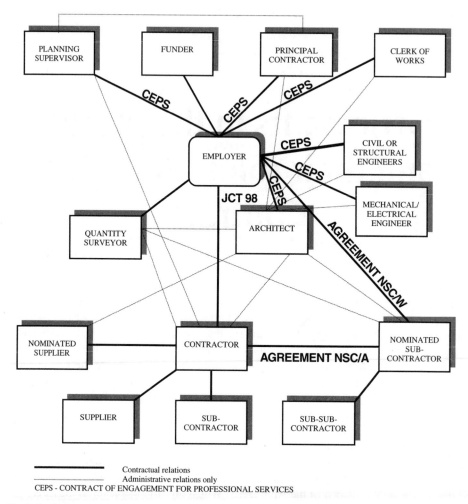

Fig. 2.1   Project participants.

2.1.1 Express duties

At the post-contract stage, the Employer generally stays in the background and allows the Architect, specialist design consultants, the Quantity Surveyor and the Clerk of Works, all of whom he employs, to act on his behalf. However, the Contract expressly requires the Employer to perform certain duties which are summarized in Table 2.1. The term 'duty', as used in this context, denotes a task that the Employer must perform in any circumstances and is generally prefaced by 'the Employer shall...'. The table does not therefore include steps in procedures that the Employer may only have to take depending upon whether the relevant procedure is invoked, e.g. notice to refer to adjudication or arbitration and notice to withhold payment. The most important of the Employer's duties is his obligation, stated in Article 2, to pay the Contractor the Contract Sum or other sum payable under the Contract at times and in a manner explained in detail in Chapter 15.

Table 2.1 Express duties of the Employer

Article/Clause:	Duty
Article 3/3A	to appoint a replacement Architect when the incumbent ceases to act as such
Article 4	to appoint a replacement Quantity Surveyor when the incumbent ceases to Act as such
Article 6.1	to appoint a replacement Planning Supervisor whenever necessary
Clause 1.6	to notify the Contractor in writing of the name and address of a replacement Planning Supervisor or Principal Contractor
Clause 5.7	not to use certain documents for any purpose other than the carrying out of the Works and not to divulge the rates and prices in the Contract Bills
Clause 6A.1	to ensure that the Planning Supervisor and the Principal Contractor carry out the duties under the CDM Regulations
Clause 6A.2	to notify the Planning Supervisor and the Architect of changes to the Health and Safety Plan notified to the Employer by the Contractor in his capacity as Principal Contractor
Clause 13A.3	to consider Clause 13A Quotations from the Contractor
Clause 19.1.1	not to assign any right under the Contract without the written consent of the Contractor
Clause 22B	where it applies, to take out and maintain insurance against damage to the Works
Clause 22C	where it applies, to take out and maintain insurance against damage to the Works and existing structures and the contents
Clause 23.1.1	to give possession of the Site to the Contractor on the Date of Possession
Clause 30.1.1.	to pay on any Interim Certificate within 14 days of its issue
Clause 30.1.1.3	to give notice of payment
Clause 30.8.2	to pay on the Final Certificate within 28 days of its issue if the amounted stated on it is payable by the Employer
Clause 35.13.2	to make direct payments to Nominated Sub-contractors if the Contractor defaults on payment obligations to them
41A.7.3	to comply with any decision of an Adjudicator and to ensure that it is given effect.

2.1.2 Implied duties

In addition to the express duties, duties normally implied into construction contracts apply unless, in respect of the particular duty, the Contract expressly provides to the contrary. Examination of most of the express duties will show that they are no more than the expression of some of the principles flowing from an employer's duty, long established by common law, to cooperate with the contractor in his carrying out of the works. Flowing from this general duty to cooperate is an implied term that neither party will do anything to prevent the other from performing his side of the contract.[1] The positive form of this term is an implied obligation to do whatever is necessary for the other to perform his obligations. However, it would appear that the positive duty only applies to contracts that specify an outcome that cannot be achieved unless both parties cooperate to bring it about. In *London Borough of Merton v. Leach*[2] Vinelott J held that all these terms were implied into the JCT 63 Conditions. As the JCT 98 does not depart from the JCT 63 in respect of general philosophy and approach, it is submitted that the implied duty to cooperate in general and its derivative duties also apply to the JCT 98.

---

1 *McKay v. Dick* (1881) 6 App. Cas. 251; *Barque Quilpé Ltd v. Brown* [1904] 2 KB 264.
2 *London Borough of Merton v. Stanley Hugh Leach* (1985) 32 BLR 51 (hereafter *Merton v Leach*); see also *Holland Hannen & Cubitts v. Welsh Hospital and Technical Services Organisation* (1983) 18 BLR 80 in which the implication of these terms was approved by Judge Newey QC.

2.1.3 Duty not to interfere with discretion of Architect or Quantity Surveyor

As explained in Section 2.4, where the Architect or Quantity Surveyor is not acting as an agent of the Employer but as an independent professional holding the balance fairly between the Employer and the Contractor, e.g. carrying out of valuation of variations or certification, the Employer does not warrant that those duties will be performed with reasonable skill and care. For example, the Employer is not liable for the financial losses of the Contractor flowing from any under-certification by the Architect.[3] The remedy is for the Contractor to invoke the appropriate dispute resolution mechanism to have the certificate revised. However, the Employer is under an implied warranty that they will act fairly independently and that he will not interfere with their free exercise of their professional judgement.[4] Indeed, case law suggests that he may be under a duty to stop any unfairness to the Contractor of which the Employer becomes aware.[5]

2.2 Employer's representative

It is common practice for the Employer's role to be delegated to an in-house project manager or an external project management firm. The JCT 98 recognizes this practice by providing in Clause 1.9 that the Employer is entitled to do this by written notice identifying the name of the individual to act for him and the extent of the delegation. The delegation takes effect from the date of the notification. It is important to note that the Employer's Representative must be a named individual and not that of the firm he may be working for. A footnote advises against the Architect or the Quantity Surveyor being named the Employer's Representative on grounds of possible confusion arising from the duality in their roles that would result.

The duties to be performed by the Employer's Representative are a matter for the contract between them. Case law suggests that, in the absence of terms to the contrary, the Employer's Representative may be considered a watchdog on behalf of the Employer, with responsibilities for reporting any failings of the Architect, the Quantity Surveyor or other consultants to the Employer. In *Chesham Properties Ltd v. Bucknall Austin Project Management Services Ltd,*[6] His Honour Judge Hicks, QC, deciding a preliminary issue, held, *inter alia*, that a project manager on a contract in the terms of the JCT 80 owed a duty in contract and in tort to advise and/or inform the Employer of actual or potential deficiencies in the performance by the Architect, the Quantity Surveyor and other consultants of their contractual duties to the Employer. It is submitted that this also applies to the Employer's Representative under the JCT 98.

---

3   See Section 2.4 for more detailed discussion on this issue.
4   *Sutcliffe v. Thackrah* [1974] AC 727; *Merton v. Leach*; *John Mowlem & Co. Ltd. v. Eagle Star Insurance Co. Ltd and Others* (1992) 62 BLR 126. Under Clause 28.2.1 interference with the issue of certificates under the Contract is a specified default by the Employer, i.e. the Contractor is entitled to terminate his own employment under the contract (see Section 16.11.2).
5   *Panamena Europea v. Leyland* [1943] 76 Lloyd's Rep. 114; *Perini Corporation v. Commonwealth of Australia* (1969) 12 BLR 82 (Supreme Court of New South Wales) see Section 15.2 for discussion of negligent certification of which the employer is aware.
6   (1996) 82 BLR 92.

2.3 Contractor

Article 1 states the Contractor's primary obligation as being to carry out and complete the Works in accordance with the Contract Documents. This is repeated in Clause 2.1 but with more detail on quality of materials and standards of workmanship to be achieved. The Conditions contain express and specific duties within this general obligation for purposes of certainty, monitoring and remedial action as appropriate.

2.3.1 Materials and workmanship

The Contractor's responsibility to the Employer for workmanship and materials at common law is next explained for the purpose of putting in context the provisions of the JCT 98 on the subject. It is to be noted that the Supply of Goods and Services Act 1982 has put the common law principles discussed in statutory form.

2.3.1.1 The common law position

The common understanding of 'workmanship' in the construction industry concerns the skill and care exercised by a contractor in the physical execution of work. However, to the extent that choice of materials is left to the contractor, it may also mean design, i.e. suitability of the materials for the purpose for which they have been used. In the absence of terms to the contrary or directing the contractor as to the detailed manner in which the work is to be done, the law has for a long time implied a term that the work will be done in a proper and workmanlike manner, i.e. with the skill and care of an ordinary competent contractor.[7]

From general principles of the law of contract, a contractor is under an obligation to ensure compliance with the specification where materials are specified in the contract. If the specification states a brand name or a particular supplier of the material, the contractor would still be under a warranty that such materials are of good quality when used. This obligation is absolute, i.e. it is no defence to liability for a defect that it was not discoverable even with the most careful examination. A warranty of fitness of purpose may also apply if the circumstances indicate that there was reliance on the contractor's skills regarding suitability of the materials, e.g. leaving to the contractor the choice of the type of material. The leading authority on quality and fitness for purpose of materials in construction contracts is *Young & Marten Ltd v. McManus Childs Ltd*:[8]

> MC, developers and the main contractor[9] under a building contract, sub-let the roofing to YM. The sub-contract called for the use of a type of tile called 'Somerset 13', which was available from only one manufacturer. Soon after construction, some of the tiles began to disintegrate. This problem was traced to a batch of the tiles which contained defects which could not have been discovered even with reasonable inspection. MC sought to recover the costs of re-roofing from the sub-contractor. YM

---

7 *Duncan v. Blundell* (1820) 3 Stark. 6; *Pearce v. Tucker* (1862) 3 F & F 136; *Test Valley Borough Council v. Greater London Council* (1979) 13 BLR 63.
8 (1969) 9 BLR 77 (hereafter *Young & Marten*).
9 The main contractors were Richard Saunders Ltd but the pleadings showed that they were acting as agents for the developers.

argued that, since they had not been relied upon regarding the type of tile and they could not reasonably have discovered the defects, they were not liable. The House of Lords decided that, unless the circumstances of a particular case are such as to exclude it, there will be implied into a contract for work and materials a term that the materials will be of good quality and a further term that the materials will be reasonably fit for the purpose for which they were used. It was held that, in this particular case, although there was no obligation for fitness for purpose, there was still a duty to ensure that the materials were of good quality because, as YM had a choice whether or not to accept the batch of defective tiles, there was reliance upon them to inspect them before acceptance. YM were therefore liable.

However, where the circumstances of the particular contract indicate that there is no reliance on the skill and care of the contractor on the issue of quality, e.g. the terms of the contract compel the contractor to accept materials from a particular supplier, the contractor will not be liable for defects in them.

Gloucestershire C. C. v. Richardson[10]: the respondents, main contractors on a building contract let on the JCT 39 (1957 revision), were instructed by architects to accept and use concrete columns from nominated suppliers. After the columns had been incorporated into the works, defects were discovered. A PC sum in the Bill of Quantities merely required the contractor to erect columns to be supplied by nominated suppliers. All the circumstances of the contract indicated that the contractor had no choice but to accept the supply even though the supplier's terms negotiated by the architect contained exclusion clauses. The House of Lords held that, in those circumstances, the contractor was not subject to a warranty that the columns would be of good quality.

As their Lordships emphasized, generally, the principles discussed above apply to the work and materials of sub-contractors and suppliers even if the employer nominated them.[11] Two reasons are normally given for this. First, main contracts often state expressly that the contractor is to be responsible for the work and materials of all sub-contractors and suppliers exactly as if they are his own. For example, Clause 19.5.1 of the JCT 98 provides that, unless otherwise stated in the Conditions, the Contractor is responsible to the Employer for the carrying out and completion of Nominated Sub-contract Works as if they were the Contractor's own direct work. Second, there is a need to maintain a chain of liability from the Employer down to the manufacturer. Without such a chain of liability, a sub-contractor or supplier can, at best, recover only nominal damages from the next party down the chain (wholesaler or manufacturer). Generally, the law follows the view that society is not well served by allowing those causing loss or damage to escape liability whilst those who suffer the loss are denied any remedy. Wherever possible, the courts therefore interpret contracts in such way that the chain of liability is maintained from the ultimate consumer right down to the manufacturer. In *Young & Marten* itself Lord Reid explained the need, wherever possible, to construe contracts so as to maintain the chain of liability in these words:

---

10 (1968) 1 AC 480; 2 All ER 1181; this decision was made by the same Law Lords and given the same day as *Young & Marten*.
11 *Rumbelows v. A. M. K.* (1980) 19 BLR 25.

There are, in my view, good reasons for implying such a warranty if it is not excluded by the terms of the contract. If the contractor's employer suffers loss by reason of the emergence of latent defect, he will generally have no redress if he cannot recover damages from the contractor. But, if he can recover damages, the contractor will generally not have to bear the loss; he will have bought the defective material from a seller who will be liable under s. 14(2) of the Sale of Goods Act, 1893, because the material was not of merchantable quality. And, if that seller had in turn bought from someone else, there will again be liability, so that there will be a chain of liability from the employer who suffers the damage back to the author of the defect.

2.3.1.2 Contractual position

On the quality of materials and workmanship, the Contract states:

8.1.1   All materials and goods shall, so far as procurable, be of the kinds and standards described in the Contract Bills, and also, in regard to any Performance Specified Work, in the Contractor's Statement, provided that materials and goods shall be to the reasonable satisfaction of the architect where and to the extent that this is required in accordance with clause 2.1.

8.1.2   All workmanship shall be of the standards described in the Contract Bills, and also, in regard to any Performance Specified Work,[12] in the Contractor's Statement, or, to the extent that no such standards are described in the Contract Bills, or, in regard to any Performance Specified Work, in the Contractor's Statement, shall be of a standard appropriate to the Works, provided that workmanship shall be to the reasonable satisfaction of the Architect where and to the extent that this is required in accordance with clause 2.1.

8.1.3   All work shall be carried out in a proper and workmanlike manner and in accordance with the Health and Safety Plan.'[13]

The requirement to work to the Contract Documents implies that, in respect of materials and workmanship specified in detail in the Contract Bills, it is no defence to liability for non-complying work or materials that the Architect or the Clerk of Works (COW) approved them or even included them in payment certificates. The same point is expressly stated in Clause 1.5. These two parties are on the site solely for the benefit of the Employer. In the performance of their duties of inspection, they have no responsibility to the Contractor to discover his mistakes.[14] They do owe that responsibility to the Employer and if they fail to discover defects that they should, with reasonable skill and care, have discovered, they will liable to him for the resulting loss.

From the general principles already explained, the Contractor is not under a warranty of fitness for purpose in respect of materials specified in the Contract Bills. Some specifications contain lists of different type of materials for the same purpose, with the

---

12 This Section covers only materials and workmanship not involving Performance Specified Work. Quality issues in respect of such work are covered in Chapter 4.
13 The construction of clause 8.1.3 is considered in Section 2.4.1, which deals with the Architect's powers in relation to the Contractor's failure to comply with it.
14 See Section 2.4.6 for a detailed discussion of the Architect's liability to the Contractor.

choice left to the Contractor as to which type to use. Such specifications raise the issue whether the Contractor is liable if his choice is not fit for purpose whilst the others on the list are. This issue came before the Court of Appeal in *Rotherham Metropolitan Borough Council v. Frank Haslam Milan & Co. Ltd and M. J. Gleeson (Northern) Ltd*[15] which arose from a contract let on the JCT 63. Clause 6(1) of that form of contract stated that 'All materials, goods and workmanship shall so far as procurable be of the respective kinds and standards described in the Contract Bills.' Hard-core to be used as fill to the underside of the ground floor slab of a building was specified in the Contract Bills as:

> Granular hardcore shall be well graded or uncrushed gravel, stone, rock fill, crushed concrete or slag or natural sand or a combination of any of these. It shall not contain organic materials susceptible to spontaneous combustion, materials in a frozen condition, clays or more than 0.25 of sulphate ions as determined by BS 1377.

The Contractor used steel slag that expanded after completion, causing heaving of the ground floor with resulting cracking in reinforced concrete floors. It was argued for the Employer that as the choice of type of hard-core was left to the Contractor, he warranted the fitness of his choice for purpose as hard-core. This was rejected unanimously in the Court of Appeal.

The Contractor's obligation to use materials of the kind and standard specified in the Contract Bills is qualified by 'so far as procurable': Clause 8.1.1. It is unfortunate that the Contract does not specify the geographic limits of procurability. This omission could give rise to arguments whether materials available only abroad are procurable. If after execution of the Contract, any item of material ceases to be procurable, the Architect may issue a variation altering the kind or standard of the item. It is debatable whether straight substitution, i.e. there is no alteration in the kind and standard of the item,[16] is within the Clause 13.1 definition of a variation. If it is not, substitution needs the agreement of the Employer. However, it is submitted that as a variation is defined by inclusive phraseology, straight substitution is likely to be within the definition. This type of problem may be avoided altogether by adding to the description of relevant items in the Contract Bills the phrase 'or equivalent approved by the Architect'. The construction of a similar phrase in *Leedsford Ltd v. Bradford Corporation*[17] indicates that such a strategy gives the Architect absolute discretion whether or not to accept materials of similar specification offered by the Contractor.

The Contractor's obligation on standards of workmanship is not qualified by 'so far as procurable'. This means that it is no defence to the obligation that the human skills or equipment to achieve that standard of workmanship are not procurable.

Where the quality of materials and standards of workmanship are left to the opinion of the Architect, they must be to his reasonable satisfaction: Clause 2.1. This is an objective standard. The Architect has no power to demand quality or workmanship of the highest standards without a variation. A decision of the Court of Appeal suggests that the Architect may be entitled to take into account the competitiveness of the Contractor's prices for the relevant materials or work in deciding whether they are to his reasonable satisfaction. In

---

15 (1996) 78 BLR 1.
16 An example is where the Contract Bills specify a brick from a particular manufacturer and there is the same kind and standard of brick obtained from another.
17 (1956) 24 BLR 45.

Cotton v. Wallis[18] the specification in the contract required all materials and workmanship to be the best of their kind and to the full satisfaction of the Architect. The contract also provided that 'the contractor shall carry out and complete the works in accordance with this contract in every respect with the direction and to the reasonable satisfaction of the architect'. It was held by a majority that it was reasonable for the Architect to accept work and materials not of the best quality because the contract price was very low. In respect of quality standards that must meet the Architect's reasonable satisfaction, Clause 8.2.2 requires the Architect to express any dissatisfaction within a reasonable time from execution of the unsatisfactory work. The Contract does not state the effect of the Architect's failure to do this. A possible construction is that the Architect cannot thereafter reject the work no matter how bad it is and that he must therefore include it in Interim Certificates. However, the Architect may advise the Employer to challenge the quality of the work before expiry of 28 days after the issue of the Final Certificate. After the 28 days, as explained under Section 3.9.3, the Contractor is no longer answerable for work expressed as subject to the reasonable satisfaction of the Architect. An alternative, and preferred, construction is that the Architect can still reject the work but the delay in expressing his dissatisfaction is a breach of Contract by the Employer for which the Contractor would be entitled to damages.

In respect of workmanship not described in the Contract Bills, Clause 8.1.2 requires a standard appropriate to the Works. This is probably a different standard from the 'good and workmanlike manner' standard normally implied at common law. For example, the standard of finishes appropriate to farm buildings intended for use by animals would not be appropriate for a five-star hotel. As the Architect must decide whether such work has been properly executed before including it in Interim Certificates under Clause 30.2.1, it is his decision whether the workmanship used by the Contractor is appropriate to the Works.

Liability in respect of quality of materials and standard of workmanship specified in detail in the Contract Documents lasts for either 6 or 12 years, depending upon whether the Contract is executed as a simple contract or a deed (limitation of actions is discussed in detail in Chapter 1). However, where quality of materials and standards of workmanship are left to the opinion of the Architect, the Employer cannot challenge any decision of the Architect that the materials or workmanship are to his reasonable satisfaction unless proceedings are commenced within 28 days after the issue of the Certificate.[19]

2.3.2 Contractor's duty to give notices

The Contractor is required to give a variety of other notices to the Architect and/or the Employer. They are summarized in Table 2.2. Where, in respect of a particular requirement for a notice or a document, the Contract prescribes the manner of giving or serving the notice or document, the Contractor must comply with the prescription. Clause 1.7 deals with situations where the Contract does not state the manner of notice. The Contractor must give the notice or serve the document by any effective means to any agreed address. If no address is agreed, the notice or document is deemed given to or served on a registered company if properly posted to its registered or principal office. In

18 [1955] 1 WLR 1168.
19 See Section 3.9.4 for detailed discussion of this issue.

Table 2.2   Notices to be served by the Contractor.

Clause	Matter to be Notified
Clause 2.3	errors, discrepancies and divergences within or between documents
Clause 2.4	any discrepancy or divergence between the Contractor's Statement in respect of Performance Specified Work and an AI (Architect's Instructions)
Clause 5.4.2	time for provision of information by the Architect of which he appears to be unaware
Clause 6.1.1	notice required by the Statutory Requirements (the appropriate body)
Clause 6.1.2	divergence between the Statutory Requirements and a Clause 2.3 document or a variation instruction
Clause 6.1.4.2	emergency work to comply with the Statutory Requirements
Clause 6A.2	amendment to the Health and Safety Plan (this applies only if the Contractor is also the Principal Contractor)
Clause 13.2.3	the Contractor's disagreement with an AI stating that Clause 13A applies to its valuation
Clauses 22A.4.1 and 22B.3.1	loss or damage to any part of work executed or Site Materials
Clauses 23.3.2 and 23.3.3	proposed use or occupation of any part of the Site or the Works by the Employer and additional premium payable for such use or occupation
Clause 25.2.1.1	material circumstances of actual or likely delay to the progress of the Works
Clauses 25.2.2 and 25.2.3	details of delays caused by a Relevant Event and additional information reasonably necessary to keep the Architect updated
Clause 27.3.2	occurrence of certain insolvency events in relation to the Contractor
Clause 28.2.1	specified default by the Employer
Clause 28.3.3	determination of the Contractor of his own employment for the Employer's insolvency
Clause 28A.1.1	determination by the Contractor of his own employment under Clause 28A
Clause 31.4	change in the Contractor's position regarding the statutory Tax Deduction Scheme
Clause 34.1.3	Fossils, antiquities and other objects of interest or value found on the site
Clause 35.8	Failure to enter into a Nominated Sub-contract within 10 days of the relevant nomination instruction
Clause 35.15.1	Failure of a Nominated Sub-contractor to complete by the completion date applicable to the Sub-contract
35.24.6	Determination of the employment of a Nominated Sub-contractor
38.4.1/39.5.1	Occurrence of events for which the Contract Sum is to be adjusted for fluctuations
42.15	that in his opinion compliance with an AI will be injurious to the efficacy of Performance Specified Work

the case of a non-registered body, there is a need to post it properly to the last known principal business address.

The notice requirements serve a number of purposes, including to:

1. enable the Architect to deal with the matter notified or to take timely and appropriate remedial action to minimize its negative effects;
2. alert the Architect to monitor the situation and to collect contemporaneous information with a view to avoiding future disagreements concerning what actually happened;
3. allow the Architect properly to keep the Employer informed about the project;

4. allow the Employer to make necessary arrangements to deal with the notified matter, e.g. making arrangements for additional funds;
5. assist the Quantity Surveyor in valuing the Works.

Failure to serve any notice in accordance with the stipulations of the Contract is technically a breach of contract for which the Employer may claim damages or even determine the Contract at common law for a fundamental breach. Also, the Contractor may not be able to enforce his contractual rights in relation in the matters he failed to notify, e.g. extension of time and recovery of loss and/or expense because the courts tend to interpret such notice requirements as preconditions for the enforcement of associated rights. Indeed, under Clause 26.1, it is a precondition for ascertainment of loss and/or expense under the Contract that the Contractor has made a written application that regular progress is being, or will be, disturbed. In any event, to the extent that the failure to notify prevented operation of the appropriate contractual machinery, he cannot enforce such operation because it is a long standing common law principle that 'no person can take advantage of the non-fulfilment of a condition the performance of which has been hindered by himself'.[20]

2.3.3 Contractor's design responsibility

This issue begs the question what is 'design'? Although it is not possible to produce a universally acceptable definition, there is no doubt that the term is wide enough to include decision-making during the drawing up of specifications for the works, determination of dimensions, determination of reinforcement needs and choice of working methods.[21] Generally, in a 'construct' contract, i.e. one in which a full design has been produced on behalf of an employer, with the contractor only implementing what has been designed, the contractor bears no design responsibility towards the employer. This proposition is based on a presumed common intention that the contractor is not to be responsible for a matter on which the employer never placed reliance upon his or her skills.

> *Lynch v. Thorne*:[22] a builder put up a house in accordance with plans and specifications annexed to the contract. The specifications required use of nine-inch bricks in an external wall without any rendering. The litigation concerned the builder's liability for water penetration through the wall. The Court of Appeal held that, since the builder had complied exactly with the specifications and drawings, he was not liable.

The wording of Article 1 and Clause 2.1 of the JCT 98 suggests that, without incorporating the Contractor's Designed Portion Supplement or properly requiring Performance Specified Work, it is a construct contract as the Contractor is simply to carry out the works in accordance with the Contract Documents. *John Mowlem & Co. Ltd v. BICC Pensions Trust Ltd & S. Jampel & Partners*[23] was decided on the unchallenged assumption that a Contractor under the JCT 63, which is indistinguishable from the JCT 98 on this issue, had no responsibility for design.

---

20 *Roberts v. Bury Commissioners* (1870) LR 5 CP 310.
21 See *Hudson's Building and Engineering Contracts*, ed. Wallace, I.N.D. (Sweet & Maxwell, 1995), paras 4.064 and 4.065 (hereafter *Hudson's*).
22 *Lynch v. Thorne* [1956] 1 WLR 303.
23 (1977) 3 Con. LR 63. For a contrasting view of the effect of Clause 2.2.1 see Section 1.4.3

A section of the Contract Bills contained a performance specification which stated: 'retaining walls forming external walls to buildings and basement slabs are to be constructed so that they are impervious to water and damp penetration, and the Contractor is responsible for maintaining these in this condition'. Clause 12(1) of the JCT 63, which is similar to Clause 2.2.1 of the JCT 98, provided that nothing in the Contract Bills was to modify or override the printed conditions. The action concerned cracking and water penetration that occurred. It was held that the performance specification imposed a design responsibility on the Contractor and that it was therefore ineffective against Clause 12(1).

Whilst the whole tenor of the JCT 98 is that the Contractor carries no design responsibility, some of the responsibilities of the Contractor require the exercise of a certain amount of design skills, e.g. workmanship and selection of materials to the extent that these matters have not been covered in the specifications, compliance with Building Regulations and other statutory obligations, and reporting of errors, departures, divergences and discrepancies.

2.3.4 Contractor's duty to warn[24]

Case law suggests that there are circumstances in which a contractor may be held liable under a construct contract for design defects on the basis of breach of an implied duty to warn the owner, or his representatives, of design defects. In the Canadian case of *Brunswick Construction v. Nowlan*,[25] the contractor was to build a house in accordance with drawings provided by the owner but without any supervision from the Architect who had carried out the design and produced the drawings. The Supreme Court of Canada decided that, in those circumstances, the owner must have relied upon the contractor regarding obvious defects in the design and that a duty to warn the owner of such defects was implied. In *Equitable Debenture Assets Corp. Ltd v. William Moss*[26] and *Victoria University of Manchester v. Hugh Wilson*[27] Judge Newey QC held that it is an implied term of a construct contract that the contractor should warn the employer of defects in the design that he believed to exist. He explained that 'belief' in this context required more than mere doubts as to the correctness of the design but less than actual knowledge of errors. These cases arose from contracts let on versions of the JCT 63 that did not contain express terms requiring the Contractor to warn the Architect of defects that he actually discovered. However, in *University of Glasgow v. William Whitfield and John Laing*,[28] which arose from a subsequent version of the JCT 63 that contained express obligations[29] to warn of discovered defects, Judge Bowsher said *obiter* (Lat. by the way) that there was no implied duty to warn of design defects. He distinguished the earlier cases on the

---

24 See Nicholls, H., 'Contractors Duty to Warn' (1989) 5 Const LJ 175; Wilson, S., and Rutherford, L., 'Design Defects in Building Contracts' (1994) 10 Const. LJ 90.
25 (1974) 49 DLR 93.
26 *Equitable Debenture Assets Corp. Ltd v. William Moss* (1984) 1 Const. LJ 131.
27 (1984) 1 Const. LJ 162.
28 (1988) 42 BLR 27; this decision was preferred by Judge Esyr Lewis, QC in *Chancellor, Masters and Scholars of the University of Oxford (trading as Oxford University Press) v. John Stedman Design Group and Others* (1991) 7 Const. LJ 102.
29 The express provisions worked against implications of additional obligations in the sense that the parties are deemed to have defined completely the contractor's duty to warn.

grounds that whilst those cases involved reliance by owners upon contractors regarding design defects, there was no evidence of such reliance in the case before him.

A contractor has a duty under the Supply of Goods and Services Act 1982 to exercise reasonable skill and care in the performance of the construction contract. It is argued in *Keating on Building Contracts*[30] that the limitation of a contractor's duty to warn to only defects actually discovered may not be effective where a defect is so glaring that any contractor complying with this duty ought reasonably to have discovered it. This argument is more applicable to undiscovered defects that present serious health and safety risks than aesthetic defects or those that only affect the pocket of the employer. Judge Newey adopted such an approach when the issue came before him again in *Edward Lindenberg v. Joe Canning and Others*.[31] In that case the defendant was engaged by the plaintiff developer to carry out preliminary work in the basement of a block of flats that the plaintiff wished to convert into flats. Drawings produced by the plaintiff's surveyor showed 9-inch internal walls, including the chimney breast wall, as non-load-bearing. The defendant demolished the walls without props or other suitable precautions, resulting in deflection of the floor of the flat above. The defendant contended that, in assuming that protective measures were unnecessary, he had simply followed the drawings provided by the plaintiff and that he was not therefore liable for the damage. Judge Newey QC held that the defendant should have had grave doubts about the correctness of the information on the drawing not only from the fact that a so obviously important structural member as the chimney breast wall was indicated as non-load-bearing but also from the fact that 9-inch walls should be required for a non-load-bearing function[32]. Following his previous decisions, he held that the defendants were in breach of an implied term of the agreement that the defendant should exercise the care expected of an ordinary competent builder.

2.3.5 Duty to comply with Architect's Instructions (AIs)

Clause 4 of the Conditions requires the Contractor to comply with all instructions of the Architect issued under the Contract. Some of the most problematical issues in contract administration arise from this duty. For that reason, AIs are treated as a separate topic in Chapters 5 and 6.

2.3.6 Duties relating to the Contractor's master programme

A construction programme is one of the most effective documents for assessing and controlling a contractor's performance on schedule. In the event of claims for extension of time and loss and /or expense (e.g. recovery of additional preliminaries) the programme is often of crucial importance. Such a programme is normally submitted with the Contractor's tender. Clause 5.3.1.2 provides that, unless he has already done so, the Contractor must provide the Architect with two copies of his master programme as soon as possible after execution of the Contract. If the Contractor amends the programme to reflect extensions of time granted or a confirmed acceptance of a 13A quotation, two copies of the revised programme must be supplied to the Architect within 14 days of the

---

30 At p. 60.
31 (1992) 62 BLR 147.
32 Usually 4.5-inch walls are used for the non-load-bearing function of partitioning rooms.

granting of extension of time or the date of confirmation of the acceptance of the quotation. The provisions on the Contractor's programme suffer from a number of shortcomings:

1. There is no indication of the nature of the programme required. It could be a simple list of activities, a bar chart, a linked bar chart or a CPM network diagram. Technically, the Contractor would be complying with Clause 5.3.1.2 if he supplied any form of programme. To avoid uncertainty about the level of detail in the programme to be provided, it is advisable to specify any programme requirements in the Contract Bills. Some sophisticated clients now even go to the extent of specifying the type of planning and scheduling software on which the programme is to be produced.
2. The programme does not have to be to the satisfaction of the Architect. This means that the Architect has no effective way of getting the Contractor to amend an unrealistic programme.
3. The Contractor is not expressly required to revise his master programme whenever there is a need to do. He is simply expected to provide the Architect with copies of amendments but only if there is an amended programme.
4. The 14 days' deadline for the provision of the revised programme is enforceable only if the Contractor actually finishes revising the programme within the 14 days. As pointed out above, there is no obligation to do this. If the programme is actually revised after the 14 days, there is the ridiculous result that the deadline is passed even before the obligation to supply the revision arises.
5. If the Contractor amends his master programme for reasons other than extension of time or a confirmed acceptance of a 13A quotation, e.g. for delay due to his own inefficiency, the Contractor is not required to give copies of the revised programme to the Architect.

The contractual status of a contractor's programme has been considered in the case law. In *Glenlion Construction Ltd v. The Guinness Trust*[33] the Contract Bills in a JCT 63 contract required the Contractor to provide a programme. The programme provided by the Contractor contained a completion date earlier than the Completion Date in the contract documents. It was held by Judge Fox-Andrews that: (a) although the Contractor was entitled to earlier completion, it was not a contractual obligation; (b) there was no obligation on the Employer or the Architect to do all the things necessary to enable completion according to the Contractor's programme. They were obliged to do only the things necessary to enable completion by the Completion Date in the contract documents.

However, a decision in *Merton v. Leach*[34] suggests that, in certain circumstances, a contractor's programme may augment an obligation provided for in the Contract Documents. This case also arose from a JCT 63 contract, which provided as a Relevant Event:

> the Contractor not having received in due time necessary instructions ... drawings, details or levels from the Architect for which he specifically applied in writing provided that such application was made on a date which having regard to the Completion Date was neither unreasonably distant from or unreasonably close to the date on which it was necessary for him to receive the same.

---

33 (1987) 39 BLR 89 (hereafter *Glenlion v. Guinness*); this was followed by Stabb J in *J. F. Finnegan Ltd v. Sheffield City Council* (1988) 43 BLR 130.
34 See n. 2.

On a preliminary issue, Vinelott J held that a programme annotated with dates by which the Architect was to supply instructions, drawings, or other information could constitute the specific application contemplated in that Relevant Event provided the two requirements of being neither unreasonably distant nor unreasonably close were met. As explained in Section 11.3.7, there is now no requirement for the Contractor to have applied for the information. The onus is rather on the Architect to determine when it is necessary for the Contractor to receive the information. The Contractor's role here is only to advise the Architect of his need for information where he is aware, *and* has reasonable grounds to believe, that the Architect is unaware of them. It is to be noted not only that the test of when the Contractor should advise the Architect is subjective but also that the advice is not required to be in writing. By analogy with the *Merton v. Leach* decision, it is submitted that the Contractor will have complied with his obligation to advise the Architect in this respect if he submits up-to-date programmes with the appropriate level of detail. The Contractor does not even have to do this because of the subjective nature of the test of when he should advise the Architect. This balance of obligation is likely to be considered unfair by many non-contractors in the construction industry. All this suggests a policy on the part of the JCT to discourage use of the form without a comprehensive Information Release Schedule.

Clause 5.3.2 provides that descriptive schedules, master programme and amendments to master programme are not to be considered as imposing any contractual obligations beyond those imposed by the Contract Documents. Apart from restating the common law position highlighted in *Glenlion v. Guinness* and *Merton v. Leach*, this provision avoids the problem highlighted by the litigation in *Yorkshire Water Authority v. Sir Alfred McAlpine & Son (Northern) Ltd.*[35] In that case, which arose from a contract in the terms of the ICE *Conditions of Contract (5th Edn)*, the contract signed incorporated the Contractor's method statement. Skinner J held that, contrary to the customary practice in the construction industry whereby the costs and schedule implications of a contractor's chosen methods of construction are matters for which the contractor is entirely responsible, the Contractor was entitled to a variation order (with the normal consequences on entitlement to recovery of extra costs and extension of time for delays) if a change in the construction methods was necessitated by impossibility.

2.3.7 Statutory requirements

By Clause 6.1.1 the Contractor must comply with, and give any notices required by any of the following:

- any Act of Parliament;
- any instrument, rule or order made under any Act of Parliament;
- any regulation or bye-law of any local authority with jurisdiction over, or in connection with the Works;
- any regulation or bye-law of any statutory undertaker with jurisdiction over, or in connection with the Works.

These obligations are referred to collectively as the 'Statutory Requirements'. Failure to comply with them constitutes an actionable breach of contract. Furthermore, such failure

35 (1985) 32 BLR 114.

may constitute some form of illegality that may affect the Contractor's right to payment because, as a general principle from common law, a contractor cannot recover payment for an illegal contract even if he did not know of the illegality.[36] This principle applies to contracts which, by the very nature, contravene statutory provisions, e.g. the Building Regulations and Town and Country Planning legislation.[37] In such a situation, not only is the Contractor not entitled to be paid, but sums already paid by the Employer would also be refundable.[38]

A distinction has to be made between inherent illegality (e.g. where there is no planning permission for the development) and illegality resulting from the way the Contractor performed his contractual obligations. Depending upon the circumstances, the Contractor may be entitled to payment even with the latter type of illegality. Factors to be taken into account include whether the illegality can be cured by remedial work and whether or not the enforcement authority is prepared to waive the contravention.

> *Townsend (Builders) Ltd v. Cinema News & Property Management Ltd:*[39] the location of water closets in a building constructed by the Contractor contravened building bye-laws. The contract was in the terms of the then current predecessor of the JCT 98. The Contractor became aware of the contravention only when the construction had reached an advanced stage. On the assurance of the Architect that he would resolve the problem with the local authority, the Contractor completed the work without a variation to deal with the contravention. Although the local authority condemned the work, they allowed it to remain subject to conditions. The Court of Appeal held that, since the local authority had allowed the work to remain, the contract was not illegal and that the Contractor was therefore entitled to payment. It was also held that failure of the Contractor to comply with a term similar to JCT 98 Clause 6.1.1 was a breach for which the Employer was entitled to damages, effectively the cost of remedial work necessary to eliminate the contravention.

Clause 6.2 provides that the Contractor shall pay and indemnify the Employer against liability for fees and charges in connection with the Statutory Requirements that are legally payable pursuant to the relevant statutory provision. However, such fees and charges are to be added to the Contract Sum unless: (i) they are priced or provided for as a provisional sum item in the Contract Bills, or (ii) they arise in respect of the statutory work or service of a statutory undertaker or authority undertaken as a Nominated Sub-contractor or Nominated Supplier.

2.3.7.1 Divergence between statutory requirements and other contractual requirements

The Conditions expressly provide for the situation where the Contractor discovers a divergence between the Statutory Requirements and the Contract Documents or any other contractual document for which the Contractor is not responsible.[40] By Clause 6.1.2, for a divergence not concerning Performance Specified Work (PSW), the Contractor must

---

36 *Re Mahmoud and Ispahani* [1921] 2 KB 716; *Bostel Bros Ltd v. Hurlock* [1949] 1 KB 74.
37 *Steven v. Gourley* (1859) 7 CB (NS) 99; *Townsend (Builders) Ltd v. Cinema News & Property Management Ltd* (1959) 20 BLR 118.
38 *Smith & Son Ltd v Walker* [1952] 2 QB 319.
39 See n. 37.
40 Divergences concerning PSW are dealt with in Chapter 4.

immediately give written notice of it to the Architect, who must then issue appropriate instructions correcting it. The Architect is also obliged to issue the instructions where he discovers the divergence independently. He must act within 7 days of the Contractor's notice or his own discovery, as the case may be. In so far as any instruction given under 6.1.3 requires a variation, it is to be considered a variation under Clause 13.2, with all the implications for entitlements to extensions of time and recovery of direct loss/expense.

The Contractor's duty to comply with the Statutory Requirements and to give notice of any divergence between them and any other requirement of the Contract is subject to Clause 6.1.5 which states:

> Provided that the Contractor complies with clause 6.1.2, the Contractor shall not be liable to the Employer under this Contract if the Works do not comply with the Statutory Requirements where and to the extent that such non-compliance of the Works results from the Contractor having carried out the work in accordance with the documents referred to in clause 2.3 or with any instruction requiring a Variation issued by the Architect in accordance with 13.2.

The intention behind this clause appears to be that, as long as the Contractor gives the notices on discovered divergences, he is not to be liable to the Employer for non-compliance with the Statutory Requirements to the extent that the non-compliance is due to the carrying out of the Works in accordance with the other requirements of the Contract. Whether the courts will give effect to this intention is debatable. The effectiveness of the same clause in the JCT 80 in exonerating the Contractor from liability was doubted by the late Dr John Parris. Citing *Street & Another v. Sibbabridge & Another*[41] as authority, he argued that it is an implied term in building contracts that the contractor must comply with building regulations and the like and that this implied term prevails over any terms requiring the contractor to comply with an architect's design and instructions. In that case, it was held that a contractor was liable for contravention of building regulations inherent in a building that he constructed to a design provided on behalf of an employer. However, that case may be distinguishable on the grounds that whilst it was about implication of terms, the issue here is one of limitation of liability for certain breaches of that term. S. 7 of the Unfair Contract Terms Act 1977 requires reasonableness if the clause is to be effective against an Employer who, for the purposes of the Contract, deals as a consumer. In respect of a similar clause in the *Intermediate Form of Contract, 1984 Edition*, (IFC 84) Jones and Bergman[42] argued that such a term is reasonable in the circumstances of a building contract in which the Contractor is required to comply with a design for which the Employer is responsible. It is submitted that the same position generally applies to the JCT 98.

However, as explained in Section 2.3.4, Clause 6.1.5 may be denied effect on the grounds that it is an implied term of building contracts that the Contractor will exercise the reasonable skill and care of an ordinarily competent contractor. The contractor would therefore be in breach of this term if he fails to notice a contravention of the Statutory Requirements that a competent contractor ought reasonably to have discovered.

---

41 (1980), reported in Parris, J., *Standard Form of Building Contract: JCT80* (Collins, 1982), at p. 92.
42 Jones, N. F., Bergman, D., *A Commentary on the JCT Intermediate Form of Building Contract*, 2nd edn, (BSP Professional Books, 1990), p. 274.

There is the related question whether the Contractor is liable to the Employer in tort in circumstances where he is exonerated by Clause 6.1.5 from liability under the Contract. There is now little doubt that the effect of the 'contract structure' defence against imposition of a duty of care is that the Contractor would not be liable in the tort of negligence. Whether there is liability for breach of statutory duty is a question of whether the relevant statute provides for civil liability. Under s. 38 of the Building Act 1984, breach of the Building Regulations gives rise to civil liability but that section is not yet in force. Where the Contract requires 'work for or in connection with the provision of a dwelling' the Contractor owes a duty to the Employer under s. 1(1) of the Defective Premises Act 1972 'to see that the work that he carries on is done in workmanlike manner, with proper materials ... and so as to be fit for the purpose required'. However, s. 1(2) exonerates from liability 'a person who takes on any such work for another on terms that he is to do it in accordance with instructions given by or behalf of that other ... to the extent to which he does it properly in accordance with those instructions'. This section is clearly Clause 6.1.5 in relation to dwellings.

2.3.7.2 Unauthorized variations to comply with statutory requirements

Under normal circumstances, the Contractor must not vary the Works without a variation order from the Architect. However, sometimes, it happens that the Contractor has to carry out variations to comply with the Statutory Requirements in circumstances where it is not practicable to wait for variation instructions. For example, the Health and Safety Executive (HSE) may make orders requiring immediate remedying of a particular hazard. The Conditions recognize the possibility of such situations arising by allowing the Contractor to supply any materials and do any work necessary without waiting for instructions from the Architect. However, the Contractor must inform the Architect of such emergency and the steps taken. Where the emergency work was made necessary by a divergence between the Statutory Requirements and Contract, it should be treated as if the Architect had given an instruction pursuant to a notice of the divergence from the Contractor.

2.3.8 Duty to indemnify the Employer

The Contractor undertakes to indemnify the Employer against any expense, liability, loss, claim or proceedings in connection with any of the following matters:

- statutory fees/charges (Clause 6.2);
- the Contractor's infringement of patent rights in the carrying out of the Works described in the Contract Bills (Clause 9.1);
- personal injury or death (Clause 20.1);[43]
- damage to property (Clause 20.2);[44]
- the Contractor's incorrect statement of his direct costs for the purposes of the Statutory Tax Deduction Scheme (Clause 31.6.2);
- the Contractor's failure without good reason to enter into a nominated sub-contract within 10 days of the Architect's nomination instruction (Clause 35.8).

---

43 See Section 7.2.1 for more details on this indemnity.
44 See Section 7.2.2.

As a cause of action under an indemnity accrues only when the Employer's loss from the relevant matter is established,[45] the Contractor may incur liability under the indemnities many years after expiry of the normal limitation period applicable to the Contract. To illustrate how onerous the indemnities are, consider a third party action in tort against the Employer for latent damage to neighbouring property caused by subsidence for which the Contractor is responsible under Clause 20.2. Under the Latent Damage Act 1986, such an action may be brought against the Employer 15 years after the date on which the damage occurred. The limitation period for the indemnity, which would be 12 years if the Contract was executed as a deed, may therefore begin to run from the fifteenth year after the damage, giving rise to potential liability for up to 27 years after commission of the relevant wrong.

2.3.9 Insurance obligations

The Contractor is required to take out and maintain insurance policies against certain risks. Details on the risks and the insurance requirements are provided in Chapter 7.

2.3.10 Duties with respect to assignment and sub-contracting

These matters are covered in Chapters 8 and 9.

2.3.11 Obligations in respect of other contractors of the Employer

It is not uncommon for the Contract to be only a part of a larger project. Clause 29 is designed to facilitate other work to be carried out by the Employer or his other contractors in parallel with the carrying out of the Works. It provides that where such other work is identified in the Contract Bills with sufficient information to enable the Contractor to carry out the Works, the Contractor must allow such other work to be carried out. Where the other work is not so identified, consent of the Contractor must first be obtained but such consent is not to be unreasonably delayed or withheld if requested.

2.4 Architect

Traditionally the Architect has been the leader of the design team. His relationship with the Employer is, in the main, governed by the contract between them. In most cases, the terms of this contract are those of the RIBA's Standard Form of Agreement for the Appointment of an Architect (SFA). However, it is not uncommon for the Architect's appointment to be informal. Where this is the case, a contract in those terms may be implied by a previous course of dealing between the particular parties. In the absence of such history of dealings, the courts may not imply an intention to contract on those terms.[46]

45 *Post Office v. Norwich Union Fire Insurance Ltd* [1967] 1 QB 363; *County & District Properties v. C. Jenner & Son Ltd* [1967] 2 Lloyd's Rep. 728; *R. & H. Green Silley Weir Ltd. v. British Railway Board* (1980) 17 BLR 94.
46 *Sidney Kaye, Eric Firmin & Partners v. Bronesky* (1973) 4 BLR 1.

The SFA covers the full diet of services offered by architects, ranging from feasibility studies, through design and supervision on site to commissioning. It was explained at the beginning of this chapter that the role of the Architect under the Contract is mainly in an administrative capacity although he may also have designed the works. This section considers the role of the Architect as the administrator of the Contract. His role as designer, if relevant, is covered in Section 2.8.

The JCT 98 expressly provides that the Architect must perform certain duties. They are summarized in Table 2.3. Such provisions, in themselves, do not impose on the Architect obligations to perform them because the Architect is not a party to the Contract. However, he would be under such obligations through express or implied terms in his contract of engagement to that effect.[47] In the performance of his duties, the Architect wears two hats but not both at the same time. With respect to some of the duties, he acts as an agent of the Employer. Any default by the Architect in the performance of such duties may therefore be treated by the Contractor as a default on the part of the Employer, for example, failure to provide necessary information or to certify interim payment. With other duties, the Architect is an independent professional and the Employer is not liable for defective performance of duties such as assessment of claims and quantification of the amount payable in certificates. In *Merton v. Leach*[48] Vinelott J described the duality of the role of the Architect under the JCT 63 in these terms:

> It is to my mind clear that under the standard conditions the Architect acts as a servant or agent of the building owner when supplying the Contractor with the necessary drawings, instructions, levels and the like and in supervising the progress of the work and in ensuring that it is properly carried out. ... To the extent that the Architect performs these duties the building owner contracts with the Contractor that the Architect will perform them with reasonable diligence and with reasonable skill and care. The contract also confers on the Architect discretionary powers which he must exercise with due regard to the interests of the Contractor and the building owner. The building owner does not undertake that the Architect will exercise his discretionary powers reasonably; he undertakes that although the Architect may be engaged or employed by him he will leave him free to exercise his discretions fairly and without improper interference by him.

2.4.1 Quality control duties

This section is to be read in conjunction with Section 2.3.1, which discusses the Contractor's obligations in respect of quality of materials and standards of workmanship. The relevant provisions in the Contract are in Clauses 2.1, 4, 6, 8, and 17.

Quality Control is easily the most important part of the Architect's supervisory role, the aim of which is to detect any non-conforming work or materials and to have the non-conformance remedied by using powers given him under the Contract. Under Clause 8, the Architect may do any of the following in relation to his quality control function:

---

47 *Townsend v. Stone Toms & Partners* (1984) 27 BLR 26 the Court of Appeal decided, *inter alia*, that it was an implied term of an architect's contract of engagement in relation to supervise that he would supervise the building contract in accordance with its terms.
48 See, n. 2, at pp. 136–37.

Table 2.3 Duties of the Architect

Clause	Duty
2.2.2.2	to correct any errors in the Contract Bills (although the clause does not state expressly that this is the duty of the Architect, it is so inferred from the deeming of the correction as a variation)
2.3.	to issue instructions to resolve discrepancies within or divergences between contractual documents
3	take any addition to or deduction from the Contract Sum into account in the computation of the next Interim Certificate
4.2	to specify the clause authorizing any instruction that is challenged by the Contractor in writing
4.3.1	to issue all instructions in writing
5.4.1	to ensure that 2 copies of the information referred to in the Information Release Schedule are released at the time stated in the Schedule
5.8	to issue any certificate due under the Contract to the Employer but with a duplicate copy to the Contractor
6.1	to issue instructions to resolve any divergence between the Statutory Requirements and any other contractual document within 7 days of the Contractor's notice or its discovery by the Architect
6.1.6	to give notice to the Contractor if he discovers any divergence between the Statutory Requirements and a Contractor's Statement
7	to determine levels and provide to the Contractor dimensioned drawings and such other information to enable the Contractor set out the Works at ground level
13.3	to issue instructions to the Contractor in regard to the expenditure of provisional sums in the Contract Bills or Nominated Sub-contracts
17.1	to issue the Certificate of Practical Completion when the conditions for its issue have been satisfied
17.2	to issue a schedule of defects not later than 14 days after expiration of the Defects Liability Period if defects, shrinkages or other faults appear within that Period
17.4	to issue a Certificate of Completion of Making Good Defects when the Contractor has made good all defects in the Works for which the Contractor is responsible
18.1	to issue a written statement to the Contractor identifying part(s) of the Works taken into possession by the Employer before issue of the Certificate of Practical Completion
18.2	to issue a Certificate of Completion of Making Good Defects in respect of any part taken over when the Contractor has made good all defects in the part for which the Contractor is responsible
22A.4.2 22B.3.2 22C.4.1	to ignore any damage to the Works from the insured risks in computing amount payable to the Contractor under the Contract
22D.1	where 22D applies: (i) to inform the Contractor whether or not insurance under the clause is required; to obtain from the Employer the information reasonably required to obtain the insurance; (ii) to instruct the Contractor whether or not to accept a quotation for the insurance
25.3	to grant extension of time in accordance with the clause
26.1	to ascertain or instruct the Quantity Surveyor to ascertain loss and/or expense for which the Contractor has applied
26.3	to state the extension of time granted in respect of a relevant matter that is also a Relevant Event
26.4	to ascertain or instruct the Quantity Surveyor to ascertain loss and/or expense due under Nominated Sub-contracts in accordance with the clause
30.1.1.1	to issue Interim Certificates in accordance with Clause 30
30.7	to issue not less than 28 days before the date of issue of the Final Certificate an Interim Certificate containing the final accounts of all Nominated Sub-contractors
30.8	to issue the Final Certificate
34.2	to issue instructions in regard to what is to be done concerning fossils, antiquities and other objects of interest of value found on the site

Table 2.3 *continued*

34.3	to ascertain or to instruct the Quantity Surveyor to ascertain loss and/or expense for disturbance to progress of the Works by discovery of fossils and the like
35.6	to issue nomination instructions in accordance with the clause
35.9	to issue instructions to deal with failure of the Contractor to enter into a Nominated Sub-contract within 10 days of receipt a nomination instruction
35.13	to direct the Contractor as to sums included in Interim Certificates as interim or final payment to Nominated Sub-contractors and to advise each Nominated Sub-contractor of such sums
35.14.2	to assess extensions of time for Nominated Sub-contract Works
35.15.1	to issue Certificates of Non-completion in respect of Nominated Sub-contract Works
35.16	to issue Certificates of Practical Completion of Nominated Sub-contract Works
35.17	on expiry of 12 months from practical completion of a Nominated Sub-contract, to issue an Interim Certificate that incorporates the final accounts of the Sub-contract
35.24.6.3 35.24.7.3 35.24.7.4 35.24.8.1 35.24.8.2	to re-nominate a sub-contractor where a Nominated Sub-contractor's employment is determined or he fails to carry out work to be re-executed or to make good defects
35.26	after determination of the employment of a Nominated Sub-contractor, to issue instructions directing the Contractor regarding financial settlement with the Sub-contractor
36.2	to issue instructions for the purpose of nominating a supplier for materials or goods covered by a prime cost sum in the Contract Bills
36.4	except where the Architect and the Contractor otherwise agree, to nominate only suppliers who will enter into a supply contract incorporating the terms itemized under the clause

- request relevant vouchers;
- carry out inspection and testing of materials or executed work;
- instruct removal of non-conforming work;
- accept non-conforming work after due consultation;
- issue appropriate variation and other instructions;
- omit non-conforming work and work when preparing Interim Certificates;
- exclude from the site any person employed on it.

In support of these powers, the Architect is entitled under Clause 11 to access at all reasonable times to the Works and workshops and other location where work is being done for the Contract. The Contractor is to ensure that the Architect has similar rights of access to the work and workshops of domestic sub-contractor and Nominated Sub-contractor. The Contractor is to include dovetailing terms in sub-contracts.[49]

2.4.1.1 Vouchers

The Architect may request the Contractor to provide him with vouchers covering materials for the Works (Clause 8.2.1). Such documents often contain some information on the materials, e.g. specifications, prices and names of suppliers, that allow the Architect to draw useful inferences on quality issues for appropriate follow up.

---

49 DOM/1 and Conditions NSC/C contain dovetailing provisions.

2.4.1.2 Inspection and testing

The Architect may issue instructions requiring the Contractor to carry out any test on executed work or materials for the Works either himself or by third parties (Clause 8.3). Where the work has already been covered up, he may instruct that it is opened up for inspection and such testing. Clause 8.4.4 is intended to deal with situations where the Architect, as a result of discovery of actual non-compliance with the Contract, has reasonable grounds to suspect other instances of further or similar non-compliance. He may issue further instructions under Clause 8.3 for opening up or testing other work as reasonably necessary to deal with those suspicions. Reasonableness of the Architect's actions pursuant to Clause 8.4.4 is governed by a Code of Practice annexed to the Conditions. For example, consider where, on excavating one of a row of column footings of identical design, it is discovered that its dimensions are wrong. The Architect may insist on excavating the other footings to check that the same or similar mistakes have not been made.

It is important to bear in mind that there are two different types of instruction involved here, those under Clause 8.3 only and those under Clause 8.4.4 and 8.3. Both may entail: (i) additional costs incurred in the tasks of opening up, inspection, testing and reinstatement; (ii) loss and/or expense incurred as a result of disturbance to the carrying out of the Works; (iii) delays to completion of the Works. The Contractor's entitlements in respect of these consequences are summarized in Table 2.4.

Table 2.4 Instruction Under Clauses 8.3 and 8.4.4

Instruction	Outcome of Inspection/Testing	Costs of Compliance	Loss and/or Expense	Extension of Time
Under Clause 8.3 only	Materials/work are in accordance with the Contract	To be added to the Contract Sum unless provided for in the Contract Bills	Entitlement	Entitlement? *See discussion*
	Materials/work are not in accordance with the Contract	To be added to the Contract Sum only if provided for in the Contract Bills *See discussion*	No entitlement	No entitlement
Under Clause 8.3 but pursuant to Clause 8.4.4	Materials/work are in accordance with the Contract	Not to be added to the Contract Sum if instruction was reasonably necessary	No entitlement	Entitlement
	Materials/work are not in accordance with the Contract	Not to be added to the Contract Sum if instruction was reasonably necessary	No entitlement	No entitlement

It is therefore to be noted that there are differences in entitlement depending upon whether an instruction is issued under Clause 8.3 alone or under Clause 8.3 but pursuant to Clause 8.4.4. To avoid disputes concerning these differences, the Architect must always state clearly in the instruction itself which type is intended where he has to issue any of them.

It is stated in Clause 8.4.4 that the Contractor is not entitled to any additional payment for compliance with an Architect's instruction issued pursuant to that clause provided that the instruction was reasonable in the circumstances. There is no entitlement even if the inspection or test shows that the work and materials are in accordance with the Contract. As this clause applies only where, as a consequence of prior discovery of non-compliance, the Architect needs to be satisfied that there is no further non-compliance of a similar nature, it is only fair that the Employer does not pay for a situation brought about by the Contractor's failure to comply with the contract. Under Clause 25.4.5.2, it is a Relevant Event if delay is caused 'in regard to the opening up for inspection of any work covered up or the testing of any of the work, materials or goods in accordance with Clause 8.3 (including making good in consequence of such opening up or testing) unless the inspection or test showed that the work, materials or goods were not in accordance with this Contract'. It is also stated in Clause 8.4.4 that, for a Clause 8.3 instruction to which Clause 8.4.4 is applicable, Clause 25.4.5.2 applies if the work or materials are found to be in accordance with the contract. This express provision was probably intended to avoid the temptation to deny the Contractor extension of time on account of the fact there is no entitlement to recovery of loss and/or expense even if the work and materials are found to be in accordance with the Contract. Why the Contractor should be given more time for dealing with a consequence of his own breach of contract is hard to justify. However, such express provision in Clause 8.4.4 for extension of time opens up a possibility for argument that the absence of such a provision under Clause 8.3 means that there is no entitlement to extension of time for delays caused by instructions under Clause 8.3 that are not pursuant to Clause 8.4.4.

Under Clause 8.3, if there is provision for opening, inspection, or testing in the Contract Bills, the costs of compliance with such instructions are not to be added to the Contract Sum regardless of the outcome of the inspection or testing. This provision was probably designed to avoid adjusting the Contract Sum twice for the same work (pursuant to the provisions in the Contract Bills and under Clause 8.3) where the work/materials are found to be in accordance with the Contract. However, it is doubtful whether the intention is to adjust the Contract Sum where work or materials are found not to be in accordance with the Contract as this would be contrary to established principles of risk allocation in construction contracts. With such an arrangement, there would be no incentive for the Contractor to avoid contravention of the Contract unless the description of the relevant item in the Contract Bills limits its application only to situations where the work/materials are found to be in accordance with the Contract.

2.4.1.3 Removal of non-conforming work and materials

If the Architect expresses his opinion that work done ought to be taken down or that specific materials should not be used, the Contractor will normally comply. However, there is always the danger that once the Architect's back is turned, the Contractor will do whatever he likes, particularly where the relevant work is to be covered up soon afterwards. Clause 8.4.1 is designed to avoid this risk by allowing the Architect to require the Contractor to remove the non-conforming work or materials from the site altogether. For the obligation to remove work or materials to arise the Architect must expressly and unequivocally require the Contractor to do so. Merely condemning the work or materials as non-conforming and requiring the Contractor to ensure compliance with the Contract would not amount to instructing their removal.[50] Here again, the Architect may advise the Employer to consider exercising his right under Clause 4.1.2 to have the work or materials removed at the Contractor's expense if the Contractor fails to comply with the AI.

2.4.1.4 Acceptance of non-conforming work

The Architect may, after consultation with the Contractor (who must also consult Nominated Sub-contractors as appropriate), allow non-conforming work or materials to be left and to make an appropriate deduction from the Contract Sum to reflect the non-conformance (Clause 8.4.2). The Architect is required to confirm such acceptance of non-conforming work to the Contractor in writing. The clause states expressly that such allowance does not amount to a variation under the contract. Such express provision avoids uncertainty of the kind that arose in *Simplex v. The Borough of St Pancras*.[51] In that case the Architect accepted an alternative proposed by the Contractor to deal with defects. The Contractor's contention that the alternative amounted to variation, with the usual entitlement to extension of time and recovery of loss and/or expense, was accepted by the court.

There is no mechanism for determining the amount of the 'appropriate deduction'. There is even no requirement for the Quantity Surveyor to be involved except to adjust the Contract Sum by the amount of the appropriate deduction although, in practice, such a role goes unchallenged by the Employer and the Contractor. The Guidance Notes to Amendment 5 to the JCT 80, which introduced Clause 8.4.2, take a different line on the role of the Quantity Surveyor. They state that the determination of the amount is a matter for him. However, there are at least two arguments against this construction. First, a duty to include an amount in the final account does not necessarily mean a duty to determine its quantum. In many similar situations, the Quantity Surveyor is expressly required to determine the quantum, e.g. valuation of variations under Clause 13.5, even though Clause 30.6.2 also requires him to take account of the amount in the final accounts. Second, and more importantly, the whole arrangement is subject to the agreement of the Employer. He is entitled to withhold agreement unless he considers the amount involved to be the 'appropriate deduction'. Generally, with matters for the Quantity Surveyor's determination, the Employer cannot interfere in this way.

It would appear, therefore, that the Employer is in the driving seat, particularly as there is no requirement for his rejection of the arrangement to be reasonable. However, the Guidance Notes to Amendment 5 caution that 'The Tribunal considers that in no circumstances could the deduction take the form of a "penalty" on the Contractor nor should it equate or relate to the cost of making the non-conforming work conform to the Contract which the Contractor has saved because the non-conforming work is being allowed to remain'. These arguments are relevant only where the Employer agrees to the Architect allowing the non-conforming work to remain. If the Employer refuses, the Contractor would have to make good the non-conformance, particularly as the Architect has several ways of ensuring this. For example, the Architect may instruct removal of the non-conforming work and advise the Employer to exercise his right under Clause 4.1.2 to have it done if the Contractor fails to comply.

---

50 *Holland Hannen & Cubitts (Northern) Ltd v. Welsh Hospital and Technical Services Organisation* (1981) 18 BLR 80.
51 (1958) 14 BLR 80.

2.4.1.5 Variations and other instructions

Under Clause 8.4.3, the Architect may issue any variation instruction made reasonably necessary by an AI to remove non-conforming work and materials or acceptance of non-conforming work. Before the Architect can do this, he must consult the Contractor and, through him, relevant Nominated Sub-contractors. Under Clause 8.5, the Architect may, after similar consultation, issue an instruction to deal with the Contractor's 'failure to comply with Clause 8.1.3 in regard to the carrying out the work in a proper and workmanlike manner' irrespective of whether he has instructed removal of the work or allowed it to remain. Amendment 14 changed Clause 8.1.3 from 'All work shall be carried out in a proper and workmanlike manner' to 'All work shall be carried out in a proper and workmanlike manner and in accordance with the Health and Safety Plan'. However, there is no reference in Clause 8.5 to the Architect's powers to issue instructions to deal with the Contractor's failure to comply with the Health and Safety Plan. This was probably inadvertent.

An instruction under Clause 8.5 may, but does not have to, be a variation. If the Contractor fails to comply with the instruction, the Employer may exercise his right under clause 4.1.2 to give effect to the instruction at the Contractor's expense. To the extent that an instruction under Clause 8.4.3 or 8.5 is made necessary by the non-conformance, it is not a ground for extension of time or addition to the Contract Sum. It is only fair that the Contractor should not gain from his own breach of contract.

There are two possible constructions of Clause 8.1.3, one wide and the other narrow. The use of the phrase 'proper workmanlike manner' suggests that it is intended to deal with workmanship. The wide construction of the clause is that it covers workmanship not only in the sense of the final quality of the work produced but also the physical methods of achieving it. Such a construction entails some overlap with Clause 8.1.2 with respect to quality matters on which the Contract is silent (quality neither covered by objective specifications nor expressed as subject to the reasonable satisfaction of the Architect).[52] This overlap may be reconciled by treating the standard in Clause 8.1.3 as the baseline when deciding what standard is appropriate for the Works. Such a wide construction has the implication that the Architect may issue an instruction under Clause 8.5 requiring the Contractor to make good defects arising from failure to comply with Clause 8.1.3. The Architect is given express powers to issue instructions requiring the Contractor to make good defects only under Clause 17. Such express powers must be exercised only after the Certificate of Practical Completion and terminate with delivery of the schedule of defects under Clause 17.2 or, where no schedule is delivered, after 14 days from the expiry of the Defects Liability Period.[53] There would be no such limitations on instructions under Clause 8.5 to make good defects arising from contravention of the wide construction of Clause 8.1.3. They may therefore be issued at any time before the Final Certificate.

The narrow construction is that 8.1.3 concerns only the Contractor's work methods, with Clause 8.1.2 alone covering the quality of the finished work. This is probably the intended construction because the clause was introduced by Amendment 9 of the JCT 80 in response to the Court of Appeal's refusal in *Greater Nottingham Co-operative Society v. Cementation Piling and Foundations Ltd*[54] to imply such term into a warranty between

---

52 Other well-known examples of overlap include Article 1 and clauses 2.1 and 8.1.
53 See Section 3.5 for details on instruction to make good defects during the Defects Liability Period.
54 (1988) 41 BLR 43.

an employer and a nominated sub-contractor. In that case negligent piling by a nominated sub-contractor caused damage to a restaurant adjoining the site. The court of first instance, who found that the collateral contract between the employer and the nominated sub-contractor was in equivalent terms to the JCT 63, declined to imply a term that the Sub-contractor would carry out the work with reasonable skill and care or without damage to surrounding buildings. This point was conceded in the appeal which concerned liability in tort.

2.4.1.6 Defective work and Interim Certificates

Under Clause 30.2.1 the Architect is to include in Interim Certificates only work properly executed. This qualification represents probably the most potent weapon by which the Architect can enforce compliance with the required quality standards. It is also an argument against a wide construction of Clause 8.1.3 on the grounds that, with such a weapon, the Architect has no need for powers to issue instructions requiring the Contractor to make good defects before Practical Completion. It is to be noted that the Architect has no discretion to include non-conforming work in Interim Certificates. In *Townsend and Another v. Stone Toms & Partners*[55] the Court of Appeal held that it was a breach of his contract for an architect to include defective work of which he was aware in interim certificates under the then current edition of the JCT's prime cost contract. It was stated that the fact that there was enough retention to cover the cost of making them good if the Contractor failed to do so was not a proper ground upon which the architect could ignore the terms of the contract to pay for only work properly executed.

2.4.1.7 Exclusion of persons

The Architect may issue instructions excluding from the site any person employed on it: Clause 8.6. This power is not to be exercised 'unreasonably and vexatiously'.[56] This power is probably designed to avoid recurrent problems with consistently sub-standard work by particular individuals or firms.

2.4.2 Certification duties

The Architect is responsible for issuing certain certificates or certifying certain matters. They are to be issued to the Employer but with duplicate copies to the Contractor: Clause 5.8. The certificates or matters are:

- the Certificate of Practical Completion (Clause 17.1);
- the Certificate of Completion of Making Good Defects (Clause 17.1);
- a certificate to the effect that frost damage occurred before Practical Completion (Clause 17.5);
- where the Employer wishes to take possession of a part of the Works, a certificate stating that defects, shrinkages or other defaults in that part have been made good (Clause 18.1.2);
- Certificates of Non-completion (Clause 24.1);
- Interim Certificates (Clause 30.1.1.1);

---

55 See n. 47.
56 For discussion of the meaning of that phrase see Section 16.18.

- certificates stating that the Contractor has failed to pay a Nominated Sub-contractor sums included in an Interim Certificate in respect of the relevant Nominated Sub-contract Works (Clause 35.13.1);
- Certificate of Completion of Nominated Sub-contract Works (Clause 35.16);
- Interim Certificate incorporating final accounts of all Nominated Sub-contracts.

It is to be noted that the certificates fall into two categories. First, there are those that the Employer warrants that the Architect shall issue, e.g. interim and final payment certificates. Failure of the Architect to issue this type of certificate in compliance with the Contract is therefore a breach of the Contract by the Employer. Generally, this type of certificate involves statements of quantum and the Employer is not vicariously liable for any negligence of the Architect in arriving at the quantum. The appropriate remedy is for the Contractor to challenge the contents of the certificate by invoking the applicable dispute resolution procedure. The second category consists of decisions of the Architect, arrived at through exercise of his professional skills that certain events have occurred, e.g. practical completion of the works or completion of making good defects. The Employer is not liable to the Contractor for issue or failure to issue this type of certificate. Here too, the appropriate remedy is through invocation of the relevant dispute resolution procedure to determine whether or not the certificate should be issued. Any negligence by the Architect in the performance of duties to certify is a breach of his contract of engagement for which the Employer would be entitled to damages.[57]

2.4.3 Duty to report on failings of other members of the design team

The Architect is often under an express duty under his contract of engagement to coordinate the work of all members of the design team not only during design but also during the construction phase.[58] Such a term may even be implied as a custom of the construction process. A first-instance decision suggests that the Architect may owe the Employer a duty in contract and in tort to advise and/or inform him of actual or potential deficiencies in the performance by the Quantity Surveyor and other consultants of their contractual duties to the Employer.[59] However, it was also held in that case that there is no duty to warn in respect of the Architect's own deficient performance or that of the Employer's project manager. It may even be argued that, depending on the type of error and all the surrounding circumstances on site, the Architect may even be under a duty to correct design errors by other designers. An example is where the Architect should reasonably be expected to possess relevant design expertise and it is clearly in the interests of the Employer that a design change is implemented immediately without inviting or waiting for a response from the original designer. However, the Architect would be prudent to follow the formal process of informing the Employer and getting his express approval to the change.

---

57 See Section 15.9.
58 See Clauses 3.3.3 and 4.2.4 of the SFA.
59 *Chesham Properties Ltd v. Bucknall Austin Project Management Services Ltd*, see n. 6.

2.4.4 Standard of skill and care

At common law, the standard of skill and care to be exercised by the Architect in the performance of his duties is that required of every professional person, i.e. the reasonable skill and care an ordinary competent architect would exercise in the circumstances of the particular job.[60] Clause 1.2.1 of SFA provides that, in the provision of the services he was engaged to deliver, 'the Architect shall exercise reasonable skill and care in conformity with the normal standards of the Architect's profession'. Where there is no such express term in the particular contract of engagement, it will be implied unless there is express provision to the contrary. Even where there is exclusion of the duty of skill and care, it will almost certainly be struck down by the Unfair Contract Terms Act 1977. Case law[61] suggests that he must possess a good general understanding of the law in relation to building contracts. In particular he would be expected to have good understanding of the terms of the Contract and relevant case law. Although he is not expected to possess the expert knowledge of a lawyer, he must be able to recognize when to advise his client to seek appropriate legal advice.

It is clear from the case law[62] that exercise of the appropriate standard of skill and care does not require the Architect to be permanently on the site. In *Corfield v. Grant and Others*[63] Judge Bowsher had this to say on the subject:

> What is adequate by way of supervision and other work is not in the end to be tested by the number of hours worked on site or elsewhere, but by asking whether it was enough. At some stages of some jobs exclusive attention may be required to do the job in question ... at other stages of the same jobs, or during most of the duration of other jobs, it will be quite sufficient to give attention to the job only from time to time. The proof of the pudding was in the eating. Was the attention given enough for this particular job.

It would appear that the experience and competence of the particular contractor is also a factor to be taken into account. In the *East Ham v. Sunley* case, Lord Upjohn said that where an architect knows the contractors sufficiently well and can rely on them to do a good job, it would be proper to relax the supervision of detail in favour of other matters. It was also suggested in *Sutcliffe v. Chippendale and Edmondson*[64] that the degree of supervision should be higher where the experience or competence of the contractor is questionable. However, where a job requires more inspection by an architect than is possible because of the Architect's other responsibilities, the Architect may still be liable.[65]

The SFA deals with the frequency of the Architect's site visits expressly. Clause 3.1.1 of that form requires him to make such visits as he reasonably expected to be necessary

---

60 *Bolam v. Friern Hospital Management Committee* [1957] 1 WLR 582; 2 All ER 118; the *Bolam* standard was applied to the supervisory duties of architects by the Court of Appeal in *West Faulkner Associates v. London Borough of Newham* (1994) 71 BLR 1.
61 *Townsend (Builders) Ltd v. Cinema News & Property Management Ltd*, see n. 37; *B. L. Holdings v. Robert J. Wood & Partners* (1979) 12 BLR 1.
62 *East Ham Corporation v. Bernard Sunley & Sons Ltd* (1966) AC 406 (hereafter *East Ham v. Sunley*); *Corfield v. Grant and Others* (1992) 59 BLR 102.
63 *Ibid.*
64 (1971) 18 BLR 149. This case went to appeal under the name of *Sutcliffe v. Thackrah*, see n. 4, but not on this issue.
65 *Corfield v. Grant and Others*, see, n. 62.

when he was appointed. This test of sufficiency of the visits is therefore subjective, a contrast with the objective standard under the common law. The Architect is under a duty to confirm this expectation to his client in writing. It is submitted that a duty to do so within a reasonable time of his engagement would be implied. Reasonableness should reflect the state of the client's brief and the complexity of the project. However, there is no need to delay the confirmation until he is certain of the frequency required because Clause 3.1.2 anticipates changes in his expectation by providing that, if he does revise them, he must so inform the Employer in writing. It is therefore clear that constant presence of the Architect on site is not the norm. Clause 3.3 of the SFA reflects this by requiring the Architect to advise the Employer in writing of any need for staff, e.g. resident architects, representatives of designers and clerks of works, to assist him in supervision and administration of the Contract. With the exception of the provision in Clause 12 for a Clerk of Works, the JCT 98 does not provide for the roles of such additional staff because, as stated in the SFA, all site staff so appointed are to be under the direction and control of the Architect.

From the above discussion, the mere fact that there are undiscovered defects does not necessarily mean that the Architect was negligent in his supervision of the Works. Indeed, in *Gray and Others v. Bennett & Sons and Others*,[66] Sir William Stabb QC, sitting on Official Referees' business, decided, on a preliminary issue, that an architect was not liable for defects that the contractor's workmen had deliberately concealed. Similarly, in *Department of National Heritage v. Syeensen Varming Mulcahy and Others*[67] Bowsher J rejected an employer's argument that defective work was itself evidence of inadequate supervision by consulting engineers who carried similar supervisory responsibility to the Architect under the JCT 98.

2.4.5 Liability of the Architect to the Employer

Where the Architect fails to discharge his role as Architect with the required standard of skill and care, he would be liable to the Employer in contract and tort. For example, the House of Lords held in *Sutcliffe v. Thackrah*[68] that the Architect was liable in contract for negligent certification. In *West Faulkner Associates v. London Borough of Newham*,[69] which arose from a JCT 80 contract, the Employer wished to determine the Contractor's employment on grounds of alleged failure to proceed regularly and diligently with the carrying out of the Works, as they were entitled to do, but could not do this because the Architect refused to issue the Contractor with a notice of the default, a precondition of the Employer's right to determine. The Court of Appeal found that, on the facts, the Contractor had committed the default and that the Architect should have issued the notice. The Architect was therefore held liable for the Employer's loss caused by not determining when he wished to do so.

Unless his contract of engagement provides to the contrary, the Architect would be under a duty to advise the Employer on the general operation of the Contract and on the

---

66 (1987) 43 BLR 63.
67 (1998) CILL 1422.
68 See n. 4.
69 See n. 60.

Employer's responsibilities within it. Where the Architect does not possess the required special knowledge, he should advise the Employer to consult a lawyer or other expert. In *Pozzolanic Lytag Ltd v. Brian Hobson Associates*[70] it was decided on a preliminary issue that a project manager appointed to administer a contract let on the JCT 80 With Contractor's Designed Portion Supplement was under a duty to check that the scope of the Contractor's insurance arrangements met the requirements of the Contract. To the project manager's contention that his role in that respect was simply to collect evidence on the insurance arrangements and to pass it to the Employer for consideration, Dyson J said:

> I cannot agree with the opinion of [the project manager]. If a project manager does not have the expertise to advise his client as to the adequacy of the insurance arrangements proposed by the contractor, he has a choice. He may obtain expert advice from an insurance broker or lawyer. Questions may arise as to who has to pay for this. Alternatively, he may inform the client that expert advice is required, and seek to persuade the client to obtain it. What he cannot do is simply act as a 'postbox' and send the evidence of the proposed arrangements to the client without comment.

It is submitted that these comments apply to the Architect under the JCT 98.

2.4.6 Liability of the Architect to the Contractor and other third parties

As there is no contract between the Architect and any of these parties, any liability would have to be found in tort. The general principles governing this type of liability are outside the scope of this book. However, three points have to be made. First, there is no legal responsibility for the Architect to protect the Contractor's interests by intervening in the manner in which the Contractor is carrying out the Works even if the Architect notices that the Contractor is working inefficiently or even dangerously. In *Oldschool v. Gleeson*[71] in which a contractor contended that supervising engineers were liable to them for the collapse of a wall, Sir William Stabb, QC, sitting on Official Referee's Business, stated:

> The duty of care of an architect or of a consulting engineer in no way extends into the area of how the work is carried out. Not only has he no duty to instruct the builder how to do the work or what safety precautions to take but he has no right to do so, nor is he under any duty to the builder to detect faults during the progress of the work. The architect, in that respect, may be in breach of his duty to his client, the building owner, but this does not excuse the builder for faulty work.

Second, as explained in some detail in Section 15.9, it is doubtful whether the Architect will incur liability to the Contractor for his loss arising from under-certification of payment. Third, there is little doubt that the Architect may be found liable in tort for personal injury and damage to property. For Example, in *Clay v. Crump & Sons Ltd*[72] an architect was found jointly liable with a contractor for injury suffered by contractor's

---

70 (1999) 15 Const. LJ 135. For a similar finding see *William Tomkinson v. The Parochial Church Council of St Michael* (1990) 6 Const. LJ 319.
71 (1974) 4 BLR 103; see similar comments in *Clayton v. Woodman & Son (Builders)* [1962] 2 All ER 33; [1962] 1 WLR 585
72 [1963] 3 WLR 866.

workmen as a result of the collapse of a wall that the architect had instructed the contractor to leave standing on the site. The architect was found negligent because had he inspected the wall, which he failed to do, he would have found that the wall was too unstable to be left standing. There may also be liability for pure economic loss under the *Hedley Byrne* principle where the Architect goes out of his way to advise the Contractor and the requirements for that type of liability are met.[73]

2.5 Quantity Surveyor

The Contract states the duties of the Quantity Surveyor as:

1. to act on variations in accordance with Clauses 13.4.1.2A2, 13.4.1.2A4, 13.3.1.2, 13.4.1.2A7.1;
2. to ascertain loss and/or expense that has been, or is being, incurred by the Contractor on account of a Relevant Matter if instructed by the Architect to do so (Clauses 26.1 and 34.3.1);
3. to carry out valuations for interim certificates (where Clause 40 applies the Quantity Surveyor must always do so but where it does not apply he must do so only if instructed by the Architect) (Clause 30.1.2.1);
4. to prepare the final accounts of Nominated Sub-contractors (Clause 35.17.2);
5. to prepare the final accounts for the Contract (Clause 30.6.1.2.2).

In the performance of his duties under the Contract, the standard of skill and care to be exercised by the Quantity Surveyor is the normal *Bolam*[74] standard applicable to all professional people. A pre-*Bolam* case suggests that occasional arithmetical or clerical errors may be consistent with exercise of the proper standard of skill and care. In *London School Board v. Northcroft*[75] it was held that a quantity surveyor was not liable for two clerical errors made by his assistant in the computation of the contract price for completed buildings. However, the learned editor of *Hudson's Building and Engineering Contracts* warns that such leniency may not apply today.[76] The Quantity Surveyor is required by s. 13 of the Supply of Goods and Services Act 1982 to exercise reasonable skill in the delivery of his professional services. It is therefore arguable that, although he has no duties with respect to quality matters, he owes the Employer a duty to draw to the attention of the Architect obvious defects that he notices in the course of measurements for valuations.

The comments about the Architect's liability to the Contractor would also apply to the corresponding liability of the Quantity Surveyor.[77]

---

73 For an example that pre-dates the development of *Hedley Byrne* liability, see *Townsend (Builders) Ltd v. Cinema News & Property Management Ltd, supra,* n. 37. in which the architect was liable to the contractor for costs of remedial work to comply with bye-laws on account of advice to the contractor during construction that he would resolve the contravention with the local authority and that the contractor should therefore ignore it.

74 *Bolam v. Friern Hospital Management Committee* [1957] 2 All ER 118.

75 (1889) *Hudson's*, 11th edn, at para. 2.230.

76 *Hudson's* para. 2.230

77 See Section 2.4.6

2.6 Planning Supervisor

Article 6.1 names the Planning Supervisor.[78] The Contract expects him to be the Architect or other named individual from a named organization. Practice Note 27 advises that having the Architect doubling up as the Planning Supervisor avoids creating an additional interface that would otherwise result in delays in managing the health and safety implications of variations and post-contract design and planning. The Employer must appoint a replacement if an incumbent Planning Supervisor dies or otherwise ceases to be the Planning Supervisor. As the Article is silent on the Contractor's right to object, a contrast to the Contractor's express rights under Articles 3 and 4 to raise reasonable objections to the appointment of replacement Architects and Quantity Surveyors, the implication has to be that the Contractor has no such right.

The Planning Supervisor's general responsibility is to oversee and coordinate health and safety aspects of design and planning right from project inception to completion. Specific duties under the CDM Regulations are summarized in Table 2.5. On the type of projects for which the JCT 98 is suitable, most of the work of the Planning Supervisor is done at the pre-tender stage. Ideally, an appointment must therefore be made to this position as early as practicable. Although we are concerned mainly with his duties during construction, some understanding of his pre-tender duties is required as well because many of them provide the necessary foundation for the duties at the post-tender stage.

Table 2.5 Specific duties of the Planning Supervisor under the CDM Regulations

Regulation	Duty
7	if the project is notifiable, to notify the HSE as soon as practicable after his appointment and again after the appointment of the Principal Contractor
11; 12	to obtain from the Employer information relevant to health and safety that is needed by designers and others
13(2); 14(a)(i)	to ensure that designers fulfil their duties under the CDM Regulations in respect of the health and safety aspects of their design
14(b)	to ensure that designers cooperate over health and safety matters
14, 14(c)(i)	if requested by the Employer, to give advice on the competencies and resources of designers and contractors
15(i)	to ensure that the Pre-tender Health and Safety Plan is prepared
14(c)(ii)	if requested by the Employer, to advise him whether the Construction Health and Safety Plan is advanced enough for the start of construction
ACOP* Paragraphs 70(c); 86; 87	to cooperate and liaise with the Principal Contractor in the continual review of the Construction Health and Safety Plan
14(d)–(f)	to ensure that the Health and Safety File is prepared and delivered to the Employer

* This is an acronym for Approved Code of Practice. The document itself has the title *Managing Construction For Health and Safety*. It contains guidance on compliance with CDM Regulations. For all intents and purposes, it is part of the CDM Regulations themselves as a court may treat failure to comply with the code as evidence of breach of relevant health and safety legislation.

2.6.1 Pre-tender duties

Specific pre-tender responsibilities include giving appropriate notices to the Health and Safety Executive, advising the Employer on competencies and resources of designers and ensuring that designers fulfil their obligations under the CDM Regulations. The end

78 See Section 1.5.2

product of his pre-tender duties is a Pre-tender Health and Safety Plan, which is a collection of information on health and safety in relation to the project, e.g. type of work, existing site conditions, access arrangements, traffic routes, security requirements and statements of risk assessments by designers. The information is acquired from many different sources such as the client and the designers. The Planning Supervisor's responsibility to produce the plan is therefore largely one of coordination of the input of the necessary information from the various sources. Each tenderer is to be provided with the plan as it has obvious pricing implications.

2.6.2 Post-contract duties

A Construction Health and Safety Plan must be developed from the Pre-tender Health and Safety Plan before construction can start. This task is to be undertaken by the Principal Contractor although the responsibility to ensure that construction does not start until this has been done remains with the Employer. However, the Planning Supervisor may be under a duty imposed by his contract of engagement to oversee the development of the plan on the Employer's behalf and to advise whether a plan submitted by the Principal Contractor has reached the stage where construction can be lawfully started.

For the purposes of the JCT 98, the Planning Supervisor has two main duties. First, he must continue to monitor and coordinate the health and safety aspects of any design and planning continuing during construction. Such design and planning may arise from variations, supply of Performance Specified Work or the appointment of sub-contractors with design responsibility. He must therefore be consulted on such matters and served with the information necessary to discharge this responsibility. Second, he must coordinate the production of a Health and Safety File for delivery to the Employer at the end of the construction phase. CDM Regulation 2 defines the Health and Safety File as a permanent record of those aspects of a construction project which might affect the health and safety of: (a) any person carrying out future construction, cleaning or demolition work upon the building; (b) any person occupying the building who may be affected by those carrying out work upon it. Typical contents of the file include 'as built' drawings, design criteria, in-built facilities for the maintenance of the building, operating and maintenance manuals for specialist plant and equipment and incoming services. Here again, the information required is scattered across the variety of participants on the project. Clause 6A.4 requires the Contractor to comply with the Planning Supervisor's reasonable timetable for the provision of information for the Health and Safety File and to ensure similar compliance by all sub-contractors. The information is to be provided to the Planning Supervisor where the Contractor is also the Principal Contractor but to the Principal Contractor if different from the Contractor.

It is important to note that, under Clause 6A.1 of the JCT 98, any failure of the Planning Supervisor properly to perform his duties under the CDM Regulations is in effect a breach of the Contract by the Employer for which the Contractor would be entitled to appropriate remedies. The Employer may in turn look to the Planning Supervisor for compensation against such liability to the Contractor. Breaches of the CDM Regulations also constitute criminal offences that may be prosecuted.

2.7 Principal Contractor

Under the CDM Regulations, the Employer must appoint, as soon as practicable, a Principal Contractor whose responsibility is to plan, manage and control health and safety matters during the construction phase of the project. Article 6.2 provides that the Contractor is also the Principal Contractor.[79] Although the Employer has a right to replace the Contractor as the Principal Contractor, Practice Note 27 cautions against this course of action for the same reasons as already given in relation to the Planning Supervisor. It is to be noted that the Principal Contractor is a creature of statute and that he must not therefore be confused with the terms 'main contractor' or 'prime contractor'.

Specific duties within the general responsibility of the Principal Contractor to coordinate health and safety during the construction phase are to:

- take over and develop the Pre-tender Health and Safety Plan into a Construction Health and Safety Plan (Reg. 15(4));
- monitor health and safety on site (Reg. 15(4)(a));
- ensure that contractors and sub-contractors are properly coordinated (Regs 15 and 16);
- obtain and examine proposals for minimizing risks from contractors (Regs 15 and 16);
- enforce site rules and site access arrangements (Reg. 16);
- ensure all workers are properly trained (Reg. 17);
- provide information for the Health and Safety File (Reg. 16(1)(e)).

The duty to develop the Pre-tender Health and Safety Plan prepared by the Planning Supervisor into a Construction Health and Safety Plan is probably the most important. This revised plan, which is that covered by the Clause 1.3 definition of 'Health and Safety Plan', must address issues such as the organization structure, site rules and safety and emergency procedures. Throughout the construction phase, the plan is to be kept under continual review so that it adequately addresses evolving health and safety issues. Such issues may arise from design changes, instructions as to the expenditure of provisional sums, appointments of sub-contractors with new design input, work methods and health and safety standards actually achieved. To make all these stringent requirements on planning meaningful, the Principal Contractor is also responsible for ensuring that the Contractor, sub-contractors and designers are competent and adequately resourced in relation to health and safety and well informed on the Health and Safety Plan.

Practice Note 27 suggests that where the Contractor is also the Principal Contractor, the costs of reviewing the Health and Safety Plan in response to variations or instruction as to the expenditure of provisional sums are to be added as part of the relevant valuation. Similarly, extensions of time considerations are to include the impact of any review work on progress. This is subject to any separate agreement reached by the Contractor with the Employer regarding his role as the Principal Contractor.

It is becoming common practice whereby the client appoints one of his professional team to take on the role of the Planning Supervisor during the pre-contract phase of the project. Upon appointment of the Contractor, the Planning Supervisor changes over to the Principal Contractor. This arrangement offers the advantage of continuity in the development of the Health and Safety Plan.

---

79 See Section 1.5.2.

2.8 Designers

This section considers the generic role of designers whose involvement with the project continues to the construction phase. Where the Architect also carried out the architectural design of the Works, his role as designer described in this section would be additional to his supervisory and administrative responsibilities already discussed in Section 2.4. Except with the simplest of buildings, it is rare for the Architect to possess all the skills required to supervise construction to designs by specialists such as structural engineers and mechanical/electrical engineers. For that reason, these specialists are often retained in a limited capacity during the construction phase to assist the Architect with supervision. As the JCT 98 recognizes only the Architect as supervisor, any instruction by these specialists has to pass through the Architect to the Contractor.

It is now settled law that a designer who continues as a supervisor is under a duty to review his design to deal with any problems that come to his attention during construction and that ought reasonably to alert him to a serious possibility of errors in his initial design. In *Brickfield Properties v. Newton*[80] Sachs LJ said:

> The architect is under a continuing duty to check that his design will work in practice and to correct any errors which may emerge. It savours of the ridiculous for the architect to be able to say, as it was here suggested that he could say: 'True, my design was faulty but, of course, I saw to it that the contractors followed it faithfully' and to be enabled on that ground to succeed in the action.

This is important for reasons of limitation of action under the contract.[81] The duty to review means that the cause of action may not accrue until completion of construction, which may be long after completion of the faulty design.

CDM Regulation 13 imposes obligations on every designer to:

- be competent to carry out the design he is commissioned to produce;
- satisfy the client that he is adequately resourced for the task;
- give adequate regard to the need to avoid foreseeable risks to health and safety of any person carrying out construction as well as others likely to affected by it;
- cooperate with the Planning Supervisor and other designers in the interest of health and safety;
- consider risks likely to affect the future maintenance and repair of the building or structure;
- provide to the Planning Supervisor or the client any information about aspects of the design that might affect health and safety in connection with the construction or use of the building.

The importance of a designer's responsibilities under the CDM Regulations cannot be over-emphasized. Indeed, the first ever prosecution under the Regulation was of a designer.[82] The firm of architects concerned were engaged to design and supervise the

---

80 [1971] 1 WLR 862, at p. 873. For application of the duty to review design see also *London Borough of Merton v. Lowe* (1981) 18 BLR 130; *Chelmsford District Council v. Evers* (1983) 25 BLR 99; *Equitable Debenture Assets Corporation v. William Moss*; see n. 26; *University of Glasgow v. Whitfield* (1988) 42 BLR 66.
81 For discussion of limitation of action under the Contract see Section 1.4.1 (dealing with Attestation).
82 See *Building* magazine, 19 January 1996 in which it was reported.

construction of an extension to a cash-and-carry store designed by them about 8 years earlier. During construction the contractor struck an 11000 volt electricity cable, causing an explosion. As original designers of the store, the architects must have known of the existence and location of the cable. It was the early days of the Regulations when procedures were still being refined. Although the architects coordinated health and safety matters on site, they were never really appointed as the Planning Supervisor. If a Planning Supervisor had been appointed, he would have been his responsibility to have identified and flagged up the hazard in the Pre-tender Health and Safety Plan. The employers avoided prosecution for failing to appoint a Planning Supervisor because they never knew that they needed to make such an appointment. Although the architects could not be prosecuted as the Planning Supervisor, the firm were found guilty of contravening the Regulations by failing to advise their client to appoint a Planning Supervisor and were fined £500.

2.9 Person-in-Charge (PIC)

Clause 10 requires the Contractor always to have on site a competent Person-in-Charge. Traditionally, this person has usually been referred to as the 'Site Agent' although, nowadays, there are other titles such as 'Contract Manager', 'Site Manager', and 'Project Manager'. In practice, he is usually the Contractor's employee with total responsibility for the carrying out of the Works on site. The Contract does not require the Contractor to notify to the Employer or the Architect the identity of the person acting in that capacity. However, the Contractor will usually do so before the commencement of site operations.

The clause states that any AI or COW direction given to the PIC is deemed issued to the Contractor. Instructions or directions issued to any other person on site are not binding on the Contractor. It is important to note that the deeming provision affects only AI and COW directions and that service of notices or other document required under the Contract is governed by Clause 1.7 as follows:

1. If there is an agreed method of service that method should be followed.
2. If there is no agreed method but there is an agreed address for service of notices and documents, they may be served by any effective means to that address.
3. If there is neither an agreed method nor address, the notice or document is considered served if it is posted, in the case of a registered company, to the registered address or, in other cases, to the last known principal business address.

2.10 Clerk of Works (COW)

It has already been explained in Section 2.4.4 that normal practice in building contracts does not require the Architect always to be on the site. Clause 12 of the JCT 98 reflects the normal practice by providing that the Employer has a right to appoint a COW whose function is 'solely as inspector on behalf of the Employer'. He is usually a very experienced foreman appointed on the recommendation of the Architect. On projects of high complexity, he will often be assisted by specialist clerks of works. The RIBA and Institute of Clerks of Works have jointly produced a Clerks of Works Manual, which describes the role and duties of a clerk of works when employed on contracts let on the JCT family of standard forms.

The Conditions are silent on the Contractor's right to raise objections to any person being appointed as a COW. As the Articles expressly provide for such a right in the case of a replacement Architect and Quantity Surveyor, the implication must be that there is no such right. The Contractor must afford the COW every reasonable facility for the performance of his duties. However, the powers of the COW to direct the Contractor in the carrying out of the Works are limited as follows:

- The direction must be given in regard to a matter in respect of which the Architect is empowered under the Contract to give instructions.
- The instruction is confirmed in writing by the Architect within two working days of the original instruction being given.

It is to be noted that the instruction becomes effective from the date of issue of the confirmation and not, as one would expect, the date of its receipt. In practice, many contractors comply with instructions of the COW without waiting for the Architect's confirmation. Although such instructions are often confirmed retrospectively, it should be borne in mind that, technically, the Contractor would be complying with an invalid instruction if he carries out an unconfirmed instruction of the COW. He may therefore lose entitlements to extra payment and extension time if compliance with the instruction entails delay. Even if the Architect includes payment in respect of such instructions, the Employer would be entitled to challenge the certificate, thereby delaying payment.

It is not uncommon for the COW to carry out some of the express duties of the Architect, some of which may go beyond inspection of matters of detail. This is most common where the Works are very simple in nature. The JCT 98 does not provide for such delegation. Where such delegation is anticipated, the Conditions should be amended accordingly, with particular attention being paid to the authority of the COW.

Clause 12 makes it clear that the COW, even if appointed on the Architect's recommendation, acts as an agent of the Employer. In any legal action against the Architect for negligent supervision, the Architect can therefore plead any negligence of the COW in the performance of his duties as the Employer's contributory negligence. The effect of such a plea is to reduce the damages recoverable against the Architect by the extent to which the Employer's loss was caused by the COW's negligence.

Kensington & Chelsea & Westminster Area Health Authority v. Wettern Composites Ltd & Adams Holden & Partners[83]: pre-cast concrete mullions in the plaintiff's hospital cracked as a result of the Contractor's poor workmanship. The plaintiff employer's action against the Architect for failure to exercise reasonable skill and care in supervising the works succeeded. However, the damages recoverable from the architect were reduced by 20 per cent because of overlooking of the defects by the plaintiff's COW.

However, the Architect is not entitled to delegate to the COW his obligation to take reasonable steps to ensure that the Works are carried out to his design. This means that the Architect must attend personally to the general scheme of the Contractor's work methods while giving clear directions to the COW on matters of detail to be followed up.

---

83 (1984) 31 BLR 57.

Leicester Board of Guardians v. Trollope[84]: this litigation raised the question of an architect's liability for dry rot discovered in a hospital four years after it had been constructed. The cause of the problem was found to be deviations from the work method specified in the contract for constructing ground floors. Although the COW was always on site during construction, he acquiesced to the deviations for corrupt reasons. The Architect, who only visited the site from time to time, admitted that he never checked the ground floors of any of the buildings. The architect had therefore left more than matters of detail to the COW. It was held at first instance that the presence of the COW on site did not eliminate the Architect's responsibility to see that the work was properly carried out to his design. However, the court suggested that if the Architect had seen to it that the work on the first block was properly carried out, and had then told the COW that the rest was to be done in the same way, he might have avoided liability.

The agency relationship between the COW and the Employer suggests that the Employer may be better off not appointing a COW as negligence on his part may reduce the Architect's liability in respect of the matter concerned. Whilst this may be true from a legal standpoint, in practice, prevention of loss through the appointment of a competent COW is better than having legal remedies against an Architect. Besides, failure to appoint a COW may itself constitute contributory negligence by the Employer although there is no direct authority on the issue.

It is not uncommon for the Clerk of Works to perform some functions of the Architect such as accepting the Contractor's confirmation of oral instructions.[85] The Contract does not provide for this type of delegation. Where such delegation is desirable, appropriate provision must therefore be added as amendments to the Contract or a separate agreement involving the Employer, Contractor and the Architect is drawn up. Another common practice that can raise problems is where the COW uses the official stationery of the Employer or the Architect to communicate with the Contractor. The problem is that the Contractor may be, entitled to treat the COW in such circumstances as agents of the Employer or the Architect, as the case may be, and treat the communication accordingly. For example, an instruction to the Contractor on the Architect's notepaper may constitute a variation without the need for further action from the Architect.

2.11 Suppliers and sub-contractors

The role of a general contractor engaged by a construction client is increasingly becoming one of only management and coordination of actual construction by third parties engaged by the contractor for the purpose. These third parties are referred to generically as 'sub-contractors' of which there are two types. 'Domestic sub-contractors' are selected and appointed by the Contractor to carry out specific work for which the Contractor has generally obtained the Architect's permission to sub-contract. Some sub-contractors are appointed by the Contractor as an obligation under the Contract because they have been selected in the Contract Bills or by the Architect. Any sub-contractor appointed in this way

---

84 (1911) 75 JP 197.
85 See Chapter 5.

is referred to as a 'Nominated Sub-contractor'. Similarly, Nominated Suppliers are selected by the Employer, or the Architect on his behalf, for appointment by the Contractor. Domestic sub-contractors, Nominated Sub-contractors and Nominated Suppliers under the JCT 98 are covered mainly in Chapters 8, 9 and 10 respectively.

2.12 Joint liability

It may be concluded from the contents of this chapter that there are certain types of loss suffered by the Employer for which more than one participant could be liable. The Employer may therefore bring action against any of the parties he considers responsible for the loss. Where two or more parties are found to have contributed to the loss the court has jurisdiction under the Civil Liability (Contribution) Act 1978 (CLCA) to apportion liability between them on whatever basis the court considers just and equitable, taking into account the extent of each person's responsibility for the loss. The CLCA also allows a party who is sued to bring into the proceedings as co-defendant any other party who would also be liable for the same loss. For example, if the Employer sues the Architect for defects arising from negligent supervision, the Architect may add the Contractor to the proceedings for a contribution towards the Employer's loss where the defects represent a breach of the Contract by the Contractor.

3

JCT 98 timetable

Certain decisions and actions required under the JCT 98 have to be made or taken at specific points in time or within specific time windows. For example, the Architect cannot give certain types of instructions after Practical Completion. Awareness of the general timetable of a JCT 98 contract is therefore necessary for effective administration of the contract. The aim in this chapter is to provide this awareness. The key events in the JCT 98 calendar are summarized in Fig. 3.1.

3.1 Possession of site

To put the JCT 98 provisions on possession of site in context, the general principles on the subject are first explained.

3.1.1 General principles

In construction contracts, the date by which the site for the works should be made available to the contractor to commence construction is referred to as the 'date of possession'. This date should normally be fixed long before there is a need to commence operations on site. However, either because of poor planning or because of circumstances compelling invitation of tenders in advance of availability of the necessary information on the site, contracts are often awarded without any firm idea as to when the site will be available. If, for these or other reasons, the date for possession is not specified, the law will imply a term that the owner will give possession of site to the contractor in sufficient time to allow completion by any agreed date for completion.[1]

It is not uncommon that on the date of possession, the owner, for a variety of reasons beyond his control, is unable to make the site available. For example, in *H. W. Nevill (Sunblest) Ltd v. William Press Ltd*[2] a demolition and site clearance contractor failed to get the site ready for the owner to hand over to the contractor for the actual building works. In *Rapid Building Co. Ltd v. Ealing Family Housing Association*[3] the defendant was

---

1   *Freeman & Son v. Hensler* (1900) 64 JP 260.
2   (1981) 20 BLR 78 (hereafter *Nevill v Press*).
3   *Rapid Building Co. Ltd v. Ealing Family Housing Association* (1985) 1 Con. LR 1.

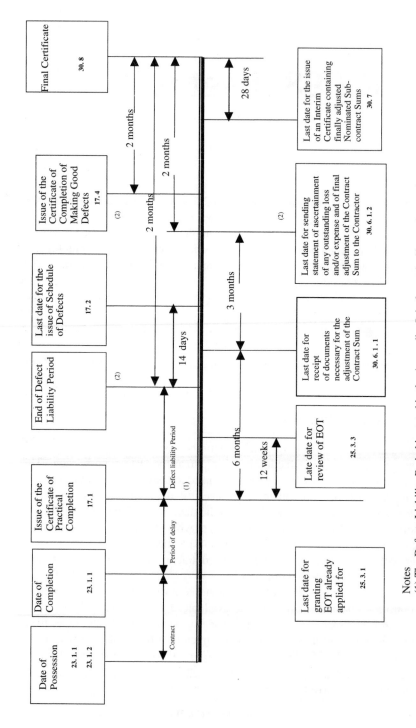

Fig. 3.1 The JCT 98 Timetable

Notes
(1) The Defects Liability Period is stated in the Appendix. It is 6 months if none is stated
(2) The period of 2 months to the Final Certificate runs from whichever of these occurs last: Clause 30.8.2

unable to give possession of site to the plaintiff because a part of the site was occupied by squatters whom the defendant could not remove.

Failure to give possession of site by the agreed date or, in the absence of an express date for possession, in sufficient time to allow completion on time, may constitute a fundamental breach of the contract. This principle was applied by the High Court of Australia in *Carr v. J. A. Berriman Property Ltd.*[4] This means that in the event of this type of breach the contractor can refuse to go on with the contract and instead sue for damages. In practice, very few responsible contractors would take such a draconian course of action. A more business-like approach would be to continue with the contract and to claim damages for breach of contract. Even where the contractor chooses to continue with the contract, on principles explained in Chapter 11, the time for completion becomes 'at large' if the contract does not allow for extension of time to cover the delay. Time for completion being at large means that the contractor is only obliged to complete within a reasonable time. Furthermore, the employer loses any rights to liquidated damages. If the contractor fails to complete within a reasonable time, the employer would have to prove any damages that he wishes to recover.

However, mere failure to grant possession in due time will not, in itself, amount to a repudiatory breach. There must be evidence either that the employer has no intention ever to grant possession or that the failure is total or very serious. In *Wardens and Commonality of the Mystery of Mercers of the City of London v. New Hampshire Insurance Co. Ltd*[5] the plaintiff (Mercers) entered into a contract with contractors to carry out works to their premises. The contract, which incorporated the JCT 80, allowed deferment of possession of site by a maximum of 6 weeks. To avoid forthcoming increases in VAT, the Employer made an advance payment of most of the Contract Sum. The defendant provided a bond guaranteeing that the advance payment would be employed towards the carrying out and completion of the Works. The Employer exceeded the maximum period of allowable deferment of possession of site by a further 4 weeks, a breach of the Contract. The issue before the Court of Appeal was whether the effect of the breach was to discharge the defendant from the obligations of the bond. The answer was in the negative. To arrive at this answer, the court held that the breach was non-repudiatory. Unfortunately the court did not explain their reasoning because both parties were in agreement that the breach was non-repudiatory.

3.1.2 Possession of site under the JCT 98

'Date of Possession' is defined in Clause 1.3 of the JCT 98 as the date stated in the Appendix under the reference to Clause 23.1. Clause 23.1.1 requires the Contractor to be given possession of the site on this date to commence the construction of the Works.

To allow for any problems with giving possession of site, the JCT 98 contains a mechanism whereby possession of site can be deferred. To avail himself of this opportunity, the Employer, or usually his professional advisers, should indicate in the Appendix that Clause 23.1.2, which allows deferment of giving possession of site, will apply. The maximum period of deferment must also be stated in the appropriate place in

---

4 (1953) 27 ALJR 273.
5 (1992) 60 BLR 26.

the Appendix. This maximum period must not exceed 6 weeks, the default maximum period if that part of the Appendix is not completed. Deferment of possession of site for up to the applicable maximum period is one of the Relevant Events, which entitle the Contractor to extension of time under Clause 25. It should be noted that, if the actual period of deferment exceeds the maximum period, the Employer would be acting beyond his contractual rights. The Architect would therefore have no authority to grant extension, with the serious consequence that time for completion would become at large.

The extent of possession can be a source of disputes, e.g. whether the Employer is entitled to grant possession in stages. If the extent of possession were provided for expressly in the contract, those provisions would govern the Contractor's entitlement. In the absence of such provisions, it was said in *Hounslow London Borough v. Twickenham Garden Developments Ltd*[6] that the contractor should be given 'such possession, occupation or use as is necessary to enable him to perform the contract'.

Under Clause 29.1 of the JCT98, where the Contract Bills indicate that other work on the site is be carried out by the Employer himself or his other contractors, the Contractor must allow the work to be done. In those circumstances, the Contractor would not be entitled to exclusive possession of the site. Even where it is not so stated in the Contract Bills, the Contractor must not unreasonably withhold consent to the carrying out of such ancillary work: Clause 29.2. The Contractor is therefore entitled to exclusive possession only in circumstances where it is necessary for satisfactory performance of his obligations.

3.2 Completion Date

Clause 23.1.1 requires the Contractor to complete the Works on or before the 'Completion Date'. This date is defined in Clause 1.3 as 'the Date for Completion as fixed and stated in the Appendix or any date fixed either under Clause 25 or in a confirmed acceptance of a 13A Quotation'. Under Clause 25, the Architect has powers to revise the applicable completion date to allow for specific delays referred to collectively as the 'Relevant Events'. These powers and the procedures governing the granting of extension of time are discussed in detail in Chapter 11.

From the use of the phrase 'shall complete same on or before the Completion Date' in Clause 23.1.1, the Contractor is clearly entitled to earlier completion if he can. However, it was pointed out in *Glenlion Construction Ltd v. The Guinness Trust*[7] that there is no corresponding duty on the part of the Employer or the Architect to go out of their way to make such completion possible, e.g. producing drawings or taking other actions required of them earlier than anticipated within the contractual programme.

3.3 Certificate of Practical Completion

Clause 17.1 provides that the Architect must forthwith issue a Certificate of Practical Completion when Practical Completion of the Works has been achieved and the following conditions have been met:

---

6   [1970] 3 All ER 326.
7   (1987) 39 BLR 89; see also *J. F. Finnegan Ltd v. Sheffield City Council* (1988) 43 BLR 130.

- The Contractor has complied sufficiently with his obligation under Clause 6A.4 to supply information requested by the Planning Supervisor/Principal Contractor for compilation of the Health and Safety File and has ensured similar compliance by his sub-contractors (Clause 6A.4).
- The Contractor has supplied the Employer with as-built drawings and other information on any Performance Specified Work (Clause 5.9).

3.3.1 Meaning of 'Practical Completion'

The Contract does not specify any factors to be considered by the Architect in deciding whether or not Practical Completion has been achieved. This is not surprising because Practical Completion, like the proverbial elephant, is more easily recognized than described. The courts have nevertheless provided some guidance. Generally, the Architect must not issue the Certificate if there are patent defects.[8] However, according to Judge Newey in *Nevill v. Press,* Practical Completion may have been achieved even where minor *de minimis* (Lat.: trifling) work remains to be done.[9] The Architect must still be slow to overlook such outstanding work because of possible disagreements on whether or not the outstanding work is actually trifling. In any case, he would be ill advised to issue the Certificate without written acknowledgements by the Contractor of such work or undertakings to carry it out without undue delay after the issue of the Certificate. This caution is necessary because, under Clause 17.2, the Architect is authorized to issue instructions requiring the Contractor to make good only 'defects, shrinkages or other faults which *shall appear* [authors' emphasis] within the Defects Liability Period'. Without such a collateral undertaking, the Contractor may therefore decline to make good defects that existed before the Defects Liability Period. Although such defects are still a breach of contract for which the Employer would be entitled to damages, exercise by the Architect of his powers to ask the Contractor back to make them good may sometimes be more advantageous.

The common understanding of the construction industry is that Practical Completion is achieved where, except for trifling outstanding work and defects, the Works are so substantially complete that the building can be put to its intended use with safety and convenience. This view is supported by an Australian decision. In *Murphy Corporation Ltd v. Acumen Design & Development (Queensland) Pty and Another*[10] the Supreme Court of Queensland stated that the concept of practical completion meant 'completion for all practical purposes, that is to say, for the purpose of allowing the principal to take possession of the works and use them as intended'. In that case, although minor works remained to be done, the Employer took possession of the building, put it to its intended use and opened it to the public. It was held that the employer's acts were sufficient proof that practical completion had been achieved. The contractor would therefore have been entitled to be put in the position he would have been in if the appropriate completion certificate had been properly issued.

---

8   *Westminster City Council v. J. Jarvis & Sons Ltd* (1970) 7 BLR 64;
9   *Nevill v. Press*, see n.2.
10 (1995) 11 BCL 274.

3.3.2 Effect of the Certificate of Practical Completion

The issue of this certificate is one of the most important events in the timetable of the contract for it triggers off a number of serious consequences:

- the commencement of the Defects Liability Period (Clause 17.3 and the Appendix);
- the end of the Contractor's responsibility for loss or damage to the Works except for frost damage traceable to a cause before Practical Completion (Clause 17.5);
- the end of the obligation of the Contractor to insure against damage to the Works (Clause 22A.1/22B.1/22C.2);
- the end of the Contractor's liability for Liquidated and Ascertained Damages (LADs) in the event of late completion (Clause 24.2.1);
- the retention percentage in subsequent interim payment certificates is reduced by a half (Clause 30.4.1.3);
- start of the period of 6 months within which the Contractor is to supply the Architect and/or Quantity Surveyor with all the necessary documentation for the preparation of final accounts (Clause 30.6.1.1).

It should be noted that the Works are at the risk of the Employer after the date of issue of the Certificate of Practical Completion. The Architect must therefore advise the Employer to ensure that his own building insurance takes effect the following day.

From the consequences listed above, whilst the advantages to the Contractor of Practical Completion are obvious, the issue of the Certificate of Practical Completion is not always in the commercial interest of the Employer. For example, the Employer may not be ready, or is otherwise unable to put the building to its intended use, e.g. occupy, rent, lease, or sell. In any such situations, he will be assuming the responsibilities of Practical Completion with no benefit. For these reasons, the Employer may be tempted either to influence the Architect into withholding the issue of the Certificate or, if it is issued, to dispute its validity. Also, there could be a genuine disagreement between the Architect and the Contractor as to whether the Works have reached Practical Completion. All such disputes between the Architect or the Employer and the Contractor as to whether or not the Works have reached practical completion must be resolved using the appropriate dispute resolution technique.

Under Clause 28.2.1.2, interference or obstruction by the Employer of the issue of any certificate under the Contract entitles the Contractor to determine his employment. Determination by the Contractor and its consequences are discussed in Sections 16.11–16.15.

3.4 Certificate of Non-completion

If the Works have not reached Practical Completion on the Completion Date, it is the Architect's duty to issue a certificate to that effect (Clause 24.1). The issue of this certificate, referred to as the 'Certificate of Non-Completion', is one of the conditions precedent to recovery of liquidated damages by the Employer. The Employer must also serve a notice to withhold or deduct payment (see Section 15.1.4).

The total amount of liquidated damages recoverable from the Contractor at any point in time is determined by multiplying the rate of liquidated damages inserted in the Appendix (e.g. £x/week or day of delay) by the total period of delay. If the Employer decides to accept reduced liquidated damages (probably because the actual consequences of the delay are less financially damaging than was anticipated at the time of entering into the contract) he must certify the reduced rate in writing. The purpose of this stipulation is probably to avoid challenges to attempted set-off by the Employer on the argument that, because the amount involved is less than the liquidated damages due, it is for something else and therefore not authorized under the contract.

The Architect must issue the Certificate of Non-Completion even if there are claims for extension of time still to be assessed. Any subsequent award of extension of time invalidates the Certificate and the Employer must return to the Contractor any liquidated damages recovered for any period before the revised Completion Date (Clause 24.2.3). It is to be noted that the Architect must issue a new Certificate of Non-Completion if on the revised Completion Date the Works are still not completed.

3.5 Defects Liability Period

The Defects Liability Period is to be stated in the Appendix. If none is stated, it is to be 6 months and the period begins to run from the day named in the Certificate of Practical Completion. Clause 17.2 provides that during this period the Contractor is responsible for the making good of defects (Clause 17.2):

- which are attributable to quality of materials and standards of workmanship being otherwise than specified in the Contract Documents;
- due to frost damage from a cause which occurred before Practical Completion.

The Architect is empowered under Clause 17.3 to issue instructions to the Contractor requiring the making good of such defects. He may carry out investigations to identify defects by using in powers to inspect and test under Clause 8.3. The Architect is not limited to any number of instructions. An instruction under this clause is subject to all the provisions of Clause 4. However, there are three differences between instructions under Clause 17 and other instructions under the Contract. First, whilst with all other instructions the Contractor is to comply forthwith, an instruction in this case is to be complied with within a reasonable time after its receipt. Second, the cost of complying with an instruction to make good defects is to be borne by the Contractor unless the Architect directs that the defects must be made good at the expense of the Employer. The clause expressly requires the Architect to get the consent of the Employer before giving any directives that the work is to be carried out at the expense of the Employer. In general, there is no such limitation on the Contractor's rights of remuneration for complying with other instructions. Third, there is an option, which requires the consent of the Employer, to ignore the defect and to make an appropriate deduction from the Contract Sum. It would appear that the Contractor does not have to give his consent to the amount of deduction. However, as the clause expressly states that the amount of deduction must be appropriate, the Contractor may be entitled to challenge any exorbitant deduction on the ground that it is not appropriate.

The last instruction of the Architect requiring the making good of defects is a special one. It should take the form of a Schedule of Defects which the Contractor must make good (Clause 17.2). This Schedule is referred to in the construction industry as the 'final snagging list'. The Architect has up to 14 days after the expiry of the Defects Liability Period to issue this Schedule. After the delivery of this schedule or after the expiry of the

14 days from the Defects Liability Period, the Architect has no further powers to order the making good of defects under Clause 17.3. It follows therefore that:

- the Architect must ensure that all defects for which the Contractor is responsible under Clause 17 are included in the schedule;
- it would be prudent on the part of the Architect to avoid issuing the Schedule too soon, having regard to the time available and the quality of the Contractor's work in general.

3.5.1 Liability for defects that appear during the Defects Liability Period

A defect that appears during the Defects Liability Period is technically a breach of contract by the Contractor for which, subject to the terms of Clause 17, the Employer is entitled to damages in accordance with the normal contractual principles. Under Clause 17, the Employer has a right to have the defects made good through exercise by the Architect of his powers under that clause. The Architect's duty to allow the Contractor to make good defects that appear during the Defects Liability Period is worded in mandatory terms in Clause 17.2. Failure of the Architect to perform this duty, i.e. produce the schedule of defects, is therefore a breach of the Contract by the Employer for which the Contractor is entitled to damages. The clause therefore also gives the Contractor a right to be asked back to make good the defects. The value of this right was judicially recognized in *Pearce and High Ltd v. Baxter and Baxter*[11], which arose from a contract let on the JCT Minor Works form. This form contains defects liability provisions similar to those in Clause 17 of the JCT 98. Comparing the right to return and repair with the situation where the Employer brought in an alternative contractor to make good the defects, Evans LJ stated at p. 104:

> The cost of employing a third party to remedy the works was likely to be higher than the cost of the original contractor remedying the work himself. The right to return to remedy defective works was therefore a valuable one to the contractor. Accordingly, when denied this right, the contractor would not liable for the full cost of repair ... the employer could not recover more than the amount which it would have cost the contractor himself to remedy the defects.'

3.5.2 Defects after the Defects Liability Period

If the defects appeared during the Defects Liability Period, as explained above, the Architect should have given the Contractor the opportunity of making them good. If he was in fact given the opportunity but he failed to comply, as a simple matter of contractual analysis, the Employer would be entitled, even after the Defects Liability Period, to have the work done by a third party and to recover the costs incurred from the Contractor.[12] Such costs may even include costs of the Employer's managerial time expended in getting the work done by the third party, e.g. time used in selecting the alternative contractor and supervising the work.[13] However, as also explained above, where the Contractor was

---

11 (1999) BLR 101 (hereafter *Pearce v. Baxter*).
12 In *Forsyth* v *Ruxley Electronics and Construction* (1995) 73 BLR 1 the House of Lords held that where cost of repair would be unreasonable, only modest damages for loss of amenity could be recovered. However, it is to be noted that in that litigation the owner had not yet carried out repairs.
13 *Tate & Lyle Food & Distribution Ltd v. GLC* [1983] 2 AC 509.

denied the chance of making good the defects, the Employer can recover as damages no more than it would have cost the Contractor if he had been given the chance.[14] In the *Pearce v. Baxter* litigation the Court of Appeal rejected emphatically an argument on behalf of the Contractor that the effect of failure to allow the Contractor to return was to absolve the Contractor completely of liability for the defects that should have been notified to him. The court expressly approved of Judge Stannard's statement in *William Tomkinson v. St Michael's PCC* that there must be clear words to that effect if the Employer is to be denied the normal remedy of damages for breach of contract on account of the contractor having been refused the chance to effect the repairs himself. There are no such clear words in the JCT 98.

A general caveat should now be entered regarding the above discussion on liability for defects after expiry of the Defects Liability Period. It is that after issue of the Final Certificate, matters of quality expressly left to the Architect's opinion and on which his approval has been received cannot be challenged unless the Employer invoked the appropriate dispute resolution method within 28 days of its issue.[15]

3.5.3 Consequential loss from defects

The Employer's right in respect of defects, whether before or after expiry of the Defects Liability Period, is not limited to having them made good at no extra costs to himself. If the process of carrying out the necessary repairs causes consequential damages, they can be recovered under normal contractual principles. Examples of consequential damages include the cost of getting alternative accommodation or cost of disruption of any business carried on in the building. On the issue of entitlement to consequential damages, in *P. & M. Kaye v. Hosier and Dickinson Ltd*[16] Lord Diplock said:

> At common law a party to a contract is entitled to recover from the other party consequential damage of this kind resulting from that other party's breach of contract, unless by the terms of the contract itself he has agreed that such damage shall not be recoverable. In the absence of express words in the contract a court should hesitate to hold that a party had surrendered any of his common law rights to damages for its breach, although it is not impossible for this to be a necessary implication from other provisions of the contract.

He went on to state that no such implication could be drawn from the defects liability provisions very similar to those in Clause 17 of the JCT 98. In *Pearce v. Baxter*, although the parties were in agreement on liability for consequential damages, Lord Diplock's statement was quoted by Evans LJ with complete approval.

3.6 Certificate of Completion of Making Good Defects

When the Contractor has finished making good the defects contained in the Schedule of Defects to the satisfaction of the Architect, the latter is to issue a certificate to that effect

---

14 (1990) 6 Const. LJ 319 (hereafter *Tomkinson v. St Michael's*).
15 See Section 3.9.3 for details on this issue.
16 [1972] 1 WLR 146.

(Clause 17.4). This certificate is referred to in the Contract as the 'Certificate of Completion of Making Good Defects'. It is worth noting that the issue of this certificate is one of the three preconditions for the issue of the Final Certificate under Clause 30.8. Clause 17.4 has the odd effect that in the rare situation where the Architect does not issue a Schedule of Defects because there are none he cannot issue a Certificate of Completion of Making Good Defects and, therefore, the Final Certificate.

3.7 Final adjustment of the Contract Sum

The JCT 98 is basically a lump sum contract. The Contractor undertakes to start and complete the Works for the Contract Sum. However, except in the smallest of projects, the Contractor cannot realistically allow for all possible eventualities in the Contract Sum. Examples of such eventualities include variations, delays and disruptions for which the Employer is responsible. The JCT 98 therefore allows for the adjustment of the Contract Sum in defined situations. Details of the adjustments allowed are provided in Section 15.7. Not later than 6 months after Practical Completion, the Contractor is to send to the Architect (or to the Quantity Surveyor if so instructed by the Architect) all documents necessary for the adjustment of the Contract Sum (Clause 30.6.1.1). Inclusion of all relevant documents relating to the accounts of Nominated Sub-contractors and Nominated Suppliers is specifically required. Within 3 months of receipt of all the necessary documents, the Architect (or the Quantity Surveyor as the case may be) has to prepare a statement of final adjustment of the Contract Sum (Clause 30.6.1.2). This final statement is referred to as the 'final accounts' of the Contract. The Contractor will be in breach of contract if he fails to submit all the necessary documentation within the 6 months. Apart from the fact the Employer would be entitled to damages that he can prove, the period of 3 months within which the Architect and the Quantity Surveyor must prepare statements of outstanding claims and the final account does not begin to run until all the necessary documents have been received. If final accounts are produced on incomplete information, the Contractor would have an exceedingly tough job challenging them, considering that any resulting inaccuracies may be attributable to his own breach of contract. In any such challenge, the Employer would be able to set off the cost of responding to it against any increase in the Contractor's entitlement.

On completion of the preparation of the final accounts, the Architect is to send copies of the statements to the Contractor forthwith. Relevant extracts are also to be sent to each Nominated Sub-contractor. The sending off of these documents by the Architect is one of the preconditions for the issue of the Final Certificate under Clause 30.8. On the assumption that the Architect (or the Quantity Surveyor) will prepare the documents promptly if all the necessary documents are made available, the control of the condition is to that extent in the hands of the Contractor. It is therefore in the Contractor's interest to send all the necessary documents and any further documents requested as quickly as possible.

3.8 Nominated Sub-contract final accounts

All Interim Certificates are to be issued at regular intervals referred to as the 'Period of Interim Certificates', the length of which is an entry in the Appendix. A default period of

1 month applies. Under Clause 30.7, the Architect is obliged to issue an Interim Certificate incorporating all final accounts of Nominated Sub-contractors. This Certificate is to be issued as soon as practicable and, in any case, not less than 28 days before the date of issue of the Final Certificate. As a special case, this Interim Certificate may be issued notwithstanding that a certification period has not elapsed since the issue of the previous Interim Certificate.

In issuing the Interim Certificate containing the final accounts of all Nominated Sub-contractors, Clause 35.13.1 requires the Architect to direct the Contractor as to the amount to be paid to each Nominated Sub- contractor and forthwith to inform each Nominated Sub-contractor of the amount due him under the Interim Certificate. As explained in more detail in Section 9.6, these steps are conditions precedent to direct payment to Nominated Sub-contractors against Contractor's sums in the Final Certificate. The rationale of Clause 30.7 is therefore to allow the Employer a final opportunity to make direct payments to the Nominated Sub-contractors before he is obliged to pay the Contractor on the Final Certificate. Provided the preconditions of direct payment are satisfied, the Employer is entitled to set off all previous certified payments that should have, but have not been made to Nominated Sub-contractors.

3.9 Final Certificate

Clause 30.8 provides that the Final Certificate is to be issued within 2 months of whichever of the following occurs last:

- the end of the Defects Liability Period,
- the date of issue of the Certificate of Completion of Making Good Defects,
- the date on which the Architect sent copies of the final accounts to the Contractor and Nominated Sub-contractors.

Clause 30.9 deals with the effects of the issue of the Final Certificate. The four areas considered are: (i) the Architect's subsequent role under the contract; (ii) adjustment of the Contract Sum and claims under Clauses 25 and 26; (iii) quality; (iv) arbitration.

3.9.1 Role of the Architect

Upon issue of the Final Certificate the Architect has no further duties under the Contract. For this reason he is said to be become *functus officio* (Lat. powers ended). This principle was applied in *H. Fairweather Ltd v. Asden Securities Ltd*,[17] which arose from a JCT 63 contract. The Architect, after having issued the Final Certificate, realized that he should have issued a Certificate of Non-Completion to allow the Employer to deduct liquidated damages. Mr Justice Stabb, QC held invalid the Certificate he then purported to issue on the grounds that, after the Final Certificate under the contract, the Architect was *functus officio*.

---

17 (1979) 12 BLR 40; this principle was also applied in *A. Bell (Paddington) & Sons Ltd v. CBF Residential Care Housing Association* (1989) 46 BLR 102.

3.9.2 Adjustment of the Contract Sum and claims

With the exception of matters on which dispute resolution proceedings have been commenced before its issue or within of 28 days from its issue, the Final Certificate is conclusive evidence that:

- save for errors, all adjustments to the Contract Sum required under the Contract have been correctly made (Clause 30.9.1.2. 1);
- all claims for extensions of time to which the Contractor is entitled have been granted (Clause 30.9.1.3);
- reimbursement made for loss and/or expense is in final settlement of claims in respect of the relevant matters regardless of whether the claim is in contract, tort or under statute (Clause 30.9.1.4).

3.9.3 Quality

In *Crown Estates Commissioners v. John Mowlem & Co. Ltd*[18] the Court of Appeal construed the then current wording of Clause 30.9. 1.1 to the effect that, in the absence of a suitable notice of arbitration, the Final Certificate was conclusive evidence that in respect of all matters of quality the Contractor had complied with the contract, i.e. he was not liable for any defects at all. This has the disastrous implication that, for all projects completed under that and earlier versions of the JCT 80, the Contractor's liability for defects ended upon issue of the Final Certificate. Any attempt to avoid this problem by bringing the proceedings in tort is unlikely to succeed. The decision was received with widespread and very scathing criticism, particularly by Mr Duncan Wallace, QC who opined that the approach adopted by the Court of Appeal to arrive at the decision was unnecessarily legalistic and lacked 'business common-sense'.[19] In *Belcher Food Products Ltd v. Miller and Black and Others*[20] the Scottish Court of Session distinguished *Crown Estates* as limited to only situations where the quality issue in dispute was left to the reasonable satisfaction of the Architect.

That the *Crown Estates* effect was never intended by the JCT is clear from the fact that the clause was very soon after the decision amended through Amendment 15, issued in July 1995. The intended effect is that on all matters of quality not specified objectively, i.e. by reference to a detailed specification clause in the Contract or an appropriate British Standard, but left to the opinion of the Architect, the Final Certificate is conclusive evidence of compliance with the Contract unless properly challenged by the Employer within the 28 days. The Contractor is not therefore liable for defects relating to matters so left to the Architect's opinion unless the Architect's approval was challenged by commencement of the appropriate dispute resolution procedure at the right time. Liability for defects relating to matters of quality specified objectively rather than left to the Architect's opinion stays with the Contractor for the appropriate limitation period. It is

---

18 (1994) 70 BLR 1 (hereafter *Crown Estates*); see also *Colbart v. Kumar* (1992) 59 BLR 89 in which His Honour Judge Thayne Forbes, QC reached the same decision with respect to identical provisions in the JCT Intermediate Form of Contract.
19 Duncan Wallace, I. QC, 'Not What the RIBA/JCT Meant: Loose Cannon in the Court of Appeal' (1995) 11 Const. LJ 185.
20 (1998) CILL 1415.

therefore in the Employer's interest that only minor matters of quality are left to the opinion of the Architect without objective specifications.

Without any remedy for such defects from the Contractor, an obvious alternative open to the Employer is to pursue the Architect who approved the relevant work or the person who drew up such specifications of quality. To succeed in such action, the Employer has to prove that the approval or the relevant approach to specifying quality amounts to negligence. A decision of the Technology and Construction Court suggests that an Architect sued for such defects would not be entitled to a penny by way a contribution from the author of the bad work, the Contractor. In *Oxford University Fixed Assets Ltd v. Architects Design Partnership and Tarmac Construction Ltd*[21] His Honour Judge Humphrey Lloyd, QC, on a preliminary issue, held that the conclusiveness of the Final Certificate acted as an evidential bar which not only precluded the Employer from establishing the Contractor's liability for the defects, but also prevented proof that the Contractor was liable to make a contribution to the Architect's liability for them. In the words of His Honour, any other conclusion 'would drive the proverbial coach and horses through the structure of the JCT Conditions which has been negotiated over many years and is thus to be taken as representing a fair balance between the competing interests'. Referring to the practice of contractors bringing contract supervisors into proceedings between them and employers for a contribution on account of alleged negligent supervision, Mr Duncan Wallace, QC drew an analogy between that conduct and a burglar seeking a contribution from the careless policeman who failed to arrest him and prevent his crime.[22] There is a similar analogy between the aftermath of *Crown Estates* as depicted in the *Oxford University* case and the careless policeman being sent to prison whilst the criminal is let out to enjoy his loot.

3.9.4 Arbitration

Another issue raised in the *Crown Estates* litigation was whether by Clauses 30.9. 1.1 and 30.9.3 arbitration or other proceedings were time-barred unless commenced within 28 days of the Final Certificate. If it is such a limitation, arbitration proceedings can be commenced after the 28 days only if the court grants an application to extend time within which to commence them. It was decided that there is no time limit other than the limitation period applicable under the general law, i.e. arbitration proceedings can still be commenced after the 28 days without any need for court intervention. However, in any proceedings commenced after the expiry of the 28 days, neither party may question the matters on which the Final Certificate is conclusive evidence. The policy underlying the provisions on the effects of the Final Certificate is to discourage payment disputes on certain issues, particularly payment, from being started long after final accounts.

3.10 Partial possession

With construction projects, it is often possible for parts of the contract works to be usable before completion of the whole project. For example, on a housing contract the completed

---

21 *Oxford University Fixed Assets Ltd v. Architects Design Partnership and Tarmac Construction Ltd* (1999) CILL 1473.
22 See, n. 19.

units may be fit for human habitation before completion of the whole contract. Clause 18 of JCT 98 allows the Employer, with the consent of the Contractor, to take possession of such parts for use. The Contractor must not withhold his consent unreasonably. It is submitted that it would not be unreasonable to withhold consent if the ground for the refusal is that use of such parts would seriously impede the carrying out of the uncompleted parts.

Upon the Employer taking over any part, the Architect is to prepare a written statement identifying the part taken over. Any part so identified is referred to as the 'relevant part' and the date from which the Employer takes possession is called the 'relevant date'. Generally, the effect of partial possession on the Contractor's responsibility for the part taken over is analogous to the effect of Practical Completion on the whole contract.

Insurance: the Contractor's (or the Employer's if Clause 22B applies) obligation to insure the relevant part against damage comes to an end from the relevant date. Where Insurance under Clause 22C (used mainly in contracts involving extension or alteration of existing structures) applies, the relevant part becomes part of the existing structures and contents for which the Employer must maintain material damage insurance under Clause 22C.1. This means that the Employer must contact his insurer to have the policy amended to include the relevant part.

Defects: the Contractor's responsibility for defects in the relevant part is the same as his responsibility for defects after the issue of the Certificate of Practical Completion. The only difference is that the Defects Liability Period for the part runs from the relevant date.

Retention: only half of the full retention applicable to work in the part is to be deducted in valuations after the relevant date, i.e. half the retention already deducted is released to the Contractor.

Liquidated Damages: there is to be a proportionate reduction in the amount of liquidated damages deductible for delayed completion. If the value of work in the part is X per cent of the total Contract Sum, the liquidated damages are reduced by X per cent.

It follows from the discussion of partial possession that if an Employer, with the consent of the Contractor uses any part of the works, the Architect must issue the statement with all the consequences described above. Sometimes, an Employer may want to make use of part of the works without incurring the attendant responsibility. This is only possible if Clause 18 had been appropriately amended before the contract was executed. In addition, such use must be drawn to the attention of insurers otherwise the relevant policies could be invalidated.

3.11 Sectional completion

The term 'sectional completion' is used to refer to the arrangement where the contract works are to be completed in phases. This arrangement requires the specification in the contract documentation of the dates by which the various phases must be completed. The JCT has published a *Sectional Completion Supplement* intended for use with the JCT 98 where the contract is to be completed in phases. Readers are referred to Sections 1.4.8 and 11.1.2 for discussion of the problems that can arise if attempts are made to impose obligations to achieve completion by phases without using it.

4

Performance Specified Work

(See also: Chapter 1, Section 1.5.2 dealing with Appendix: Clause 42.1.1;
Chapter 6, Section 6.12.4–6 dealing with Variations.)

4.1 General

The general nature of JCT 98 is that of a 'build only' contract,[1] in which the Employer specifies the work to be done in exact terms. All the design work is then carried out either by the Architect, or in his name; the Contractor is not expressly required to do any.[2] However, there are times when the Employer wishes the Contractor to do all or part of the design. If the majority of the design is to be done by the Contractor, JCT 98 is not appropriate, and the form of contract which should be used is the JCT Standard Form of Building Contract With Contractor's Design 1998 Edition (WCD 98). If, the Employer wishes the Contractor's design input to be substantial in an identifiable portion of the Works, but not for the whole of the Works, the Contractor's Design Portion Supplement should be used. The Supplement introduces amendments into JCT 98 which, in respect of the identifiable portion, copy the principles of the 'Design and Build' form WCD 98.

Sometimes the Employer will instruct the Contractor to carry out specified work to meet a defined performance, which necessitates some design input by the Contractor. There are many instances in a building project where a proprietary product may be specified which then requires specialist design in adapting to meet the particular circumstances of the surroundings. A typical example would be a composite unit such as a uPVC window. Whilst the window is made from standard extrusions, the choice of extrusions and the manner in which they are put together is determined by wind loadings, security and maintenance needs, or any other factor which may affect the manufacturer's detailed design. These are criteria which determine the performance required of the product. Other examples would include roof trusses, proprietary flooring systems, curtain walling, high bay storage systems, all of which require 'fine tuning' to fit into the project, and all of which require design input by the manufacturer or installer. However, the use of any of these examples in isolation would not normally suggest that the Architect would no longer

---

1 But see Chapter 1, Section 1.4.3, dealing with bills of quantities.
2 The Contractor may have some implied design responsibility arising out of his obligation in respect of workmanship and materials, and his duty to conform to Building Regulations.

be designing the Works. He would simply be choosing a product to be built into the Works, albeit specialist design would be required to be completed by someone else.

Whilst design obligation may be placed on the Contractor by specification in the Contract Bills,[3] in the absence of express provisions in the contract to deal with design by the Contractor, there is no obligation on the Contractor to fulfil his duties in any particular manner. Likewise, in the absence of provisions dealing with liability, the Contractor who designs and installs work will probably take onto himself, albeit unintentionally, an obligation to design and install work which is fit for its intended purpose.[4] This is a higher standard than that of reasonable care and skill which applies to the Architect,[5] and one which is difficult to cover with professional indemnity insurance.[6] Clearly there are benefits for both the Architect and Contractor if the contract deals with limited design when required to be carried out by the Contractor. That is the purpose of Part 5 of the Conditions: Performance Specified Work, Clause 42.

The intention of the JCT is that Performance Specified Work should be limited to simple or easily identifiable types of work. Practice Note 25[7] suggests 'The Provisions of Performance Specified Work should not be used for items which will materially affect the appearance of the building, or may result in changes in the design of other work (otherwise than at the interface ...), or will affect the use of the finished building, so that it would be essential to examine and accept the Contractor's proposals for the work before acceptance of the tender.' In an example the JCT suggest suitable works would be trussed rafters, or a whole roof timber construction incorporating trussed rafters, a simple package of electrical works, or a simple gas central heating installation.

4.2 Meaning of Performance Specified Work

Clause 42.1 defines Performance Specified work as being work to be carried out by the Contractor, for which requirements are shown on the Contract Drawings, and in respect of which the required performance together with sufficient information for pricing, or a provisional sum, is stated in the Contract Bills. Such work must be identified in the Appendix and, by Clause 42.18, cannot be carried out by a Nominated Sub-contractor or Nominated Supplier.

The extent of work to be carried out under this clause is limited to that specified in the Appendix, a point which is emphasized in Practice Note 25. Under Clause 42.12, the Architect has no authority under the Contract to introduce new Performance Specified Work during the course of the Contract, unless otherwise agreed between the Employer and the Contractor. Indeed the Architect could also be in breach of his terms of

---

3   In *Haulfryn Estate Co. Ltd v. Leonard J. Multon & Ptnrs and Frontwide Ltd*, ORB; 4 April 1990, Case No 87-H-2794, it was held design obligations placed on the Contractor in Specifications under the JCT Minor Works form was not inconsistent but added to the obligations in the Conditions.
4   See *Independent Broadcasting Authority v. EMI Electronics & BICC Construction* (1980) 14 BLR 1, and *Viking Grain Storage v. White (T. H.) Installations* (1985) 33 BLR 103.
5   See Section 4.9 dealing with reasonable care and skill.
6   Note there is no contractual obligation to take out Professional Indemnity insurance, although most Contractors involved in design and build contracting would do so; however, most insurance companies resist offering to cover fitness for purpose.
7   Contained in Amendment 12 to JCT 80, issued July 1993.

engagement with his Client if he did so,[8] although this point would obviously depend on his Client's previous instructions.

4.3 Contractor's statement

Before carrying out the work, Clause 42.2 obliges the Contractor to notify the Architect of his proposals. Prior to submitting his Statement, the Contractor must first provide a draft copy to the Planning Supervisor, and incorporate any amendments necessary to take account of the Planning Supervisor's comments. There is a commercial risk for the Contractor in this procedure. The Planning Supervisor is placed in the position where he can, in effect, instruct the Contractor under the guise of making comment which the Contractor must incorporate. Since the Planning Supervisor is often the Architect, he needs to be careful that he does not confuse his two roles, and the Contractor needs to be wary that he is not coerced into changing his design to incorporate a Variation at his own cost. It is the Planning Supervisor's normal role to comment, not to instruct; the choice of method to comply with health and safety best practice remains with the Contractor.

The Contractor's Statement has to be in sufficient detail in the form of drawings, specifications and calculations to explain to the Architect how the Contractor intends to do the work. It must include any information required by the Contract Bills (Clause 42.3) and must also be given to the Architect in sufficient time for him to provide further information if necessary. In any event the Statement must be provided by any time specified in the Bills. If no time is specified, the Statement is required to be provided within a reasonable time before the work is intended to be done. A reasonable time will have to make allowance both for the time needed by the Architect to respond, and for the Contractor to deal with possible valid criticism. Any delay to the progress, or the completion, of the Works caused by a delay in providing the Statement or amended Statement within the specified time limits, expressly will not entitle the Contractor to extensions of time or to loss and expense (Clause 42.16).

4.4 Architect's response to Contractor's statement

If, within 14 days following receipt of a Contractor's Statement, the Architect considers it to be deficient, Clause 42.5 gives him power to require the Contractor to supply more information and amend the Statement. However, 'deficiency' in Clause 42.5 refers to the sufficiency of the Contractor's Statement to explain his proposal, not the credibility of the Contractor's design. Thus the power to require amendment is limited to form and detail. It does not extend to the substantive content of the proposals; that is dealt with in Clause 42.6.

If the Architect considers the proposals in the Contractor's Statement will not achieve the required performance, he has a duty under Clause 42.6 to notify the Contractor immediately. Such notification is not a vehicle for the Architect to impose his own

---

8 In *Moresk Cleaners v. Thos Henwood Hicks* [1966] 2 Lloyd's Rep. 338. It was held that the Architect who had required the Contractor to design a concrete floor had no implied authority from the Employer to allow others to carry out that which he had been employed to do.

preferred method on the Contractor. It is simply an opportunity, and an express obligation, to warn the Contractor of deficiency in the design proposed. Provided the Contractor's proposal meets the specified requirement, any insistence by the Architect that the Contractor should use a particular method, or that the design should be 'improved' in some way, would probably amount to a deprivation of choice, ie. a Variation.[9]

There is no requirement for the Contractor to await 'approval' or comment, other than that implied in the 14-day period in which the Architect may respond to the Statement, so there should be no cause for the Contractor to be delayed by lack of response by the Architect. If however, as sometimes happens, the Contract is amended to require the Contractor to obtain clearance before starting work, then a delay by the Architect would be a breach of contract giving rise to damages for which no corresponding Event appears in Clause 25,[10] and no corresponding 'Matter' appears in Clause 26.

4.5 Provisional sums

In order that a provisional sum for Performance Specified Work may create an obligation on the Contractor, the Contract Bills must provide the information set out in Clause 42.7:

- the required performance,
- the location in the building,
- sufficient information to have enabled the Contractor to make allowance in his programming, and in his tender for pricing relevant preliminaries items.

However, if either the location or the information provided in the Bills contains an error or omission, it must be corrected and treated as though it were a Variation under Clause 13.2.

The Architect is expressly prevented from requiring Performance Specified Work against any provisional sum in the Contract Bills other than a sum identified as being for that purpose.

4.6 Variations

Clause 42.11 allows the Architect to issue instructions under Clause 13.2 requiring a Variation in Performance Specified Work, although he is not entitled to create such work by an instruction. Variations to Performance Specified Work and related problems are dealt with in Chapter 6, (Sections 6.12.4 to 6.12.6).

4.7 The Analysis

The work stated in the Bills as being Performance Specified Work is unlikely to be measured out in full in the Bills or in accordance with the general detailed rules of the

---

9   In *English Industrial Estates v. Kier Construction* (1991) 56 BLR 93, the Engineer's instruction to use crushed arisings from the excavation in preference to imported fill, when the specification provided for either, was held to be a deprivation of the Contractor's choice and a variation.

10 See Chapter 11 for commentary on 'time at large'.

Standard Method of Measurement (SMM 7). The JCT explain in Practice Note 25 that SMM 7 makes no reference to Performance Specified work, but that rule 11.1 provides for stating in the Bills any rules of measurement adopted for work not covered by the detailed rules. It is suggested by the JCT that rule 11.1 will cover the measurement of contractor-designed work.

Clause 42.13 requires the Contractor to provide an analysis of the portion of the Contract Sum related to Performance Specified Work, if the Bills do not contain such an analysis. The analysis is required for evaluation purposes, and is similar in concept to the 'Contract Sum Analysis' used in the With Contractor's Design form of contract. Practice Note 25 refers to Practice Note 23 for an example of a Contract Sum Analysis. The amount of detail provided by the Contractor in his analysis is a matter for the Contractor to decide, unless it is specified in the Bills. The prudent Contractor will ensure that allowance for design is clearly identified for calculating Variations in accordance with Clause 13.5.6.

4.8 Execution

4.8.1 Integration of Contractor's statement with Architect's design

It may be that the Contractor's design proposal in his Statement is incompatible with the overall design prepared by the Architect. In those circumstances the Architect must decide how to integrate the Performance Specified Work into the surrounding work, and give instructions accordingly (Clause 42.14). If the Contractor considers compliance with such instructions would 'injuriously affect' his proposals he must notify the Architect within 7 days. Unless the Contractor gives written consent, the instruction is of no effect, although consent cannot be unreasonably withheld or delayed (Clause 42.15). The Architect may then amend his instruction to remove the 'injurious affection', and the contractor is obliged to comply. Such an instruction will be a Variation instruction under Clause 13.5, varying surrounding work not being Performance Specified Work.

4.8.2 Divergence between Contractor's statement and statutory requirements

Compliance with Statutory Requirements in respect of Performance Specified Work is the Contractor's responsibility. Clause 6.1.6 provides that the Contractor or the Architect, if either finds a divergence, shall immediately notify the other; the Contractor is then obliged to inform the Architect of his proposal to remove the divergence. The Architect "*shall issue instructions in regard thereto*", and the Contractor's compliance is at no cost to the Employer unless caused by a change in the regulations after the Base Date of the contract; it is however expressly subject to the Contractor's right to notify an 'injurious effect' under Clause 42.15 (see Section 4.8.1). Thus the Architect cannot simply impose his own design on the Contractor at the Contractor's cost.

4.9 Liability for design and installation

4.9.1 Common law standard: fitness for purpose

Where a contractor designs and installs work under the same contract the common law obligation, in the absence of terms to the contrary, is one of fitness for purpose. The Contractor has the same liability in respect of his design as that in respect of workmanship and materials; it must be reasonably fit for the intended purpose previously made known to him by the Employer. In *Greaves v. Baynham Meikle*,[11] Lord Denning, when referring to the relationship between the contractor and the building owner, said (*obiter*):

> Now, as between the building owner and the contractors, it is plain that the owners made known ... the purpose for which the building was required. ... It was, therefore, the duty of the contractors to see that the finished work was reasonably fit for the purpose. ... It was not merely an obligation to use reasonable care. The contractors were obliged to ensure that the finished work was reasonably fit for the purpose.

In *IBA v. EMI & BICC*,[12] Lord Scarman underlined the position:

> I see no reason why one who ... contracts to design, supply and erect a television mast is not under an obligation to ensure that it is reasonably fit for the purpose for which he knows it is intended to be used.

The design must work reasonably; if it does not, the Contractor is liable. That would be the general position if the contract did not limit design liability.

4.9.2 Express obligation (Clause 42.17)

In line with other JCT design and build terms,[13] the liability for Performance Specified Work is expressly reduced in Clause 42.17 to the use of reasonable care and skill. This obligation is the same as that of the Architect or any other person providing services only.[14] To avoid any doubt, and to ensure that the common law standard is not expected by the Employer in respect of Performance Specified Work, Clause 42.17 states that nothing in the Contract shall operate as a guarantee of fitness for purpose. The limitation of liability applies only to design, so Sub-clauses 42.17.1.2 and .1 respectively also make clear that the provision is not to be construed to reduce the general obligation as regards workmanship and materials. In summary, materials supplied by the Contractor must be both fit for their general purpose and in accordance with Clause 8, workmanship must be to the standards described in Clause 8, and design must be carried out using reasonable care and skill.

However, the express limitation on design liability contained in Clause 42.17 seems to breach the provisions of the Defective Premises Act 1972. S. 1 of the Act provides 'A

---

11 *Greaves & Co. Ltd v. Baynham Meikle & Ptnrs.* [1975] 3 All ER 99, CA.
12 See n. 4.
13 See JCT Standard Form With Contractor's Design (WCD 98), Clause 2.5.1, and Contractor's Designed Portion Supplement for use with the Standard Form Clause 2.7.1, both of which state the Contractor '*shall have ... the like liability ... as would an Architect or ...*'.
14 Supply of Goods and Services Act 1982, s. 13.

person taking on work for the provision of a dwelling ... owes a duty ... to see that the which he takes on ... will be done in a ... manner ... so that as regards that work dwelling will be fit for habitation when completed.' Thus, if the contract is for the provision, conversion or enlargement of a dwelling, the design element in Performance Specified Work will be subject to a fitness for purpose requirement, notwithstanding apparent limitation in Clause 42.17.

4.9.3 Reasonable care and skill

The express obligation to use reasonable care and skill is the same as that implied by statute in a contract for services. If a contract for the provision of services is silent as to liability, a duty to use reasonable care and skill is implied by the Supply of Goods and Services Act 1982, Part II, s. 13. The Act does not define the standard to be applied, but the courts have considered the matter in relation to negligence. The extent of liability was described in *Bolam v. Friern Hospital Management Committee*:[15]

> Where you get a situation which involves the use of some special skill or competence, then the test as to whether there has been negligence or not ... is the standard of the ordinary skilled man exercising and professing to have that special skill. A man need not possess the highest expert skill ... it is sufficient if he exercises the ordinary skill of an ordinary competent man exercising that particular art.

Thus where an architect or designer-contractor has an obligation to use reasonable care and skill, he will not have failed if he does what his peers would have done in the same circumstances.[16] In the case of design and build under the provisions for Performance Specified Work a Contractor's peers are not contractors experienced in building work, but instead are contractors holding themselves out[17] to be experienced in similar design and build work. This can be compared with the position under the JCT design and build forms,[18] in which the Contractor's liability is expressed as 'the like liability ... as ... an Architect or ... other appropriate professional designer who acting independently ... had supplied such design ... to be carried out ... by a building contractor not being the supplier of the design.' It seems under the full design and build options the Contractor's 'peers' are to be found amongst Architects and other professionals, rather than designer-contractors.

15 [1957] 1 WLR 582 at 586.
16 Eg. in *The London Borough of Merton v. Lowe* (1982) 18 BLR 130, the Court of Appeal held the Architect was not negligent when the ceiling was found to be unsuitable, because it was a new product and he had made the sort of enquiries which a competent architect would be expected to make.
17 See *Tharsis Sulphur & Copper Co. v. McElroy & Sons* (1878) App.Cas. 1040: '*a contractor who expressly or impliedly undertakes to complete the work ... impliedly warrants that he can do so*'.
18 JCT With Contractor's Design, WCD 98, Clause 2.5.2; JCT Contractor's Design Portion Supplement, CDPS (WQ) 98, Clause 2.7.1.

5

itect's Instructions

5.1 Power to issue instructions

The Architect's authority comes from his terms of engagement with his Client. However, irrespective of the Architect's actual authority, the Contractor is entitled to assume it is the same as that set out in the building contract;[1] if it is not, the Employer will be in breach of his obligation to maintain a duty-holder to fulfil the duties of the Architect.[2]

The Architect's power to issue instructions is limited to the type of instruction expressly identified in the contract,[3] the most important being:

- Clause 2.3 (discrepancies in documents);
- Clause 2.4 (discrepancy or divergence);
- Clause 6.1.3 (statutory requirements);
- Clause 7 (levels and setting out);
- Clause 8.3 & 8.4.4 (opening up and testing);
- Clause 8.4.1 (removal of work not in accordance with contract);
- Clause 8.4.3 (Variations resulting from removal);
- Clause 8.5 (carry out Works in proper workmanlike manner);
- Clause 8.6 (exclusion from site of any persons employed thereon);
- Clause 12 (confirmation of clerk of works directions);
- Clause 13.2 (Variations, subject to the right of reasonable objection in Clause 4.1.1);
- Clause 13.3 (instructions on provisional sums);
- Clause 13A.4.1 (Contractor's '13A quotation' not accepted, instruction to proceed with Variation and to be valued pursuant to Clause 13.4.1);
- Clause 17.2 (schedule of defects delivered to Contractor as an instruction);
- Clause 17.3 (instruction for defects);
- Clause 23.2 (postponement);
- Clause 34.2 (antiquities);
- Clause 35.5 (removal of objection from Contractor to Nominated Sub-contractor);

---

1 The use of the RIBA Terms of Engagement assists compatibility, but it is still possible for the Employer to employ the Architect to carry out only limited duties, even under that form.
2 See *Croudace Ltd v. London Borough of Lambeth* (1986) 33 BLR 20, CA, where a Local Authority Employer failed to re-appoint an Architect and attempted to rely on the absence of an Architect to avoid paying the Contractor.
3 See also Chapter 6: Section 6.4 dealing with Variations.

- Clause 35.6 (instructions on nomination);
- Clause 35.18 (nomination of persons to carry out rectification work);
- Clause 35.24.6.1 (notice specifying default of Nominated Sub-contractor);
- Clause 36.2 (nominating a supplier for any materials or goods);
- Clause 42.5 (Architect's notice to amend Contractor's Statements);
- Clause 42.11 (Variations in respect of performance specified work);
- Clause 42.14 (instruction for the integration of performance specified work with the design of the works).

5.2 Form of instruction

5.2.1 Instructions to be in writing

Clause 4.3.1 states all instructions must be in writing. This does not mean that an instruction is only issued validly if it is headed 'Architect's Instruction'. There is no proscribed form for an instruction. Nevertheless it is in both parties' interests that wherever possible a form such as the pro forma published by the RIBA is used, which is immediately identifiable as an instruction. This is particularly important in instances where an instruction gives rise to a Variation, or where it may be unclear as to whether the communication is an instruction or a permission.[4] However, despite best intentions, instructions are not always reduced to a standard pro forma. The issue of a drawing may be an instruction in writing, and so too may a handwritten note on a drawing. Communication by e-mail may include instructions in writing if the Supplemental Provisions for EDI apply.[5] The issue of minutes of a site meeting may contain instructions in writing, providing they are issued by the Architect, although site meeting minutes produced by the Contractor may constitute only confirmation of oral instructions.

5.2.2 Instructions 'other than in writing'

Whilst Clause 4.3.1 requires all instructions to be in writing, the Contract does not bar the giving of oral instructions. Clause 4.3.2 allows instructions 'otherwise than in writing' to be confirmed in writing by the Contractor, although such instructions are stated to be of no immediate effect. Such instructions may be given under any of the provisions listed in Section 5.1, but the main source of dissatisfaction amongst contractors tends to be the issue orally of instructions giving rise to Variations. This is because an oral instruction has an air of immediacy about it, yet it is of no immediate effect so the Contractor acts on it at his risk.

The general procedure for confirming oral instructions is explained, together with commentary, in the chapter on Variations (see Chapter 6: Section 6.8, Oral Variation Instructions).

---

4   E.g. permission to sub-let part of the Works under Clause 19.2.2.
5   See Chapter 1: Section 1.4.1, Annex 2 to the Conditions: Supplemental Provisions for EDI (Electronic Data Interchange).

5.2.3 Instructions given to the Contractor

Throughout JCT 98 power is given to the Architect to issue instructions to the Contractor. Clause 1.7 provides that notices and other documents will be effectively served if given or delivered by pre-paid post to an agreed address, or to the last known business address or registered or principal office. This does not include electronic interchange such as e-mail unless an EDI agreement[6] has been signed.

One common difficulty experienced by contractors is the issue of instructions to site personnel. The Contractor is required by Clause 10 to constantly keep on site a person-in-charge, and it is expressly provided that instructions given to that person are deemed to be given to the Contractor. This can create difficulties on some sites, where the contract contains some contractor design, including Performance Specified Work, or where Clause 13A Quotation instructions are common. Unless the site establishment includes all the necessary disciplines, the person-in-charge at the site may not be the right point of contact. In such cases, the issue of variation instructions can introduce an unnecessary risk of delay in communication. Whilst the Contract does not require instructions to be issued to the site person-in-charge, it does sanction them. Clearly there is a good case for issuing building related instructions direct to site, such as those related to emergency work; but arguably it is in both parties' interests to ensure all variation instructions go to the Contractor's project leader, wherever he might be.

5.3 Compliance and query

5.3.1 Contractor's obligation to comply and entitlement to query

Clause 4.1.1 requires the Contractor to comply forthwith with any *bona fide* instruction given to him by the Architect, ie. instructions of the type which the Architect is expressly empowered to issue by the Conditions. There are two express exceptions, both related to Variations. The Contractor is not obliged to comply with a Variation if he makes reasonable objection in writing (see Chapter 6: Section 6.6); nor is he obliged, or indeed entitled, to start work against a 13A Quotation until expressly empowered to do so under the Clause 13A procedures (see Chapter 6: Sections 6.13.2 to 6.13.5).

The Contractor has one other entitlement to veto an instruction. Clause 4.2 allows the Contractor to request which provision of the Conditions empowers the Architect to issue any specified instruction. If so requested by the Contractor the Architect is obliged to comply 'forthwith'. Whilst there is no express provision for the Contractor to delay compliance with the instruction, the Contractor may put himself at risk if he were to carry out an instruction given by the Architect outside his authority.[7] However, if the Architect delays in identifying the particular provision obliging the Contractor to comply with the instruction, he will put the Employer in breach, and any effect of the Contractor's failure to put the work in hand will entitle the Contractor to damages. One important point here

---

6   Ibid.
7   See Chapter 6: Section 6.5.2., Objection to Purported Variations, regarding Employer's obligation to pay for work ordered by the Architect outside his authority.

is that the Architect must identify the empowering provision in the 'Conditions'; thus an express power or authority purported to be given under a clause in the Contract Bills (or any other part of the Contract) would not be sufficient, and the Contractor would not be entitled or obliged to comply.

If, after receiving a response from the Architect, the Contractor proceeds to comply with the instruction without either party first commencing one of the dispute resolution procedures, it is deemed to be a valid instruction.

5.3.2 Non-compliance by Contractor

Clause 4.1.2 provides a remedy for the Employer in the event of the Contractor failing to comply with a valid instruction from the Architect. The Architect may give notice to the Contractor requiring compliance, and if the Contractor fails to comply within seven days of receiving such a notice, the Employer may employ others to do the work. All the costs associated with such action may be deducted from the amounts due to the Contractor, or may be recovered as a debt. The use of the word 'costs' limits the Employer's recovery to actual additional cost incurred, including fees, but excludes losses that might arise as a result of delay. Losses suffered through delay would normally be recovered through the deduction of liquidated damages.

6

Variations and provisional sums

6.1 Variations: the general position

In a perfect world a building owner would know exactly what he wanted in advance of detailed design and part construction of his building. Likewise the designer would produce both the entire conceptual and detailed design before construction work started. There are circumstances when this can be achieved regularly, but they normally relate to projects involving repetition, for example, a standard design fuel filling station, retail store, or fast food restaurant.

However, the nature of construction work and the permanence of the finished product are such that it is rare for the building owner and his professional team to know exactly what the final requirement will be. Even the standard model may need adjustment to meet the peculiarities of the site. Many building owners procuring building work are strangers to the building industry; a new building is often a one-off experience and needs are dictated by circumstances which themselves change with the passage of time. The length of the total construction process enables a building procurer to have second thoughts, to take into account changed fortunes, to fine tune what at the outset were vague ideas; this often happens only as the substantive construction work proceeds, and rightly so if the procurer is to receive the product he needs and wants.

The secret to delivering the required product is flexibility, and whilst the contractor may view changes of mind as disruptive, it is unlikely that building owners as a class would wish to be deprived of the opportunity to develop ideas. Similarly the designer needs to be able to correct errors in the design, and to translate the procurer's developing requirements into design instructions for the contractor. However, the flexibility expected by the procurer and the designer is not an automatic right.

If the contract for the construction work does not provide for change, flexibility may not be available to the procurer and his designer. In a contract for a specified scope of work, the contractor is bound to provide that scope, no more and no less; he is obliged to produce the work he contracted to do. To do anything else would be a breach of the contract. The contractor is not only obliged to perform the specified work, but he is also entitled to perform it; to prevent him is also a breach of the contract.[1] In short, neither party can

---

1 See *Tancred Arrol & Co. v. The Steel Company of Scotland Ltd* (1890) 15 App. Cas. 125.

change the obligations of the other without the other's agreement. There is no flexibility. Changes in the work can still be introduced, but only by changing the initial agreement, or by entering into a separate contract; in either case there is no obligation to enter such an arrangement. This can create difficulties. For example, if the building owner wished to change the colour of the sanitary fittings during the course of the project before they were ordered by the contractor, and the contractor did not want the hassle of changing, the owner would have only two options. He could either accept the colour provided, or he could wait until the work was completed and pay the contractor for the fittings within the contract price. It would then be for the owner to arrange for the new fittings to be replaced, and pay again.

Even if the nature of required changes were such as to enable the parties to enter into a separate contract the position would be fraught with potential difficulty. Procedures would be duplicated, and work under each contract could materially affect and disrupt performance of the work under each of the other contracts. This would be so, even though the parties were the same.

It is for this reason that most forms of construction contract[2] contain express provisions enabling the procurer to change the obligations of the contractor. Such provisions normally allow wide change to the scope of work and the circumstances or conditions under which it is to be carried out. However, in setting parameters such provisions also create the limits, albeit wide limits, of the parties' entitlement to introduce change.

6.2 Variations 'to' or 'under' the contract?

Variations to the contract are changes to the contract itself, ie. changes to the agreement including terms and conditions, or changes to a party's performance which fall outside the types of change provided for in the contract.

Such changes can only be made by consensus between the parties. A typical variation to a contract would be an agreement to accelerate the works where no acceleration provision is made in the contract.[3]

Variations under the contract are changes to the scope of the work or conditions under which the work is carried out to the extent that terms in the contract provide for such changes. The term 'variation' is often used generically in the construction industry but contracts differ in their definition of the term. Indeed some contracts use other terms such as 'change'.[4]

Such widespread generic usage of the term 'variation' sometimes leads to confusion, and often leads both to abuse of the power to vary works, and to unwarranted high expectation of payment amongst contractors. Some of the resulting problems are dealt with later in this chapter.

---

2 But see *Form of Contract For Use Where The Contractor is To Design and Build, January 1974 Revision*, issued by the Construction Confederation, which expressly requires varied work to be carried out under a separate contract.
3 See *John Barker Construction v. London Portman Hotels* (1996) 83 BLR 31; (1996) 12 Const. LJ 277; [1996] 50 Con. LR 43.
4 See Clause 12 of JCT WCD 81.

6.3 Variations under JCT 98: definition

Clause 13 is an enabling provision, introducing into the contract the ability to change the parties' performance obligations. In this contract such changes are described as 'Variations' and the extent to which Variations are permitted is expressly set out in Clause 13.

Variations are defined in Clause 13.1 as:

a) The alteration, modification of design, quality or quantity of the Works. (This includes the addition, omission or modification of any work, alteration of materials specifications, and removal of work or materials.)

An important point here is that Variations can be wide ranging. Architects may sometimes give instructions which, to the contractor, seem to change the nature of the work; for example, an instruction to build a swimming pool as a Variation under a contract to build a sheltered housing scheme may invite objection from a contractor who only builds houses. In *Blue Circle Industries plc v. Holland Dredging Co. (UK) Ltd*[5] an instruction was given under a dredging contract to dump the dredged material to form an island bird sanctuary; it was held to be outside the scope of the original contract and therefore the work had been carried out under a separate contract. However, the courts sometimes appear to strain to bring a change within the parameters of the Contract. In *McAlpine Humberoak Ltd v. McDermott International Inc.*[6], the Court of Appeal were influenced by the purpose of a variation clause. The contract was for the construction of four pallets (shown on 22 drawings) for an oil rig tension leg. By instruction this was reduced to two pallets (but shown on 161 drawings). The contractor argued that the whole nature of the work had changed, effectively making a new contract. It was held that the contract was still a contract for the construction of pallets, and the purpose of the variations clause was to deal with such circumstances.

b) Imposition, removal or changing by the Employer of obligations or restrictions in regard to site access or use of parts of the site, limitation of working space, working hours, the order in which work is done or finished.

In the interests of flexibility this entitlement is significant for the Employer. For example, during building work on a factory refurbishment and extension, the Employer may need to change its own working practice to cope with a large order, delays to which could put the Employer in breach of its own supply contract. It is of great benefit to be able to change the sequence of building work on the extension, or to change working hours to free areas of the premises at critical operational times. However, in the absence of terms enabling such changes, it would be a breach of contract by the Employer[7] if he were to prevent the Contractor from working as agreed, or to deprive the Contractor of working areas. In return for the benefit of

---

5   (1987) 37 BLR 40.
6   CA; 5 March 1992; (1992) 58 BLR 1.
7   In *John Barker Construction v. London Portman Hotels*, it was held the Employer was in breach of an implied non-hindrance clause in a bonus agreement when he instructed variations preventing the Contractor from finishing on time.

flexibility, the Employer has to pay. The payment is not compensation; as a Variation it is treated as adjusted sales turnover, thus entitling the Contractor to overhead and profit contributions.

c) 'Variation', under Clause 13.1.3 does not include nomination of a sub-contractor to do work which in the Contract Bills has been priced as the Contractor's work.

The Architect cannot unilaterally change the provider of work, although he can take work away from the Contractor simply by omitting it, if the work is not required.

The general position is that, whilst work can be omitted (since the contract provides for it), work cannot be taken away from the Contractor to be given to someone else; to do so is a breach of the contract.[8] The nominated sub-contractor, although still a sub-contractor to the Contractor, is regarded here in the same way as a separate contractor, since otherwise the Contractor would be deprived of the right to do the work for which he contracted.

6.4 Architect's power to order Variations

Clause 13.2.1 states the Architect may issue instructions requiring a Variation.

The Architect is acting here as an agent of the Employer. The clause gives him authority to change certain of the parties' obligations under the contract; but his power is limited to Variations as specified. Thus the Contractor has no obligation to carry out an instruction purporting to be a Variation, if the subject matter of the instruction falls outside the definition of a Variation. An example of an invalid instruction would be one requiring the Contractor to accelerate the works in order to finish before the completion date or the proper extended date; the power to order acceleration is not included in the definition of a Variation.

6.5 Contractor's right of objection

The Contractor's obligation to carry out Variations is not absolute. Clause 13.2.2 provides the Contractor with rights to object to a Variation instruction; those rights are in addition to the right to challenge the Architect's authority to issue a particular instruction.

6.5.1 Objection to *bona fide* Variations

Clause 4.1.1.1 states the Contractor need not comply with a *bona fide* Variation instruction to the extent that he makes reasonable objection in writing if it varies obligations or restrictions regarding access, working space, hours or sequence.

The Contract offers no clues as what reasonable objection may be. However, it is suggested that personal circumstances may be reasonable grounds. For example a

---

8   See *Carr v. J. A. Berriman Pty Ltd* (1953) 879 Con. LR327; *Amec Building Ltd v. Cadmus Investments Co. Ltd* [1997] 51 Con. LR 105.

Variation may affect sequence on specialist work extending the Contractor's involvement into a period when his specialist resources are already fully committed elsewhere; in other words stretching his organization beyond its capability. But it may not be a reasonable objection that the Variation would itself be less profitable than the rest of the job, or that the Contractor suspects the Employer has insufficient funds to meet the extra cost.

Prior to 1 May 1998[9] and prior to Amendment 18, a dispute over the objection to a Variation instruction was a matter which the parties could refer to immediate arbitration. The arbitrator, in the absence of clear guide lines would then be left to apply such instinct of reasonableness as he possesses, and to try to justify his decision if the arbitration rules or the parties required a reasoned award. Under the adjudication provisions introduced by Amendment 18, the dispute is now more likely to be referred to an adjudicator. The adjudication rules in Clause 41A.5.4 state the adjudicator is not required to give reasons for his decision. This together with the uncertainty of what constitutes a Contractor's reasonable objection, the need for robustness by the adjudicator, and his immunity from liability, means the outcome of any objection referred to adjudication is likely to be a lottery. However, in the absence of strict rules of evidence, the Contractor's chance of success is probably greater. Nevertheless, it is still a risky business for the objecting Contractor; it is a brave contractor who challenges the Architect's authority, particularly towards the beginning of the job.

6.5.2 Objection to purported variations

Clause 4.2 gives the Contractor the opportunity to challenge the validity of any purported instruction issued by the Architect (see Chapter 5). In relation to Variations the Contractor may use the right to challenge an instruction which he considers is outside the definition of a Variation, and thus outside the power and authority of the Architect to instruct.

The types of instruction likely to be most commonly challenged are for work which could qualify as a separate contract, and administrative matters outside the apparent authority of the Architect. The latter would include an instruction to provide an invoice prior to an interim valuation, or an instruction to complete the work of a nominated sub-contractor.[10]

The decision to challenge or accept a Variation instruction when its validity is in doubt should not be taken lightly. If the instruction is carried out, and the Employer subsequently refuses to pay on the grounds that the Architect had exceeded his authority, the Contractor may be left with no remedy. Whilst the Architect in giving instructions generally acts as an agent of the Employer, he does so only to the extent that he is authorized. If he acts outside his authority the Employer is probably not liable. In *Stockport Metropolitan Borough Council v. O'Reilly,*[11] Judge Edgar Fay, QC said:

> An Architect's ultra vires acts do not saddle the employer with liability. The Architect is not the employer's agent in that respect. He has no authority to vary the contract. ... he cannot saddle the employer with responsibility for them.

9   Introduction of Housing Grants Regeneration and Construction Act 1996, Part II.
10   See Court of Appeal in *Fairclough Building Ltd v. Rhuddlan Borough Council* [1985] 3 Con. LR 368.
11   [1978] 1 Lloyd's Rep. 595 at 601.

That leaves only the Architect in the Contractor's sights. The Architect is not in contract with the Contractor, leaving no contractual remedy. The claim could only be in tort, but the Court of Appeal[12] has restricted that route where the parties to the construction contract have made an arbitration agreement. It has been suggested[13] that if the Employer were impecunious the Contractor could get around the Employer to the Architect, but the point is yet to be tested.

6.6 Architect's power to sanction Variations

The Contractor's obligation is to carry out and complete the Works in compliance with the Contract Documents. It is not for the Contractor to provide something different; to do so is a breach of contract.

However, there may be times when the Contractor is in a position to suggest an alternative to the specification, timing, or method of construction. It may be that the Architect takes the suggestion and converts it into an instruction for a Variation. However, if the Architect is indifferent about the suggestion, because it may benefit only the Contractor, it would be understandable if the Architect took the matter no further, requiring the Contractor to comply with the contract as it stood. That would be the Architect's obligation in the absence of authority from the Employer, since the terms of his own contract with his client usually include administration of the building contract, but not the authority to allow the Contractor to provide something different. The position is dealt with in Clause 13.2.4.

Clause 13.2.4 states the Architect may sanction in writing any Variation made by the Contractor. There is no requirement here for the Contractor to obtain prior permission, although the prudent Contractor will do so to avoid risk of having to remove work. The main benefit of the clause is to give the Architect the ability to allow retrospectively, a change made without prior permission, without breaching his own terms of engagement, and at the same time avoiding disruptive and perhaps unnecessary correction of the Contractor's work. Clearly, if the Architect sees the change as a financial benefit, such as an omission from the Contract Sum, there is nothing to prevent him from giving an instruction for a Variation in the normal manner; but in so doing he would be transferring risk from the Contractor to the Employer who would be paying less for the Works.[14]

A difficulty arises when the Variation would result in an addition to the Contract Sum. Whilst it may seem strange that the Contractor should be entitled to payment for a Variation which he has introduced without authority, that is the effect of Clause 13.4.1.1, which states 'all Variations required by an instruction of the Architect or subsequently sanctioned by him .. shall, unless otherwise agreed by the Employer and the Contractor, be valued .. (under Alternative A or B)'. If the Architect wishes to sanction a Variation without committing the Employer to payment, he must ensure he has the authority of the

---

12 *Pacific Associates Inc. v. Baxter* [1989] 2 All ER 159.
13 See discussion by Duncan Miller, 'The Certifier's Duty of Care to the Contractor – Pacific Associates v. Baxter Reconsidered' [1993] 10 ICLR 172.
14 See *Simplex Concrete Piling Ltd v. The Mayor and Aldermen and Councillors of the Metropolitan Borough of St Pancras*; (1980) 14 BLR 80, in which a letter assenting to a Variation was held to be an Architect's instruction for a Variation and the Employer was liable to the Contractor.

Employer to act as an agent, and then make an agreement in the name of the Employer with the Contractor. The Employer is unlikely, in these circumstances, to withhold authority. However, if the Contractor is unwilling to enter into such an agreement, then the Architect may exercise his discretion to refuse sanction, thereby keeping the Contractor in breach for not complying with the Contract specification.[15]

6.7 Variations after Practical Completion

6.7.1 After due date for Practical Completion

In *Balfour Beatty Building Ltd v. Chestermount Properties Ltd*[16] it was held that the Architect is entitled to issue Variation instructions after the date on which the Works should have reached Practical Completion; this applies whether or not the delay is one for which the Contractor is entitled to an extension of time. In short, so long as the Contractor is still in possession of the site and the Works are incomplete, the Architect can keep him there *ad infinitum*, although the Contractor will be entitled to relevant extensions of time.

6.7.2 After actual Practical Completion

The process of changing of mind or fine tuning requirements by the building owner, and correcting defects in the design by the Architect, often continues after the Contractor has achieved practical completion of the Works. The practice of attaching lists of 'snagging items'[17] to the Certificate of Practical Completion often leads to the presence of the Contractor on site after 'actual' Practical Completion. He is there to finish off minor works, but is often drawn into carrying out extra work, which falls into two broad categories.

The first category is overtly extra work instructed by the Architect taking advantage of the Contractor's continuing presence.

The second category is work done under instructions which have the appearance of requiring the Contractor to put right a defect. A common example is filling plaster cracks; the crack could be the result of the Contractor's poor workmanship, or it could be poor design of the building creating excessive vibration when over-specified door closers slam the doors. If it is the latter, the work is an extra.

Whichever of the two categories the additional work falls into, the Architect in issuing the instruction at this time acts outside his authority and power, unless the contract expressly provides.[18] The work is finished (that is finished within the meaning of the contract), and the final calculations of time and value have started. The Contractor is not then obliged to accept the instruction. He may if he wishes refuse outright, or he may

---

15 But note the decision in *Howard de Walden Estates Ltd v. Costain Management Design Ltd* (1992) 55 BLR 124, in which an architect's instruction letter was qualified to make it at the Contractor's cost; it was held that the variation was at the Contractor's cost because it resulted from the Contractor's defective work; it was said to be the sort of agreement made by sensible parties who wanted to get on with the work.

16 (1993) 62 BLR 1.

17 See discussion on Practical Completion in Chapter 11.

18 Compare with ICE *Conditions* (6th edn), cl. 51 (1) which provides for variations to be ordered during the Defects Correction Period).

propose a separate contract. Alternatively he may, by express agreement with the Employer, amend the contract to incorporate the extras; but in any event he is not obliged to do the work at the Contract rates.

In practice, instructions coming within the second category are often carried out, then disputed later. This may be because the cause is not immediately apparent, or because the list of work is handed to operational supervisors at site who may not be familiar with the detailed provisions of the contract. However the Contractor needs to take care, for if an item is later identified as an extra, it will have been the subject of an instruction which the Architect had no power to give, and the Employer may decide not to pay for it.[19]

6.8 Oral Variation instructions

Under Clause 4.3.1 all instructions issued by the Architect must be in writing. Ideally Clause 4.3 should end there; but unfortunately, it does not. Clause 4.3.2 goes on to provide a mandatory system of confirmation, to be initiated by the Contractor in the event that the Architect gives an instruction orally.

Clause 4.3.2 provides:

a) *'(an oral instruction) shall be of no immediate effect'*. The Contractor cannot start to carry out the instruction, even though he could at the time be building work which under the instruction would have to be taken down later. This negates any intended benefit to be derived from an oral (and, thus immediate) instruction.

b) *'but it shall be confirmed in writing by the Contractor ... within 7 days'*. The Contractor has no choice in the matter, except under provisos in Clauses 4.3.2.1 and 4.3.2.2 dealt with below. He must act, although the Contract is silent as to any sanction if he does not. However, the failure to confirm is a breach of the term and could affect the Contractor's entitlement to payment for removal of excessive work caused by late execution of an instruction. Under the provisos, the obligation to confirm is lifted if (i) the Architect himself confirms the instruction within 7 days, or (ii) in the event of the Contractor having complied with the instruction, but not having confirmed it, the Architect himself confirms it before the issue of the Final Certificate. The Architect's confirmation under the latter proviso is not obligatory; the Architect *'may confirm'* the instruction. The second proviso applies only when the Contractor has complied with the instruction so, if the Architect chooses not to confirm it, the Contractor will be in breach, having deviated from the scope of work in the contract.

c) If the Contractor's confirmation is not dissented from, within 7 days after the Architect's receipt, the confirmation is effective; the oral instruction is then as binding from that point as if it had been issued by the Architect.

Unfortunately, the system of confirming instructions other than in writing is open to abuse, and can be treated as an invitation to issue oral instructions. This has the effect of leaving the administration to the Contractor; with it goes the risk of delaying work or carrying out an unauthorized variation potentially in breach of the contract, in the hope that the Architect will confirm it. Clearly, the prudent Contractor will do nothing until the instruction is ratified expressly or by silence; that is both his right and obligation. That

---

19 See Section 6.5.2, Objection to Purported Variations.

being the position, the confirmation provisions in Clause 4.3.2 seem, at best, to be of doubtful benefit, and at worst conducive to poor practice and a spawning ground for dispute. This is particularly so when there are many oral instructions, many of which may impact upon each other.

Some contractors supply their site staff with a pro forma confirmation sheet; it is commonly headed 'Confirmation of Verbal Instruction'. Such contractors should be applauded for their recognition of the procedural requirement, but unfortunately they may suffer from making oral instructions too convenient. A better contractor's pro forma, it is suggested, would be one addressed to himself, headed 'Instruction', for completion by him, but to be signed by the Architect. The advantages are clear. It avoids the need for retrospective confirmation by the Architect which invites the Contractor's compliance at risk, and it removes the possible 2-week waiting period before the Contractor can be sure of his position.

6.9 Instructions in regard to provisional sums

The Architect is given power and duty in Clause 13.3 to issue instructions in regard to provisional sums in the Contract Bills, and expenditure of provisional sums in a Nominated Sub-contract.

In practice, express instructions dealing with work covered by provisional sums are unusual, except where the work content is capable of being separated; an example would be a provisional sum for an entire work section like a gatehouse. It is more common to see an instruction for the omission of a provisional sum, and the corresponding additions, being lost in other measured variations.

Clause 13.4.2 clarifies the position when the Contractor tenders for work covered by a prime cost sum created by the Architect out of a provisional sum. In those circumstances the Contractor's tender, if accepted by the Employer, is the basis of valuation. It is not treated as valuation of an instruction in regard to a provisional sum.

6.10 Variations instruction and evaluation: a choice of method

(NOTE: for convenience only as a means of describing the method of valuing variations, the valuation provisions of Clause 13.5, ie the valuation provisions of JCT 80 preceding Amendment 18, are referred to here as the 'basic method'. This is not a term used by the JCT.)

6.10.1 Introduction

Until recently the long-standing method in JCT contracts of instructing and calculating the value of variations was, in simple terms, for the Architect to instruct, for the Contractor to carry out, and for the Quantity Surveyor then to value what had been done. There was no provision for quotations, and the Architect had no authority to accept a quotation if it were provided. The contract set out a list of detailed rules, and it was by those rules that a variation had to be valued unless the Employer and Contractor expressly agreed otherwise. The Architect and Quantity Surveyor had no choice in the matter.

In recent years a choice has been introduced to ratify various growing or habitual practices in the industry; these include:

- design input by the Contractor;
- the general preference of many Employers to know the building cost with some certainty, usually by the use of advance prices;
- the tendency for contractors to measure and value Variations, either for their own purposes or to assist the Quantity Surveyor.

JCT 98 still sets out detailed rules as a basic method of evaluation;[20] these are to be found in Clause 13.5. Variations to work where the contractor designs to meet a performance is valued as a variant of the 'basic method' and is dealt with under that heading below.

The concept of advance prices was introduced in Clause 13A, by Amendment 13 to JCT 80; it is an alternative system of both valuation and of administration. Clause 13A enables the Architect to require a quotation from the Contractor for acceptance by the Employer. This is carried out under a specific regime of instruction, quotation and acceptance. A quotation, called a '13A Quotation', under this procedure must be accepted before the work described in the instruction is implemented by the Contractor.

Amendment 18 to JCT 80 introduced a means by which the Contractor may submit price calculations for a traditional instruction. This is not a facility for providing quotations. It is simply an alternative to mandatory unilateral evaluation by the Quantity Surveyor; it gives the Contractor the opportunity to put his price to the Quantity Surveyor and to invite early agreement on the value of a Variation.

6.10.2 Directions as to choice of method

Clause 13.2.3 directs that a Variation instructed by the Architect is to be valued in accordance with Clause 13.4.1 unless it is dealt with under the Quotation provisions of Clause 13A.

In turn, Clause 13.4.1.1 directs that all Variations instructed or sanctioned by the Architect including provisional sums, all work treated as though it were a Variation, and work covered by approximate quantities shall be valued under Alternative A of Clause 13.4.1.2, i.e. by submission of a Contractor's Price Statement.

There are several caveats. The direction does not apply:

1   if the Employer and Contractor have agreed that any Variation is to valued by some other method, or
2   if the Contractor does not implement the provisions by failing to submit a statement, or
3   if a Price Statement or amended Price Statement is not accepted.

If Alternative A does not apply, other than by reason of agreement between Employer and Contractor, then Alternative B of Clause 13.4.1.2 applies, i.e. the 'basic method'.

Thus the choice of Variation rules is:

1.   *'Traditional instruction method' (Clause 13.4.1.2) (see Fig. 6.1)*
- Pricing option Alternative A: Contractor's Price Statement – Contractor's choice.

---

20 See note to section heading.

- Pricing option Alternative B: basic method – default provisions for all Variations, Provisional sums approximate quantities and Performance Specified Work.

2. *'Quotation Method' (Clause 13A)*
    - 13A Quotation procedure – Architect's choice.

6.11 Variation rules – Alternative A: Contractor's Price Statement (Clause 13.4.1.2) (see Fig. 6.1)

Contractors traditionally have contributed to the valuation process, often by providing their version of what the Quantity Surveyor should produce. The practice is ratified in Alternative A by encouraging Contractors to value an instruction and to present their valuation in a Price Statement. The option is entirely the Contractor's choice. But under Clause 13.4.1.2, Alternative A is stated to be the means of evaluation and the traditional 'basic method' is to be used only if the provisos referred to in the previous section apply. In making the Contractor's Price Statement the first option the JCT clearly expect the option to be used.

The drafting form of Alternative A differs from Alternative B. Alternative A is described as 'Contractor's Price Statement' and is drafted as paragraphs A1 to A7 within Clause 13.4.1.2 containing the detailed Alternative A rules. Alternative B is also contained in Clause 13.4.1.2 but has no descriptive heading and introduces the evaluation provisions by reference to Clauses 13.5.1 to 13.5.7.

6.11.1 Instructions

Evaluation under Alternative A: Contractor's Price Statement is no more than an alternative system of valuing Architect's Variation instructions or work which is required by the Contract Conditions to be treated as though it were a Variation resulting from an instruction. It is not for the Architect to state that Alternative A shall apply; in this it differs from Clause 13A Quotation procedures. There is no special instruction required to trigger a Contractor's Price Statement, and the only bars to the choice of Alternative A are instances where the Conditions specifically provide otherwise. Typical examples are instances where a 13A Quotation is instructed, and where work the subject of an Architect's instruction affects other work which was itself included in an accepted 13A Quotation.

6.11.2 Price Statement: submission

Contractors should be wary of relying on receiving the amount included in their Price Statement. The JCT have chosen their terms carefully; a Price Statement is not a quotation. It is not a vehicle for deciding whether work will or will not be done. The Price Statement is simply a valuation of a Variation instruction which the Contractor has already received and which he is obliged to carry out. He has no choice in whether or not to do the work; but he does have a choice in influencing the evaluation of the Variation.

Under paragraph A1, within 21 days of receiving an instruction the Contractor may

CPS: Contractor's Price Statement
APS : Amended Price Statement
Q/S: Quantity Surveyor

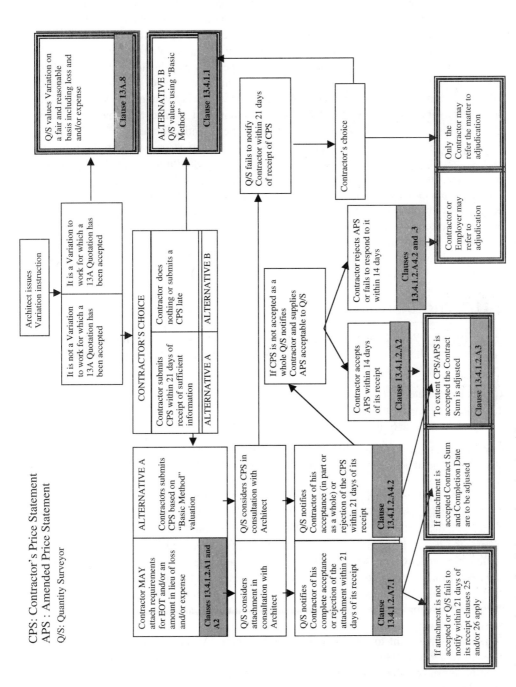

Fig. 6.1 Valuation of Variations – alternatives A and B

submit his price for the work in a Price Statement. The Price Statement is submitted to the Quantity Surveyor. If there is insufficient information provided with the instruction to enable the Contractor to prepare a price, he may submit the Price Statement within 21 days of receiving the necessary information. Alternative A contains no express provision for the Contractor to request more information needed solely for the purpose of pricing.

Thus it is arguable that the Contractor can wait, perhaps months, until he feels he has sufficient information, then submit his Price Statement. That is the Contractor's right, and in such circumstances a cautious Quantity Surveyor will delay valuing the work himself until he is sure his work will not be wasted; although it is also arguable that if the Contractor has insufficient information, so too has the Quantity Surveyor!

The Price Statement states the price for the work and must be based on the rules in Clause 13.5. It is clear that the Price Statement is not an opportunity for the Contractor to ignore the Contract rules and to price on a more advantageous basis.

The Contractor may also, if he wishes, attach to the Price Statement a separate statement of his requirements for extension of time and amounts in lieu of loss and/or expense. The only proviso to his entitlement is that he must not request time or amounts which have already been awarded or ascertained under Clauses 25 or 26, or in an accepted 13A Quotation. However, he may duplicate requests made in other Price Statements or 13A Quotations which have not yet been accepted.

6.11.3 Price Statement: Response

On receipt of a Price Statement the Quantity Surveyor must consult with the Architect, then write to the Contractor (para. A2) within 21 days of receiving the statement, stating:

1. the Price Statement is accepted, or
2. the Price Statement is not accepted, or
3. part of the Price Statement is not accepted.

Practitioners need to remember that para. A2 deals only with processing the Price Statement. The attachments to the Price Statement are dealt with separately in paragraph A7.

If the Quantity Surveyor fails to write to the Contractor, under para. A5 the Price Statement is deemed to be not accepted. The Contractor is then expressly able to refer the matter as a dispute to the Adjudicator. The Contractor would in any event have the right to refer the deemed rejection (for which the Quantity Surveyor has given no reasons) to adjudication under Clause 41A of the Contract and under the Housing Grants Construction and Regeneration Act 1996, Part II, s.108. The express right to refer to adjudication seems to underline the importance of the Price Statement option; it is a constant reminder to the Quantity Surveyor that he cannot ignore the Contractor's submission, and also that he must act promptly and speedily in checking the Price Statement within a tight timescale. Significantly the right to refer to adjudication in para. A5 is limited to the Contractor, whereas under the 1996 Act either party is entitled to refer any dispute or difference. It must be presumed that the drafters of this clause intended that the Quantity Surveyor should be seen overtly as an agent of the Employer in this matter, thus depriving the Employer of the right to refer the matter, since the Employer cannot dispute his own actions.

The Quantity Surveyor's consultation with the Architect is necessary to ensure the Price Statement relates to matters which are Variations in principle. If he fails to consult and then subsequently agrees a Price Statement which should not be considered as a Variation, the Employer may be bound to pay the Contractor, leaving the Quantity Surveyor exposed to a claim from the Employer.

Price Statement Accepted: If the Contractor's Price Statement is accepted in whole or in part, the price or part of the price accepted becomes the amount of the Variation for adjustment of the Contract Sum under Clause 13.7.

Price Statement Not Accepted: If the Price Statement is not accepted, either in whole or in part, the Quantity Surveyor must write to the Contractor giving reasons in similar detail to that in the Price Statement. He must also supply an amended Price Statement to the Contractor.

6.11.4 Amended Price Statement: submission by Quantity Surveyor

To the extent that a Price Statement is not accepted, para. A4.1 requires the Quantity Surveyor to provide to the Contractor an amended Statement. The amendment must be acceptable to the Quantity Surveyor. At first sight this latter requirement may appear unnecessary, since arguably the Quantity Surveyor would not put forward an amendment which was not acceptable to him. However, it emphasizes the nature of the amended Price Statement as a route towards agreeing the value of the Variation, rather than a means of abusing the system to reject the Contractor's price.

6.11.5 Amended Price Statement: response

On receipt of the Quantity Surveyor's amended Price Statement, the Contractor has 14 days in which to state that he accepts all or part. Significantly, unlike the Quantity Surveyor's acceptance in para. A2, para. A4.2 does not require such statement to be in writing, although the prudent Contractor will write.

To the extent that the Contractor accepts the amendment, the accepted amount becomes the relevant amount to be the adjustment of the Contract Sum in accordance with Clause 13.7.

If the Contractor does not make a statement within the 14-day period, the Contractor is deemed not to have accepted all or part of the amended Price Statement.

At this point the submission of statements back and forth between the Contractor and Quantity Surveyor ceases. To the extent that agreement has not been reached, the parts of the Price Statement and of the amended Price Statement that are not agreed may be referred by either the Employer or the Contractor to the Adjudicator under Clause 41A.

If after following the Alternative A procedures neither the Price Statement nor the amended Price Statement is accepted, and if the non-acceptance has not been referred to an Adjudicator, para. A6 states that Alternative B (the 'basic method') applies. However, there is no timescale suggested, and dispute may arise when a Quantity Surveyor ignores his own amended Price Statement in favour of a valuation under the basic method. In theory there should be little difference, since the Price Statement, and consequently the amended Price Statement, are supposed to be calculated in accordance with the rules of Clause 13.5, which is Alternative B. However, it is possible that Contractors will be tempted to move away from the strict rules of Clause 13.5 in preparing a Price Statement, possibly confusing Alternative A with Clause 13A, and the Quantity Surveyor may be tempted to respond, not only in similar detail but also in the same manner. If that occurs then clearly the Quantity Surveyor runs the risk of a claim from the Employer, but in the event of dispute the Variation should be valued under Alternative B.

6.11.6 Response to attachments to Price Statement (request for extra time/loss and expense)

It is emphasized elsewhere in this section[21] that Alternative A is not an alternative option to Clause 13A for giving quotations. Unlike acceptance of a 13A Quotation, the acceptance of a Contractor's Price Statement or a Quantity Surveyor's amended Price Statement does not include acceptance of requirements for loss and/or expense or for extensions of time. Claims for extra time and money are dealt with separately in para. A7.

On receipt of attachments to a Price Statement the Quantity Surveyor must consult with the Architect, and within 21 days of receiving the Price Statement must notify the Contractor:

1. Either that the amount requested in lieu of loss and/or expense is accepted, or is not accepted. If not accepted, any loss and/or expense due will be ascertained under Clause 26.1.

    There is no provision for an amended response.
2. Either that the request for extension of time is accepted, or is not accepted. If not accepted, adjustment to the time for completion, if any, will be made under Clause 25.

    There is no provision for an amended response.

If the Quantity Surveyor does not respond within the 21 day period, Clause 25 and Clause 26 apply as though the Contractor had attached no time or loss and expense requirements to his Price Statement. Although the Contractor may choose to submit details, there is no absolute obligation on the Quantity Surveyor to respond; neither is there any express remedy for the Contractor, other than the right to refer any dispute at any time to adjudication under Clause 41A.

6.12 Variation rules – Alternative B: 'basic method' (Clause 13.4.1.2) (see Fig. 6.1)

The first point, which many contractors forget (and unfortunately some quantity surveyors too), is that the duty to value the work under the basic method lies squarely with the Quantity Surveyor. It is not the role of the Contractor, albeit the Contractor is entitled to witness measurement on site if he wishes (Clause 13.6), and is obliged to provide information necessary for the calculation of the Final Account (Clause 30.6.1.1). Too often

---

21 See Section 6.11.2 dealing with Contractor's Price Statement.

in the past it has been left to the Contractor to present calculations for checking by the Quantity Surveyor; that may still occur, particularly if Alternative B becomes applicable as a result of missing Alternative A deadlines. Contractors probably see the effort as rewarding, driving the timing and content of calculation to their advantage. However, it has often led to disillusionment and dispute when the Quantity Surveyor has ignored the Contractor's submission, or used the Contractor's calculation for interim payment and budgeting purposes, and has then 'done it properly' for the Final Account, reverting to the strict rules.

The valuation rules set out in Clause 13.5 apply to several types of work:

- variations required or acceded to by the Architect,
- work covered by Approximate Quantities in the Contract Bills,
- work covered by Provisional Sums in the Contract Bills,
- work described as Performance Specified Work in the Contract Appendix.

6.12.1 Variations required or acceded to by the Architect including work covered by Approximate Quantities

Clauses 13.5.1 and 13.5.2 contain rules for evaluation where work can be measured. Sub-clause 5.1 sets out a list of criteria to be applied to the description of varied work in order to link variation prices with the contract price. Thus work closely resembling that in the Contract in all respects must be valued at the Bill rates, but the Bill rates become progressively less relevant to any additional work which bears little or no similarity to that described in the Bills. Sub-clause 13.5.2 deals with omitted work. The criteria and consequential rules are:

Clause 13.5.1
1. work of a similar character, carried out under similar conditions with no significant change of quantity: Bill rates apply;
2. work of a similar character, but not carried out under similar conditions and/or significant changes in the quantity: Bill rates form the basis of evaluation, adjusted to make fair allowance for differences;
3. work not of a similar character: Bill rates do not apply, and evaluation shall be by use of fair rates and prices;
4. where an Approximate Quantity is a reasonably accurate forecast and provided only the quantity has changed: Bill rates apply;
5. where an Approximate Quantity is not a reasonably accurate forecast and provided only the quantity has changed: Bill rates form the basis of evaluation, adjusted to make fair allowance for difference in quantity.

Clause 13.5.2
6. Omitted work: Bill rates apply.

The criteria continue in Clause 13.5.4:

work which cannot be properly valued by measurement must be valued by application of daywork rates. Daywork rates are those calculated in accordance with the Royal Institution of Chartered Surveyors' 'Definition of Prime cost of Daywork carried out under a Building Contract' plus the relevant percentage additions stated in the Contract

Bills. There is a caveat: detailed work records described as vouchers are required to be delivered to the Architect or his representative not later than the end of the week following that in which the work was carried out.

The effect of the Variation or changed Approximate Quantity on other work is also covered. If there is a substantial change to conditions under which other work is carried out, as a result of the Variation or changed Approximate Quantity, such other work is to be treated as though it were itself a Variation and must be valued under the rules in Clause 13.

6.12.2 Some typical problems

Whilst the variation rules cover many situations, there are a number of typical recurring problems:

1. Identification of work which can properly be valued by measurement

Contractors will often claim work cannot be measured by reason of a change to conditions under which it is carried out, or that it is of a different character. Clearly, such claims must fail since those are the precise circumstances covered by Clauses 13.5.1.1 to 3 for application to measured work. The only logical test can be application of the measurement rules of the contract, ie. the Standard Method of Measurement. If the rules can be applied, then the work can properly be measured. In practice the difficulty often lies in identifiable parameters. For example, a variation may relate to damage to completed plasterwork. If the boundaries are clear, such as a replacement of a whole wall, then measurement will be possible; but if the damage is in undefined patches, requiring cutting back to firm bonded plaster wherever that might be, the areas may not be measurable in the practical sense.

Quantity Surveyors, on the other hand will sometimes strain to achieve measurement to avoid resorting to daywork. The result can be measurement by approximation, or assertion that measurement can be made simply because some form of measurement has been made! It is submitted the meaning of properly in this context means measurement by the rules, using descriptions categorized in the rules, but applied only where the nature of the work would allow accurate physical measurement.

2. Not of similar character or conditions

Where work is not of a similar character to that described in the Bills, the Bill rate does not apply and a fair valuation must be made. In practice the concept of similar or dissimilar character seems to cause little difficulty, since detailed specification including performance determines character.

What comprises similar conditions is more likely to create problems. The conditions under which the work is carried out includes such factors as difficulty of access to the work face, distance from stores, difference in height, whether work is carried out in natural summer light or artificial light in winter. In *Wates Construction v. Brodero Fleet*[22] it was held such conditions are those to be derived from the express provisions of contract

---

22 (1993) 63 BLR 128.

documents, and do not include the constructive knowledge and expectations of the parties gained during pre-contract negotiations.

3. Bill Rates apply to measured variations – even when relevant rates are commercially low or high.

A frequent cause of dispute is the rate in the Bill which the Contractor realizes during the contract, was entered in error. Sometimes the rate is patently inadequate; sometimes it is simply a bit high or low.

Contractors frequently argue that they are entitled to have a low rate corrected, and that the Architect is not entitled to order greater quantities than those in the Bills, in order to take advantage of a bargain, albeit the error is discovered after conclusion of the contract. Quantity Surveyors will sometimes take pity on the Contractor by valuing increased quantities using a fair rate, at the same time reminding the Contractor that he is obliged to bear the loss for which he contracted , i.e. the quantity in the Bills at the Bill rate.

The problem was considered by the Court of Appeal in *The Mayor Aldermen and Burgesses of the Borough of Dudley v. Parsons and Morrin Ltd.*[23] The Contractor had inserted in the Tender Bill a rate of 2 shillings per cubic yard extra over for excavation in rock. A fair price was £2. The Contractor had been allowed a fair rate by the Architect for all the quantity over the provisional quantity in the Bill. The Contractor sought the whole quantity at a fair rate. Lord Justice Pearce said:

> Naturally one sympathises with the contractor in the circumstances, but one must assume that he chose to take the risk of greatly under-pricing an item which might not arise, whereby he lowered the tender by £1,425. He may well have thought it worth while to take that risk in order to increase his chances of securing the contract.

Clearly the Court was influenced in this case by the provisional nature of the quantities in the Bill, but the outcome was that the Bill rates were held to apply to the total quantity. The whole principle of using pricing levels in the contract for variations would be undermined if it were otherwise. Once the contract is concluded, and provided the Employer was not aware of and taking advantage of a patent error in accepting the erroneous tender, the rate is accepted at risk. The risk is borne by both parties.

Consequently a Bill rate which is erroneously high must also stand, and the Contractor would have no right of objection if the Architect omitted quantities in order to save money for the Employer. This point was confirmed in the case of *Henry Boot Construction Ltd v. Alstom Combined Cycles Ltd*[24] where it was held that the parties had agreed the Bill rates, and it was immaterial whether they appeared high or low.

4. Evaluation based on Bill Rates, adjusted to make fair allowance for differences

This is often misread by quantity surveyors and contractors alike to mean 'evaluation shall be pro rata the Bill rates'. It may be that pro rata has a part to play in some instances where a factor such as increase in thickness may be proportionate for the materials part of a rate; but it is rare for a change in thickness, height, colour etc. to affect the cost of work entirely

---

23 8 April 1959: see 'Building and Civil Engineering Claims', 1984, unreported, cited by Wood, R. D., at p. 527.
24 (1999) BLR 123.

proportionately. Only when a new description falls squarely between two Bill rates is there any excuse for a simple pro rata calculation, and even then care should be taken to ensure the element of difficulty does not in reality increase or decrease exponentially.

5. Evaluation by Fair Rates and Prices

The concept of 'fair rates' can be confusing. Normal market rates can be 'fair rates'; so too, depending on the context, can 'not unreasonable costs excluding profit'.[25] The context in Clause 13.5 suggests equal fairness to both parties. Quantity Surveyors sometimes see this as meaning the best price the Employer could have obtained by approaching local firms, or the price from a pricing book; whereas contractors sometimes see it as being all their actual costs plus an arbitrary percentage for overheads and profit. However, fairness, for the purpose of Clause 13.5, is tinged with subjectivity. What is to be considered here is the fair price to be paid to this contractor (not to some other contractor) by this Employer under this contract (not some other contract). In short the Contractor's actual net costs must be taken into account, provided he did not waste costs. Likewise the profit margins built into the Tender must be included; but so too must any actual commercial concessions which make up the market level of his Tender, although likely profitability or unprofitability should be ignored. Contractors sometimes object to the Quantity Surveyor's request to see the estimator's Tender workings on the grounds that they are confidential; but without them the Quantity Surveyor is unable to fulfil his duty to properly value the Variation.

6. The status of daywork vouchers

The daywork voucher supplied by the Contractor under Clause 13.5.4 is no more than evidence of the time, materials and plant expended on the described task. The existence of a voucher, whether signed or not, does not give rise to entitlement in principle for work to be valued on a daywork basis. Entitlement follows the test of whether work can be measured, and it is commonly left to the Contractor to guess whether the Quantity Surveyor will be able properly to measure the work. This can lead to a glut of vouchers as the Contractor plays safe and submits a voucher for anything which he considers to be out of the ordinary. The result is often a pile of unsigned vouchers when the Architect resists signing for work he considers is not a variation, or is measurable, or was done when he was not present.

This last reason is one frequently faced by Contractors, even in circumstances where the work is clearly not measurable. The Contractor should not worry – he is entitled to rely on his vouchers whether signed or not. The purpose of giving the Architect (or, as is often the case in practice – the clerk of works) an opportunity to verify a voucher is to protect the Employer's interest, and the Architect who avoids signing vouchers on the grounds of absence from site runs the risk of putting himself in breach of his contract with the Employer.[26]

A view sometimes held by Architects and Quantity Surveyors is that before signing a daywork voucher, there is a right or even a duty to correct the contents of the voucher to what they consider is a fair amount. For example, a voucher may show 10 hours' labour

---

25 *Semco Salvage & Marine Pte Ltd v. Lancer Navigation Co. Ltd*: HL; 6 February 1997.
26 See *Chartered Quantity Surveyor*, September 1986, 'Questions & Answers', p. 15 for correspondence on this point.

carrying out what should in the Architect's opinion have been completed easily within 3 hours. However, it is not for the Architect, or later the Quantity Surveyor, to change the hours or any other entry on the voucher, unless the entry is not a correct representation of what actually occurred. This point was considered recently by the Court of Appeal[27] in which it was said, after the trial judge had gone behind daywork timesheets:

> [The trial judge was] ... wrong to go behind the timesheets. The timesheets were not suggested to be fake ... . The most that could be said ... was that perhaps the workmen did not work as expeditiously as they might have done. That is the danger of day-work contracts. It is a danger which is often dealt with by the Architect, making sure that the men are on site and working.

Clearly, the Quantity Surveyor's duty in these circumstances (if daywork is the appropriate method of evaluation) is to apply the Contract daywork rates to the hours on the voucher, provided it is an accurate record of the time taken. This will be so even if he considers the hours are unreasonably high. If the hours recorded are not a true record of the time expended, then the Contractor may be guilty of a criminal offence under the Theft Act 1968.[28]

6.12.3 Work covered by provisional sums in the Contract Bills

A provisional sum is simply an amount of money established by the Employer, which he requires the Contractor to include in his Tender to cover the cost of designated work. When the work can be properly valued the sum is omitted and replaced by the proper calculated value. The method of calculation is the same in principle as for any other Variation.

6.12.4 Variations in respect of Performance Specified Work: Generally

Amendment 12 introduced the concept of 'performance specified work', by incorporation of an optional Part 5. Part 5 is dealt with in Chapter 4, but the drafters of JCT 98 saw fit to deal with variations in Performance Specified Work in the general variations clause.

Despite the original general 'build only' nature of JCT 98, there are circumstances where a contractor's specialist building knowledge includes design. Frequently the Contractor supplies a specialist product, often by use of a sub-contractor, which requires detailed design knowledge of the product. A typical example would be a uPVC window system designer who decides what profile of extrusion is necessary for the circumstances. If a Variation is required which affects the influences on design, (e.g. a requirement for windows to match those on the ground floor, but instead to be fixed at the twentieth floor), then the best person to determine the detailed design and suitability is the system designer. In these circumstances the 'traditional' system of valuing a Variation is inadequate, and open to abuse when the Parties attempt to make it fit. Clause 13.5.6 provides separate valuation rules.

---

27 *Clusky (t/a Damian Construction) v. Chamberlain*, CA: 24 November 1994 unrep.
28 S.15(1) – Obtaining property by deception, s.17(1) – False accounting.

Whilst Performance Specified Work is incorporated in Clause 13.5, described here as the 'basic method', the Employer's professional team and the Contractor need to be wary. The general principles are more akin to those of 'design and build', and misunderstanding as to what is, or is not, a variation is commonplace.

The concept, in general terms, is that the Employer requires a specified result. It is for the Contractor to achieve that result, and how he achieves it is his choice, limited only by any parameters set out in the Contract Bills or within the Contractor's Statement (see Chapter 4 dealing with Part 5, Performance Specified Work). Some examples of typical problems frequently encountered are described in Section 6.12.6 below.

6.12.5 Variations in respect of Performance Specified Work: evaluation

The rules for evaluation are set out in Clause 13.5:

1. The Valuation must include for relevant design related work such as preparation of drawings.
2. The Valuation 'shall be consistent with rates and prices of work of a similar character set out in the Contract Bills or the Analysis'.

   The Analysis (see Chapter 4) is used here like a schedule of rates, or if in insufficient detail, to set the market level of the price for this work.
3. Where rates and prices of work of a similar character are used due allowance must be made for 'any changes in the conditions under which the work is carried out and/or any significant change in the quantity'.
4. Where the Contract Bills or the Contractor's Statement contain no work of a similar character, a fair valuation shall be made.
5. Valuation of omission of work shall be in accordance with the rates and prices in the Contract Bills or the Analysis.

   There is often difficulty in valuing omissions, particularly if the relevant work is not adequately described or priced in detail. The Quantity Surveyor may then be left with the task of determining a fair value for work which will not be done and for which detailed design is not available. Although this may not cause difficulty if the variation is simply one of quantity, in an instance such as that in the uPVC windows example cited above,[29] there may be need for a new design before the original is prepared. If the new design appears to involve no more than a change in extrusion profile, it is tempting to apply some form of pro rata calculation, whereas the changes may in fact be complicated, involving a different manufacturing technique. For this reason, Variations involving specification change in Performance Specified Work should always prompt the Quantity Surveyor to ask the Contractor for a price analysis from the system designer.
6. The valuation must include allowance for any adjustment necessary in preliminaries items. Preliminaries items are identified as those referred to in SMM7.
7. Where the basis of a fair valuation is daywork, the provisions for daywork in Clause 13.5.4 apply.

---

29 Section 6.12.4.

8. If a Variation instruction regarding Performance Specified Work, including an instruction in respect of a provisional sum, affects other work by changing the conditions under which that other work is carried out, such other work is to be treated as though it were itself the subject of a Variation. The appropriate valuation rules are those applying to the other work.

6.12.6 Variations in respect of Performance Specified Work: a few typical problems

Difficulties in identification and evaluation of Performance Specified Work are often problems normally associated with the principles applying to 'design and build' and lump sum contracts, albeit there may be a unit rate for the work in the Contract Bill.

1. Variations introduced by the Contractor

The Contractor's general obligation under JCT 98 is to build what he is told to build by the Architect. When the Architect's instructions relate to Performance Specified Work, the Contractor has some discretion in how the performance will be achieved; the amount of discretion will depend on the parameters set out in the description of the work required and the Contractor's statement of how he intends to achieve it. Such discretion is not only an entitlement; it is also an obligation.

Many contractors, particularly those new to Performance Specified Work, wrongly assume when there is a bill of quantities that they are entitled to be paid for any change to the work. However, there are a number of situations in which the Contractor will be required to carry out work at his expense whether or not it is expressly referred to in any of the contract or working documents. Such work is implied either by terms in the contract, or by common law.

A common example of necessary work implied by terms is the obligation to comply with statutory requirements, including building regulations. This type of work seems to create few problems; maybe this is because contractors are generally familiar with regulatory control, and the extent and nature of the work is defined and foreseeable. However, work implied by common law is more vague, although the principle behind the concept is far from vague. In *Tharsis Sulphur and Copper Co. v. McElroy & Sons*,[30] it was said:

> a contractor who expressly or impliedly undertakes to complete the work ... impliedly warrants that he can do so ... . In consequence any additional work necessary to achieve completion must be carried out by him at his own expense if he is to discharge his liability under the contract.

Contractors will often claim extra for work which, although necessary is not expressly identified in the Contractor's Statement. The acid test of whether or not particular work must be done is often no more stringent than the simple question 'Will the installation operate as required, or meet the specified performance without it?' If the answer is 'No', the work is necessary, and must be performed without any adjustment of price.

---

30 (1878) 3 App. Cas. 1040.

2. Work omitted by the Contractor

The Contractor may wish to leave out work which he considers unnecessary.

The general position is that the Contractor is in breach of the contract if he provides something other than that which is described in the contract documents. Despite the Architect's power in Clause 13.2.4 to sanction a variation the Contractor does not have an unfettered right to vary the Works. However, provided the relevant work forms part of the Works within his discretion and is not outside the parameters limiting that discretion, it is submitted he can omit work, on the principles described in (1) above applying to necessary work. The extent to which the Contractor is limited in his choice depends entirely on the amount of detail in his Contractor's Statement provided under Clause 42.2 (see Chapter 4).

3. Performance which the Contractor cannot achieve

The rapid development of technology in recent years has increased the risk of contracting to do the impossible. This applies particularly in the field of electronics, where the impossible today may be commonplace in the near future. Construction contracts often include items such as the supply of computerized components within bespoke equipment designed by a domestic sub-contractor. The task is sometimes to solve a problem using or developing knowledge at the frontiers of science; but there are times when technological advance does not match expectation. There are occasions when the Contractor cannot achieve the performance required, but unless the contract provides otherwise,[31] impossibility is no excuse.

When the Contractor enters into the contract and provides his Contractor's Statement for Performance Specified work, he impliedly warrants that he is capable of doing what he has said he will do. If he fails he is in breach of the contract, and the Employer is entitled to damages.

The difficulty in this situation is in identifying whether the work is merely outside the capability of the Contractor, or really is impossible. If it is the former, the Employer would be entitled to go elsewhere by giving notice under Clause 4.1.2 and employing others. However, if it is the latter, the work cannot be done and another solution must be found. This will probably be by way of an instruction varying the Works, but the Contractor is still in breach and the Employer is still entitled to damages. The damages are likely to be the extra cost incurred by the Employer in achieving the required performance by other means, or extra cost finding a satisfactory alternative and some allowance for disappointment.

4. Evaluation of design

Clause 13.5.6.1 provides for 'allowance' to be paid for the 'addition or omission of ... (design work)'. There are several ways of valuing design effort. The principal methods used in practice are by lump sum, a *quantum meruit*, and by percentage of the building cost. It is for the parties to ensure an adequate means of calculation is incorporated in the contract, sometimes by setting out the basis in the Contract Bills but normally by insertion in the Analysis provided under Clause 42.13.

The difficulty in agreeing a satisfactory basis for 'allowance' lies in the nature of design work. In the case of a Variation requiring innovative design the resource expenditure may

---

31 See ICE *Conditions 5th Edition*, Clause 13.

be considerable, yet the building cost may not vary. Likewise the building cost may be reduced, by omission or variation, but the cost of achieving that omission may involve positive design costs. The Contractor who agrees a percentage design allowance runs the risk of a shortfall.

However, under Clause 13.5.6.2 it is clear that the Contractor is entitled to allowance for the varied design work carried out. This is not the same as design allowance on the varied building work. In other words allowance must be made for greater or less design work irrespective of the effect on the building cost. The variation to building work is no more than a trigger to the Contractor's entitlement. The method of evaluation which seems to meet the requirements of the contract best is some form of *quantum meruit*, either based on proven actual cost, or on an agreed labour rate applied to time records. In either case the Contractor may have difficulty establishing the exact nature of work being carried out, and it is not unusual for Contractors to under-recover on design costs.

6.13 Variation rules: quotations (Clause 13A)

6.13.1 Clause 13A Quotation: overview (see Fig. 6.2)

Clause 13A is an alternative to Clause 13 for use by the Architect, the aim being to allow the parties to agree the value of a Variation including effect on time, before the work is executed, and without allowing the Contractor to hold the Employer to ransom over the price.

The concept of Clause 13A is admirable. The Employer knows his expenditure with certainty, thus avoiding the need to lock up reserves needlessly; likewise the Contractor knows what his income (turnover) will be, enabling him to plan his financial year. Such an arrangement is common in bespoke contracts, particularly where the contractor is also the designer, but until recently was not a feature of many of the construction industry standard forms.[32]

However, the Clause 13A procedure is not mandatory; in practice the Architect may use it selectively. The calculation of the Final Account is then eased and accelerated, but possibly not sufficiently to give either party the full benefit.

The heading to Clause 13A sets the scene: 'Variation instruction – Contractor's quotation in compliance with the instruction'. Under the Clause 13A procedure, it is for the Architect to initiate and for the Contractor to respond.

The Architect first instructs the Contractor to supply a '13A Quotation'. The Contractor, unless he disagrees with the principle of a quotation, responds and by submitting a quotation to the Quantity Surveyor starts a somewhat tortuous procedure towards an agreed price. The quotation is then open for acceptance by the Employer. The Employer's wish to accept is sent to the Contractor followed by acceptance communicated to the Contractor by the Architect in a 'confirmed acceptance'.

The Contractor is barred from starting work which is the subject of a 13A Quotation until he has received a confirmed acceptance from the Architect, or an instruction to proceed otherwise (under the 'basic method').

---

32 See Institute of Chemical Engineers 'Red Book'; JCT *With Contractor's Design Form* WCD/98.

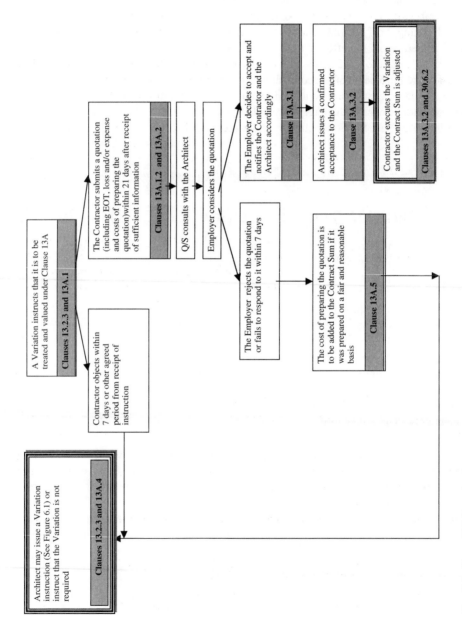

Fig. 6.2 Clause 13A Procedure

A Variation instructs that it is to be treated and valued under Clause 13A

Clauses 13.2.3 and 13A.1

The Contractor submits a quotation (including EOT, loss and/or expense and costs of preparing the quotation) within 21 days after receipt of sufficient information

Clauses 13A.1.2 and 13A.2

Q/S consults with the Architect

Employer considers the quotation

The Employer decides to accept and notifies the Contractor and the Architect accordingly

Clause 13A.3.1

Architect issues a confirmed acceptance to the Contractor

Clause 13A.3.2

Contractor executes the Variation and the Contract Sum is adjusted

Clauses 13A.3.2 and 30.6.2

Contractor objects within 7 days or other agreed period from receipt of instruction

The Employer rejects the quotation or fails to respond to it within 7 days

The cost of preparing the quotation is to be added to the Contract Sum if it was prepared on a fair and reasonable basis

Clause 13A.5

Architect may issue a Variation instruction (See Figure 6.1) or instruct that the Variation is not required

Clauses 13.2.3 and 13A.4

6.13.2 Clause 13A Quotation: the instruction

If the Architect wishes to deal with a Variation under the quotation procedure, Clause 13.2.3 states he must first issue an instruction stating that Clause 13A shall apply. Clause 13.2.3 further provides that the Contractor may within 7 days disagree in writing that Clause 13A should apply to that particular instruction. If the Contractor disagrees, Clause 13A will not apply. The Architect must then revert to the basic method if he requires the Variation to proceed.

It is significant that the Contractor's right to disagree is unfettered; it is not qualified by an express obligation to be reasonable. The Contractor is not obliged in any way to proceed down the quotation route, although most contractors would probably see the procedure as a potential benefit in principle.

The 7-day period for the Contractor's statement of disagreement may be changed by agreement. This is a necessary provision if the procedure is to be seen as credible, particularly in instances where the difficulty of producing a quotation is not immediately apparent, or where the commercial risk of committal to a price cannot be evaluated readily. However, the means of agreeing an alternative period are not described. It must be presumed unless a global agreement is reached, that the Contractor will give notice of the need for a longer period for disagreement during the first 7 days after receipt of a 13A instruction, since otherwise the 13A procedure will apply by default. Although Clause 13.2.3 provides for varying the period for disagreement 'within 7 days (or such other period as may be agreed)', it is not expressly stated with whom the agreement must be made. Since the Architect is the contract administrator and is best placed to decide whether changing the period for disagreement is in the best interests of contract progress, it seems sensible for the Architect to have authority to make such agreements. The problem is that the Architect is not expressly given that authority, and guidance from Clause 13A.7, relating to agreements changing Quotation and acceptance periods, suggests it is only the Employer who may make such agreements.

6.13.3 Clause 13A Quotation: the quotation

If the Contractor does not disagree within the proscribed notice period, then either:

1. if he reasonably considers the information provided is insufficient to provide a quotation, he must within 7 days of receiving the instruction request further information; he must then provide a quotation within 21 days of receiving the further information; or
2. if satisfied with the information, within 21 days after receipt of the instruction he must submit a quotation.

The Contractor's 13A Quotation is submitted to the Quantity Surveyor (not the Architect) and remains open for acceptance by the Employer (not the Quantity Surveyor or the Architect) for 7 days after receipt of the quotation by the Quantity Surveyor.

The periods for requesting further information, submission of the Quotation and acceptance by the Employer referred to in Clauses 13A.1.1 and 13A.1.2 may be changed by agreement. However, Clause 13A.7 makes clear that such agreement must be with the Employer (not the Architect or the Quantity Surveyor) who must confirm his agreement in writing to the Contractor. This procedure lacks practicality, and it may assist smooth running if the Employer were to notify the Contractor that the Architect or the Quantity Surveyor had limited powers of agency to agree different timescales to match the circumstances of individual 13A Instructions.

The content of the quotation is proscribed in Clause 13A.2. It must be split to show the following elements separately:

1. *The value of adjustment to the Contract Sum:* this excludes loss and/or expense but includes the effect on preliminaries items and on other work, together with an effect on Nominated Sub-contractors' work. The value must be supported by calculations with reference where relevant to the rates and prices in the Contract Bills.

    It is clear that the quotation is not a means for the Contractor to charge what he likes; the pricing level must be the market level of the Contractor's accepted tender which produced the Contract Sum, adjusted only in the manner which the Contract provides.

2. *The effect on time:* the Contractor must state his requirement for any adjustment to the Completion Date, including if relevant, an earlier date than that stated in the Appendix. However, the quotation must not duplicate any allowance for revision of the Completion Date already made by the Architect either under Clause 25, or included in any other 13A Quotation which has been accepted.

    There is opportunity here for confusion, particularly if the Contractor has previously given notice of delay for other events and the Architect has not yet awarded an extension. The proviso in Clause 13A.2.2 referred to here does not prevent the Contractor duplicating a request; it simply prevents duplication of a request for an extension where the relevant period has been granted already. Thus the Contractor may be obliged to qualify his request depending not only on extensions awarded, but also on other quotations which may be awaiting acceptance, which if accepted could affect the most recent quotation.

3. *The amount to be paid in lieu of loss and/or expense:* again the Contractor is required to avoid claiming amounts already ascertained under Clause 26 or included in other accepted quotations.

4. *A fair and reasonable amount in respect of preparing the 13A Quotation:*

    In this provision the JCT have gone further than might be expected. Whilst it would be natural for the Contractor to build the cost of preparation into his price, the cost will almost inevitably be overhead costs, and any allowance will have to be taken into account in ascertaining loss and/or expense if and in so far as it relates to office overheads. The identification of this cost is sometimes ignored by Contractors but is of particular importance for the Contractor when the quotation is not accepted. This aspect is considered below.

5. *A statement of any additional resources required to carry out the Variation, if the 13A instruction specifically requires:*

    The provision of a resources statement is an option for the Architect, but is a useful tool in determining the effect on time as well as cost. It is a pity the JCT did not make the requirement mandatory. The production and subsequent consideration of such a statement forces both Contractor and Employer's team to address their minds to the real and often disruptive nature of Variations before, rather than after, the event.

6. *A method statement for carrying out the Variation, if the 13A instruction*

specifically requires: The comments made under (5) above apply equally to method statements.

A number of the problems which can arise for the Contractor in producing a 13A Quotation are considered at the end of this section.

6.13.4 Clause 13A Quotation: acceptance

It is important to remember that the Architect does not have general authority under the Contract to accept quotations from the Contractor or otherwise to make agreements on behalf of the Employer. In order for the Architect to make agreements on the Employer's behalf, the Architect must first be given authority. In addition that authority must be communicated to the Contractor; otherwise the Contractor may refuse to comply with such an agreement. That position is not altered by the Clause 13A procedure. It is the Employer who makes it known to the Contractor under Clause 13A.3.1 that he wishes to accept the 13A Quotation. This must be done by notification within the 7-day period for acceptance.

Under Clause 13A.3.2 the Architect then simply confirms the Employer's wish by immediately writing to the Contractor. The notification, referred to as 'a confirmed acceptance', must state:

1. that the Contractor is to carry out the Variation,
2. the value of the Variation including any amount in lieu of loss and/or expense and preparation costs,
3. any adjustment of the Completion Date, and relevant adjustment of any Nominated Sub-contractors' contract periods,
4. where relevant, that the Contractor is to accept any Nominated Sub-contractor's 3.3A Quotation contained in his 13A Quotation.

6.13.5 Clause 13A Quotation: not accepted

If the Employer does not accept the 13A Quotation within the 7-day period for acceptance, it is rejected by default. Under Clause 13A.4 the Employer is not required to state expressly that he does not accept the quotation.

However, the Architect must act to tie up loose ends. The Contractor is neither entitled nor obliged to start work against the Architect's instruction which triggered the 13A procedure in the absence of acceptance. Consequently, if notwithstanding the Employer's rejection of the quotation, the Architect still requires the work to be done, he would, in the absence of an express procedure, have to regenerate the instruction in some way. Clause 13A.4 provides the procedure. If the quotation is not accepted by the Employer, the Architect must instruct that the Variation is to be carried out, or that it is not to be carried out.

If the Variation is required to be executed, albeit the Contractor's quotation is not accepted, then the Architect must instruct that the Variation is to be executed and valued under the procedures described above as the 'basic' procedure.

If the Variation is not required at all, the Architect must instruct accordingly. Strictly, the JCT could have drafted Clause 13A to allow the originating instruction and the resulting quotation to die in the absence of acceptance, since there is no obligation on either party. The general position in common law is that a tenderer incurs costs of preparing a quotation[33] at his own risk in the hope that he will recover in earnings from successful bids.[34] The enquirer is only liable for tendering costs if he asks the tenderer to expend resources, either deceitfully,[35] or to obtain something of benefit[36] when the tenderer reasonably expects payment.

The JCT apparently took the view that the instruction of the Architect to expend the Contractor's resources on preparing a quotation warranted reimbursement of preparation costs if the quotation was rejected. Thus, under Clause 13A.5 a fair and reasonable amount in respect of preparation costs must be added to the Contract Sum. It is not clear whether the fair and reasonable amount to be paid to the Contractor for preparation costs is the same fair and reasonable amount which the Contractor included in the 13A Quotation pursuant to Clause 13A.2.4. Since there is no reference to Clause 13A.2.4 in Clause 13A.5, it may be assumed that the fair and reasonable amount in Clause 13A.5 must be objective and need not be based entirely on the amount stated in the Contractor's quotation.

There is a caveat to the entitlement to preparation costs; the quotation must have been prepared on a fair and reasonable basis. This is consistent with the often voiced objection that, in giving a quotation for a Variation, a Contractor can hold the Employer to ransom if he does not want to do the work; he can simply price the work too high in the knowledge that it would be impractical for the Employer to get others to carry out the change. Clause 13A gets around the problem, first by providing the Employer with the right to revert to the 'basic' procedure, and second with the threat of a sanction on the Contractor. The sanction is deprivation of preparation costs if the Contractor's quotation is not prepared on a fair and reasonable basis. Whether this sanction is sufficient to induce a reluctant Contractor to prepare a price in a reasonable manner is questionable; in any event it is always open to the reluctant Contractor to refuse to provide a quotation on receipt of the originating instruction, by stating his disagreement to the application of the Clause 13A procedure.

The JCT clearly foresaw reluctance in the industry to pay wasted preparation costs and have dealt with what might have been an excuse used by Architects. Clause 13A.5 states that non-acceptance by the Employer is not of itself evidence that the Quotation was not prepared on a fair and reasonable basis. In a situation where an entitlement to preparation costs is triggered expressly by non-acceptance of a Quotation, it is difficult to conceive that the mere fact of non-acceptance could be considered, or even proposed, to be evidence to deprive the Contractor of those preparation costs. The fact that the JCT contemplated, and found it necessary to deal with, the proposition seems a sad reflection on the culture of the construction industry in 1994 when Amendment 13 was introduced.

---

33 The terms 'tender', 'quotation', 'bid', 'offer', 'estimate' may each be construed as meaning 'offer' in law if on a true construction it fulfils the requirements for an offer and is capable of being accepted. In *Crowshaw v. Pritchard and Renwick* (1899) 16 TLR 45, it was held that the defendant contractor's 'estimate' had no special customary meaning and was an offer which had been accepted to create a contract.
34 See *William Lacey (Hounslow) Ltd v. Davis* [1957] 2 All ER 712.
35 In *Richardson v. Sylvester* (1873) LR 9 QB 34, property was advertised for sale by auction with no intention of selling. The costs incurred by a potential bidder in valuing the property were recovered in the tort of deceit.
36 See discussion on *Marston Construction Co Ltd v. Kigass Ltd* in Chapter 1: Section 1.7.2, Letters of Intent, sub-section '*Quasi Contract or Restitution?*'

6.13.6   Clause 13A Quotation: further variation

There is always the possibility that a Variation instruction, whether issued under the 'basic procedure' or under the 13A procedure, may affect work which itself forms part of another previously accepted 13A Quotation.

Clause 13A.8 requires the Quantity Surveyor to value such later Variation, not under Clause 13.5 (ie. the 'basic method') but on a fair and reasonable basis, having regard to the content of the relevant 13A Quotation. The Quantity Surveyor must also include any direct loss and/or expense incurred by the Contractor as a result of the Variation.

This clause seems to have been added almost as an afterthought. Why variations on Variations should be treated differently from any other Variation is unclear; arguably it can only lead to confusion and dispute. The rules for Variations in Clause 13 as a whole may be applied just as well to sequential variations as they do to work which is introduced by variation and is not changed further. This provision can have far-reaching effect on the value of the Works, for it is common for varied work to be further varied. If on a project, a large number of Variations are instructed under the 13A Quotation procedure, the result could be a high proportion of the total work in variations calculated on a fair and reasonable basis, rather than under the Clause 13.5 rules or even under the quotation procedure.

The obligation to include loss and/or expense simply adds more difficulty. Since the further Variation may not be subject to a 13A Quotation instruction, the Contractor is required to comply with the Clause 26 procedures if he requires loss and/or expense to be ascertained. The Quantity Surveyor then is obliged to separate out, as best he can, the portion applicable to the Variation. Clearly the task is easier if the further variation is instructed under the 13A procedure, since then the Contractor would provide the necessary information related to the individual Variation.

6.13.7   Clause 13A Quotation: some typical problems

1. Misuse of Clause 13A procedure

The timetable for dealing with Quotations usually requires an element of estimating rather than historical surveying. The Contractor is obliged to foresee possible risks and to price those risks. The result is sometimes an amount higher than the Employer would like to pay. Architects in that situation may reject the Quotation and proceed under the 'basic method'. Later when the work is done, if the value of the Variation exceeds the Contractor's 13A Quotation the Architect may be tempted to hold the Contractor to his earlier price. The Architect is then using the rejected 13A Quotation as a form of cost advice.

This situation is common whenever a contract provides for quotations against Variations. However, under JCT 98 the Contractor is protected in that Clause 13A.6 provides that neither party shall use an unaccepted 13A Quotation for any purpose whatsoever. In short, mimicking general principles of offer and acceptance, the rejected Quotation is dead.

2. Employer does not express wish to accept 13A Quotation

The Architect is not empowered to accept quotations; under Clause 13A the Employer accepts and the Architect confirms acceptance after the Employer has expressed a wish to accept in writing to the Contractor. It sometimes happens that the Architect accepts a quotation without the Employer having first written to the Contractor. In that situation the Contractor is not obliged to comply with the instruction since he knows that the Architect is acting outside his authority. This contrasts with Clause 13.4.1.2 Alternative A. Under that clause an acceptance by the Quantity Surveyor follows consultation with the Architect, but such consultation is carried out in secret so far as the Contractor is concerned, and acceptance of a Price Statement is not subject to the Contractor having first been notified either by the Employer or by the Architect.

3. Multiple Quotation Instructions

When a quotation is requested it is not unusual for the Contractor to be instructed to provide prices for alternative schemes. The Contractor is then obliged to provide several quotations in the knowledge that only one (if any) will be accepted. Under Clause 13A.5 the Contractor is entitled to recover preparation costs of a quotation which is not accepted, so he is protected against the cost of preparing each alternative which is rejected.

The difficulty for the Contractor is in establishing the base from which the Quotation is calculated. In circumstances where there are few Variations and no other 13A Quotations pending, the task is comparatively easy. But when the Contractor is aware that the Quotation he is preparing may be affected if another 13A Quotation previously submitted is accepted, he can do no more than qualify his Quotation accordingly. When there are a number of Quotations awaiting acceptance the Contractor may need to provide an alternative price for each permutation of accepted Quotations. This is hardly likely to endear the Contractor to the Architect, but it seems there is little choice if confusion is to be avoided.

Strictly under the rules of Clause 13A the Quotations could be allowed to stack up sequentially, then depending on the order in which they are accepted, could be identified as Variations on earlier Variations. The valuation rule would then be a fair and reasonable basis under Clause 13A.8.

There is no simple solution. The important issue is recognizing that the situation can occur, and drawing it to attention before havoc sets in!

6.14 Errors, discrepancy and divergence in and between documents

6.14.1 Generally

A common category of Variation arises from errors and inconsistencies in the Contract Documents. If a contract contains an error such an error in quantity, description of work, or even in a contract term, the general position is that it must stand,[37] unless either the parties agree when it is discovered that it should be rectified, or the contract expressly sets out what is to be done, or unless a court can be persuaded that the contract does not express what the parties intended.

---

37 *Ewing & Lawson v. Hanbury & Co.* (1990) 16 Times Law Reports 140.

Sometimes an error appears as an inconsistency, such as a specification on a drawing conflicting with a description in the bills of quantities. In these cases the parties, or a dispute resolution forum, have to choose which of the specifications describes the parties' obligations and entitlements. In the absence of provisions in the contract, the general rules of contract construction will apply. So, for example, the written word will normally prevail over the typed word, and the typed word will prevail over the standard printed word.[38] If the inconsistency creates ambiguity in the documents as a whole, or if a single document contains conflicting provisions or an ambiguity, then the ambiguity may be construed *contra proferentem*,[39] that is the option least favourable to the party whose document contains the ambiguity.[40]

In some contracts the priority of the various documents is listed,[41] while in others the choice of priority to suit the circumstances is left either to one of the parties,[42] or to a third party such as the administrator of the contract. JCT 98, Clauses 2.2 to 2.4 deal expressly with specified types of errors and conflict.

6.14.2 Priority of documents and status of Contract Bills

The Contract does not list the priority of the various Contract Documents, although it deals with the status of the Contract Bills in relation to the Agreement, Conditions and Appendix.

Clause 2.2.1 states 'Nothing in the Contract Bills shall override or modify the application or interpretation of that which is contained in the Articles of Agreement, the Conditions or the Appendix.' This provision reverses the general rule of construction that the specially prepared conditions prevail where they conflict with the standard printed conditions. It protects the Contractor against terms inserted in the Bills which may otherwise change the allocation of risk agreed by the members of the JCT. However, the clause has attracted some criticism and judicial opinion that it should be removed[43] to allow all terms to be read together and not be shut out and left in the Bills only (see also Chapter 1: Section 1.4.3, Bills of Quantities, and Section 1.5.5, Amending the Standard Form).

The effect of Clause 2.2.1 is theoretically wide, but in practice it tends to be limited to dealing with conflict. The difficult area is where an obligation imposed in the Contract Bills adds something new, and it is then a matter of interpretation as to whether the entry in the Bills modifies the application or interpretation of the provisions in the printed form. Each case must be taken on its merits, but the courts have demonstrated a degree of robustness. For example, the imposition of design obligations by reference in the specification has been held 'added to but were consistent with obligations imposed by the

---

38 For full commentary on construction of contracts see *Keating*, 6th edn, Chapter 3.
39 This doctrine normally applies where the contract is based on one party's unilateral terms; even though the particular terms used have been chosen by the Employer, JCT contracts terms are not unilateral since they are negotiated by representatives of different sectors of the industry.
40 This doctrine can be seen applied in the JCT With Contractor's Design Form WCD 98, Clause 2.4 dealing with errors in the Employer's Requirements and the Contractor's Proposals.
41 For example see Clause 2.2 of Domestic Sub-contract DOM/1 published by the Construction Confederation.
42 See JCT WCD 98 Clause 2.4, which operates on the *contra proferentem* (Lat. against the one who puts forward) principle.
43 See *dicta* of Lord Justice Stephenson in *English Industrial Estates Corp. v. Geo. Wimpey and Co. Ltd* (1972) 7 BLR 122.

conditions';[44] similarly the obligation to submit a programme in a specified form was held not to conflict with the Conditions.[45] If, however, a paragraph in the Bills purported to place an obligation in conflict head on with the JCT 98 Form, such as for example stating a period for payment of an interim certificate different from that in the Appendix, then the Appendix would prevail over the Bills.

6.14.3 Errors in the Contract Bills

The Contractor's obligations include carrying out the works described in the First Recital, i.e. as shown on the drawings and described in the Bills. A difficulty can arise if the Bills contain errors. In the absence of provisions in the Contract, the descriptions in the Bills would be at the Contractor's risk; if work were shown or implied on the drawings, it would form part of the Contract even if it were missing from the Bills. An example can be seen in *Williams v. Fitzmaurice*,[46] where it was held that floorboards missed from a schedule of works were necessarily part of the Contract to complete a house. Following this judgment, in *Patman & Fotheringham v. Pilditch*[47] it was said that items omitted from the bills of quantities did not necessarily give rise to extra payment; the employer was not bound to pay for things that everybody must have understood are to be done, but which happen to be omitted from the quantities. However, it was also held in the *Patman* case that the bills formed part of the contract, and whilst the Contractor was obliged to do the omitted work, he was entitled to have the price adjusted to account for it.[48] This general position is formalized in JCT 98 where quantities form part of the contract. It is in the Employer's interests to have the Bills prepared by his professional team,[49] and in return JCT 98 Clause 2.2.2.1 places the risk in the Bills with the Employer.

Clause 2.2.2.1 states that the Bills have been prepared in accordance with The Standard Method of Measurement of Building Works,[50] unless the Bills contain specific qualification in respect of each specified item not complying. It is important for the tenderer, and eventually the Contractor, to know what he is pricing, and he is entitled to take the content of the Bills at face value.[51] Thus, to avoid misleading the tenderer, any identification of deviation from the Standard Method of Measurement must be specific. It must be sufficiently specific for the tenderer (and indeed the Employer) to know what the non-compliant description entails; otherwise such a qualification will amount to no more than a denial of warranty as to accuracy.

Clause 2.2.2.2 contains the rules for correcting an omission or an error of quantity or description in the Bills. This includes insufficient information about the content of

---

44 *Haulfryn Estate Co. Ltd v. Leonard J. Multon & Ptnrs and Frontwide Ltd*: ORB; 4 April 1990, Case No 87-H-2794. Note this case dealt with the JCT Minor Works Form, but Clause 4.1 of that form is similar to Clause 2.2.1 of the JCT Standard Form.
45 *Glenlion Construction Ltd v. Guinness Trust Ltd* (1987) 11 Con. LR 126.
46 (1858), *Hudson's*, 11th edn (1995), vol. 1, p. 498.
47 (1904), *Hudson's*, 4th edn (1914), vol. 2, p. 368.
48 See also *Meigh & Green v. Stockingford Colliery Co. Ltd* (1922); *Hudson's*, 11th edn, vol. 1, p. 975, in which builders were held entitled to be paid for brickwork missed from the bill of quantities.
49 See Chapter 1: Sections 1.4.3 and 1.5.4, Bills of Quantities.
50 See Chapter 1: Section 1.4.5, the Standard Method of Measurement.
51 In *C. Bryant & Son Ltd v. Birmingham Hospital Saturday Fund* [1938] 1 All ER 503, the equivalent of Rule 1.1 of the SMM7 when read with the conditions stating the Standard Method had been used, was construed as a warranty by the Employer that the information provided to the Contractor was both accurate and sufficient to identify the nature and extent of the Works; see also Chapter 1, Section 1.4.5.

Provisional Sums for defined work.[52] The error or omission must be corrected by treating it as though it were a Variation required by an Architect's instruction given under Clause 13.2. Since an Architect's instruction is not required, it is not clear who carries out the correction, although the error or omission, being within the expertise of the Quantity Surveyor would normally be best performed by the Quantity Surveyor.

The correction of errors, particularly errors in quantity, together with the correction of omissions, is necessary to maintain the concept of a lump sum contract which is adjusted only in respect of Variations and other specified changes, thus avoiding the need to remeasure parts of the Works. Clause 2.2.2.2 simply provides machinery for rectifying the Contract to what it should have been from the outset.

Unfortunately, the provision is difficult to apply in practice. It is often left to the Contractor to identify errors, which he can do only by comparing records of work done with the Bills, so errors in the Bills often overlap, and become confused with, Variations. Since the Contractor is usually more concerned with being paid for what he has done, than identifying the theoretical classification, he is likely to press for remeasurement of suspect quantities (this is particularly so if the Contractor opts to use evaluation Alternative A, under which the Contractor puts forward his Price Statement for agreement by the Quantity Surveyor). However, the Quantity Surveyor's duty is to apply the rules in the Contract, even if it means identifying his own errors to the Employer.

The requirement to treat Clause 2.2.2.2 rectification as if it were a Variation enables such adjustment to be considered as a potential 'event' for the purposes of extensions of time, and a potential 'matter' for the purposes of loss and expense.

6.14.4 Discrepancies in or divergence between documents

In and between documents generally
Clause 2.3 deals, somewhat inadequately, with a discrepancy in, or a divergence between, two or more documents other than the Agreement, Conditions and Appendix. The documents referred to are: the Contract Drawings, the Contract Bills, Architect's instructions other than Variations, drawings or documents issued by the Architect to further describe the Works, and the Numbered Documents annexed to a Nominated Sub-contract Agreement.

The purpose of Clause 2.3 is to correct ambiguity, yet the action of the Architect having being made aware of a problem is simply to 'issue instructions in regard thereto'. There is no guidance as to the principles to be considered in issuing instructions, unlike the similar provisions of Clause 2.4 in respect of Performance Specified Work. In the absence of express direction, the general principles of the *contra proferentem* doctrine are likely to apply. Where two documents conflict, or where there are conflicting provisions within a document, the Architect may choose which provision he wishes to apply. If the Architect's choice is not the same as that of the Contractor, the instruction will give rise in effect to a rectification of the Contract, and to corresponding adjustment of the Contract Price.

The significant feature of Clause 2.3 is the requirement for the Contractor to find the discrepancy or divergence, and then immediately to give written notice to the Architect. Whilst the word 'find' suggests a search, this does not impose an obligation on the

---

52 See SMM7, Rules 10.1 to 10.6.

Contractor to search for discrepancies; he is entitled to stumble across them. It may be that a divergence or ambiguity does not occur to the Contractor until (say during a dispute) an interpretation counter to his own is put to him. In those circumstances it is suggested the point at which the Contractor 'finds' the problem is when he realizes there is a discrepancy or divergence. Similarly, it is not for the Architect to search out, or even spot, discrepancies or ambiguities in the documents, but even if he happens upon them, he is not obliged to do anything until he is notified by the Contractor. Whilst the Contractor is required to give notice '*immediately*', there is no express timescale imposed on the Architect. Nevertheless, it has been held[53] that under a similar provision in JCT 63 a reasonable period in which the Architect should act would be implied if a substantial delay might disrupt the contract.

In and between documents relating to Performance Specified Work
Clause 2.4.1 deals with discrepancy between the Contractor's Statement[54] and the Architect's instructions issued after receiving the Statement. The procedure and action is the same as for documents generally, dealt with in the foregoing sub-section.

Clause 2.4.2 provides for discrepancies in the Contractor's Statement. In contrast with Clauses 2.3 and 2.4.1, the procedure may be triggered either by the Architect, or by the Contractor: 'If the Contractor or the Architect shall find any discrepancy'. On finding or becoming aware of a discrepancy, the Contractor must correct the Statement and inform the Architect of the correction. Such correction shall be at no cost to the Employer. If, for example, one part of the Contractor's Statement describes an expensive specification, contradicting the description of a cheap specification elsewhere in the Statement, the Contractor may choose the cheaper option, and the Architect has no grounds for challenge. If the Architect would prefer the alternative, the only course available to him is to instruct a Variation. Perversely this procedure reverses the *contra proferentem* doctrine. Why the JCT should give the Contractor the advantage is not clear, particularly since it goes against commonsense and conflicts starkly with their own procedures for similar circumstances in the With Contractor's Design form.[55]

---

53 *R. M. Douglas Construction Ltd v. CED Building Services* (1985) 1 Const. LJ 232; referring to Clause 1(2) of JCT 63.
54 See Chapter 4, Performance Specified Work.
55 See JCT WCD 98, Clause 2.4.2 dealing with discrepancies in the Contractor's Proposals, under which the Employer is able to make the final choice at the Contractor's expense.

7

Risks to health and safety: allocation and insurance

Construction work has always involved three main categories of risks relating to health and safety. First, there is the risk of injury or death to people on or passing by the site. Second, the carrying out of the Works may cause damage to property other than the Works themselves or unlawful interference with personal rights associated with ownership or occupation of such property, e.g. noise, dust, vibrations and smells. Third, there is the possibility of damage or loss to the Works themselves or to Site Materials. The best strategy against these risks is, of course, one of prevention through appointments of competent designers, contractors, sub-contractors, and suppliers with good health and safety records. Even where this is done, prudence still demands that construction contracts contain clear provisions on these risks because it is impossible to eliminate them completely. The aim in this chapter is to examine how the JCT 98 deals with these risks. Adequate appreciation of the nature of the problem that they present is a prerequisite to developing sufficient understanding of the provisions. For this reason, the chapter begins with an explanation of the general nature of the problem.

7.1 General nature of the problem

A particular feature of the risks is that they affect a large number of people. The usual reaction of anybody injured or who suffers damage to his property is to sue whoever he thinks responsible, either in part or wholly, for the injury or damage. This could be the employer, contractor, a sub-contractor, the architect or other designer. The ways in which liability may attach to any of these potential defendants include:

- negligence
- nuisance
- trespass
- strict liability
- liability under *Rylands v. Fletcher* principle
- liability in contract
- breach of statutory duty.

In deciding whom to sue for his injury or loss, the strategy of the typical plaintiff is often that of the proverbial 'duck shoot' or the so-called 'scatter gun' approach, i.e. he sues everybody involved on the site. Under the Civil Liability (Contribution) Act 1978, a defendant may also bring in as joint defendants others whom he considers to have contributed to the injury, damage or loss. With more than one defendant, the tendency is for each of them to blame the others. It is then the court's duty to find out who is responsible for what and to what extent. This can be a long drawn out and very expensive process. Most construction contracts seek to avoid this situation by spelling out clearly in the contract itself who, as between the employer and the contractor, should be responsible for specific categories of the risks. Such statements of liability are referred to as 'risk allocation clauses' or, simply, 'risk clauses'.

It is to be noted that the mere fact that a particular risk is allocated to a party does not mean that that party caused or is otherwise to be blamed for its occurrence. In some cases the employer and the contractor may both be blameless but nonetheless one of them must bear the financial consequences, which may or may not be able to pass on to the third party ultimately responsible for the cause of the loss. For example, a member of the public who is injured by structural collapse due to faulty design by a structural engineer may sue the contractor because of lack of knowledge of who is really responsible. Although the contractor should usually win in such an action, there is no guarantee that he will get all the costs of defending the action from the victim. In a construction contract, as between the employer and the contractor, this risk will usually be allocated to the employer, i.e. the contractor can claim his unrecovered costs from the employer who ought to bear the consequences of mistakes of designers that he appoints. The employer may in turn pursue the engineer for compensation.

In a contract between parties A and B, allocation of a risk to party A carries the implication that if party B incurs liability as a consequence of occurrence of that risk, A must compensate B. However, many construction contracts contain express statements to the effect that 'A shall be indemnify B if ...'. Such statements are referred to as 'indemnity clauses' or, simply, 'indemnities'. As explained in Section 2.3.8, indemnities are very onerous obligations because they can still be enforceable many years after expiry of the limitation period applicable to the contract containing them.

Risk and indemnity clauses in themselves are of limited value because it would be a fruitless exercise to determine who should shoulder the consequences of an accident if the responsible party has no funds to meet the liability. Although knowledge of potential liability will often spur a prudent party to obtain appropriate insurance cover, it would be ill advised to leave the availability of such insurance to the party's discretion. To guarantee availability of funds, construction contracts complement risk allocation clauses with additional provisions requiring a party to take out insurances against identified risks for which he is responsible. Referred to as 'insurance clauses', such provisions often state that if a party fails to maintain the specified insurance cover, the other party may take out or pay the premiums and recover the cost of so doing from the defaulting party.

In summary, the contractual approach to dealing with the problem of risks from construction work involves clear risk allocation clauses, requirements for certain risks to be covered by appropriate insurance policies and clauses designed to allow easy and effective enforcement of the insurance obligations.

7.2 Allocation of risks under the JCT 98

For the purposes of risk allocation, the following categories of risk are identified in the Contract:

- risk of personal injury/death;
- risk of damage to property other the Works and Site Materials;
- risk of damage to the Works and Site Materials;
- damage from the Excepted Risks.

7.2.1 Risk of personal injury/death

Clause 20.1 deals with liability for personal injury and death in the following terms:

> The Contractor shall be liable for, and shall indemnify the Employer against, any expense, liability, loss, claim or proceedings whatsoever arising under any statute or at common law in respect of personal injury to or the death of any person whomsoever arising out of or in the course of or caused by the carrying out of the Works, except to the extent that the same is due to any act or neglect of the Employer or of any person for whom the Employer is responsible including the persons employed or otherwise engaged by the Employer to whom clause 29 refers.

The above wording shows very clearly that Clause 20.1 is a combined risk and indemnity clause and that the Contractor is not responsible for every incident of personal injury or death on or near the site. For the liability and indemnity to apply, two conditions must be met. First, the injury or death must arise 'out of or in the course of or by reason of the carrying out of the Works'. *Richardson v. Buckinghamshire County Council and Others,*[1] in which the plaintiff was injured when riding a scooter by a construction site, illustrates this qualification. His claim against the employer, incidentally the Council, failed. However, the Council had to pay the plaintiff's legal costs because local authorities were then responsible for the legal aid of their residents. The Council sought to recover its expenses from the Contractor under a similarly worded indemnity clause. The court ruled that the loss did not fall under the indemnity because the Council suffered the loss not by reason of the carrying out of the works but because it was responsible for paying legal aid for its residents. Second, the liability and indemnity will be reduced to the extent that the death or injury is caused by the act or neglect of the Employer or of any person for whom the Employer is responsible under the Contract or the general law. Such persons include:

- the Architect (Article 3),
- the Quantity Surveyor (Article 4),
- the Planning Supervisor, if different from the Architect (Article 6.1),
- the Principal Contractor if different from the Contractor (Article 6.2),
- the Employer's Representative,
- Clerks of Works (Clause 12),
- other consultants of the Employer,
- employees of the Employer,

---

1   (1971) 69 LGR 527 (*hereafter Richardson v. Buckinghamshire*).

- other contractors employed by the Employer under Clause 29,
- third parties invited by the Architect to examine fossils, antiquities, etc. found on the site (Clause 34.2),
- statutory undertakers carrying out work outside their statutory duties.[2]

The burden of proof that death or personal injury is caused by the act or neglect of the Employer or of people for whom the Employer is responsible rests on the Contractor.

7.2.2 Damage to property

Clause 20.2 states the extent of the Contractor's liability for loss or damage to property in the following terms:

> The Contractor shall be liable for, and shall indemnify the Employer against, any expense, liability, loss, claim or proceedings in respect of any loss, injury or damage whatsoever to any property real or personal in so far as such loss, injury or damage arises out of or in the course or by reason of the carrying out of the Works and to the extent that the same is due to any negligence, breach of statutory duty, omission or default of the Contractor, his servants or agents or of any person who may properly be on the site upon or in connection with the Works or any part thereof, his servants or agents, other than the Employer or any person employed, engaged or authorised by him or by any local authority or statutory undertaker executing work solely in pursuance of its statutory rights or obligations. This liability and indemnity is subject to clause 20.3 and, where 22C.1 is applicable, excludes loss or damage to any property required to be insured thereunder caused by a Specified Peril.

The use of the term 'property real or personal' is significant because, in law, property has a wider meaning than physical property. In addition to damage to physical (real) property, the wording of the clause appears wide enough to include infringements of intangible property rights such as easement, rights to light and air and the rights of people not to have enjoyment of their property detracted from by unreasonable levels of smells, noise or vibrations.

Specified Perils

The Specified Perils are defined in Clause 1.3 as consisting of: 'fire, lightning, explosion, storm, tempest, flood, bursting or overflowing of water tanks, apparatus or pipes, earthquake, aircraft and other aerial devices or articles dropped therefrom, riot and civil commotion, but excluding Excepted Risks'.

The scope of each peril must be seen in the light of the interpretation given similar words in *Computer & Systems Engineering plc v. John Lelliott (Ilford) Ltd and Another*[3] which arose from a contract let on an earlier revision of the JCT80. In that case, a pipe in a sprinkler system was sheared off through the negligence of a sub-contractor. This caused discharge of 1600 gallons of water with resultant damage to the plaintiff Employer's property. The litigation concerned whether the damage arose from the equivalent of the Specified Perils. The Court of Appeal held that the word 'flood' used in a similar context

2   See *Henry Boot Construction Ltd v. Central Lancashire New Town Development Corporation* (1980) 15 BLR 1.
3   (1990) 54 BLR 1.

suggests rapid accumulation of larger volumes of water from an external source and that it did not cover discharge from the sprinkler system. The court also confined 'bursting of pipes' to damage from internal stresses.

Examination of the terms of Clause 20.2 identifies four qualifications to the Contractor's liability for damage to property that need to be examined in some detail. The first three are that liability applies: (i) 'so far as such loss, injury or damage arises out of or in the course or by reason of the carrying out of the Works'; (ii) 'to the extent that the same is due to any negligence'; (iii) subject to Clause 20.3. The fourth qualification is that, where clause 22C.1 applies, the Contractor is not answerable for damage against which the Employer is required to insure under that clause. As Clause 22C.1 requires the existing structures and their contents to be insured against damage by the Specified Perils, the general intention appears to be that the Contractor is not liable to the Employer for this type of damage.

First Qualification

The effect of the first qualification is illustrated by the facts in *Richardson v. Buckinghamshire*.

Second Qualification

It is illustrated by *Gold v. Patman & Fotheringham Ltd*[4] in which piling work damaged adjoining property and the Employer was found liable for the damage in nuisance. His claim under an indemnity clause similar to Clause 20.2 failed because, as the piling had been carried out according to drawings and instructions of the Architect, there had been no 'negligence, omission or default' of the Contractor. The use of the terms 'negligence, breach of statutory duty, omission or default' in this qualification gives rise to the problem that whilst the terms 'negligence' and 'breach of statutory duty' have precise legal meanings, different meanings have been attached to 'default' in the case law. According to Parker J in *Re Bayley Worthington & Cohen's Contract*,[5] it refers to personal conduct raising questions of either not having done what one ought to have done or having done what one ought not to have done. Breach of contract is therefore just an instance of such conduct. This principle was followed in *City of Manchester v. Fram Gerard*,[6] another case involving an indemnity, and *Northwood Development Co Ltd v. Aegon Insurance Co (UK)*,[7] which concerned the construction of a performance bond. However, the Court of Appeal in *Perar BV v. General Surety and Guarantee Co. Ltd*[8] rejected the reasoning in the earlier cases and equated 'default', as a precondition for the enforcement of a performance bond, with breach of contract. Considering the inherent uncertainty of 'not having done what one ought to have and having done what one ought not to have done' the Court of Appeal's decision is much to be preferred in the context of indemnities.

Third Qualification

The 'subject to Clause 20.3' qualification has the general effect that before practical completion or determination of the Contractor's employment under the Contract, the

---

4   [1958] 1 WLR 697; [1958] 2 All ER 497 (hereafter *Gold v. Patman*).
5   [1909] 1 Ch. 648.
6   (1974) 6 BLR 70.
7   (1994) 10 Const. LJ 157.
8   (1994) 66 BLR 72.

Works and Site Materials (defined under Clause 22.2) are not part of 'property' for the purpose of Clause 20.2. However, where under Clause 18 the Employer takes possession of parts of the Works before practical completion, the indemnity applies to such parts. As in Article 1 and Clause 2.1 the Contractor promises to carry out and complete the Works, he is under an obligation to reinstate any loss or damage to the Works and Site Materials at his own expense. This is an obligation for which insurance under Clauses 22A/22B/22C is required. As explained later, the Contractor is required under Clause 21.1.1.1 also to insure against his liability for damage to property. The reason for this qualification must therefore be a need to avoid the duplication that would otherwise result.

Fourth Qualification

Qualifications in the JCT family of contracts similar to the fourth qualification have been the subject of much litigation. At the centre of the controversy is the question whether the effect of the qualification is to exonerate the Contractor and his sub-contractors completely from liability for loss or damage to existing structures and their contents caused by a Specified Peril even if that risk, usually fire, was created by the negligence of the Contractor or of his sub-contractor. The answer to this question is very important for two reasons. First, it is possible that even if the Employer effects the required insurance, the insurance proceeds may prove inadequate for the cost of their repair or replacement, e.g. price increases between payment on the claim and time of repair/replacement. Second, the insurance required is of the material damage variety, i.e. it only covers the costs of repair or replacement of the thing insured. There is no requirement for cover against consequential losses arising from the physical damage, e.g. loss of business for the period within which the building is unavailable or is used inefficiently. The court decisions on the issue have arisen from previous editions or versions of the JCT 80 or other contracts in the JCT family. They therefore have to be treated with considerable caution because minor differences in wording or even the use of commas at different places in contracts may give rise to major differences of interpretation. However, most of them are very useful in that they throw some light on the approach followed by the courts in resolving disputes of that kind.

Before examining the cases, it is important to re-emphasize a fundamental conceptual difference between risk and insurance clauses. A risk clause specifies who shall be liable for what risk, thus avoiding disagreements after occurrence of the risk as to who shall shoulder its burden. An insurance clause states who is to take out what insurance against what risk. Its function is to ensure availability of funds to deal with the occurrence of the relevant risk. It is therefore possible that an insurance clause requires one party to take out insurance against risk for which the other party is responsible. The courts have pointed out that it is a misconception to conclude that a party is responsible for a risk simply because the contract requires him to take out insurance against it. For example, on such arguments advanced by sub-contractors in *The National Trust v. Haden Young Ltd*,[9] Nourse LJ said at page 11A-B:

> The essential fallacy in them is the assumption that an obligation to insure against loss can, without more, throw that loss on the insured when, by another provision of the contract, it is to be borne by the other party. It is misleading to speak of the apportionment or allocation of risk in this context.

9   (1994) 72 BLR 1.

The main source of the problem has been that some risk clauses make references to insurance clauses, thus raising the question of whether the intention is to modify the risk allocation structure that would apply without the reference. This issue is essentially a question of the construction of the particular risk clause. In *Surrey Heath Borough Council v. Lovell Construction Ltd and Another*[10] Dillon LJ said at p. 121:

> The effect of the contract agreement must always be a matter of construction. People are free to contract as they like. It may be the true construction that a provision for insurance is to be taken as satisfying or curtailing a contractual obligation, or it may be the true construction is that a contractual obligation is to be backed by insurance with the result that the contractual obligation stands or is enforceable even if for some reason the insurance fails or proves inadequate.

In a line of cases involving contractors and employers, the contracts, in addition to requiring insurance by the employer, stated that the risk was solely the employer's. In such cases the decisions were that the contractor was not liable for the risk even where the risk was created by the negligence of the contractor or his sub-contractors.

> *James Archdale & Co. Ltd v. Comservices Ltd*:[11] this arose from a contract let on the predecessor of the JCT 98 current in 1952. The Employer was required by the equivalent of today's clause 22B to take out All Risk insurance against damage to the Works. The contract also provided that damage to the Works by fire was the sole risk of the Employer. The Court of Appeal held that the Contractor was not liable for destruction of the Works by fire caused by his own negligence.

In another line of cases, the contracts required the employer to take out insurance against the risk concerned but it was not stated that the risk was the employer's. In those cases, the contractor was held liable for fire damage caused by his own negligence or that of his sub-contractors.

> *The National Trust v. Haden Young Ltd*:[12] this arose from a contract for works of repair to Uppark House let on the JCT Minor Works form. Lead work was sub-contracted to the defendants. Clause 6.2 provided that the Contractor should be liable for loss to any property in so far as it arose out of or in the course of the works and to the extent that it was caused by the negligence of the Contractor or of people for whom the Contractor was responsible in law. Clause 6.3(B) required the Employer to insure, in the joint names of the Contractor and the Employer, against loss or damage to the existing structures and their contents. During the course of the sub-contract works, two employees of the sub-contractor went for a tea break leaving a lit oxyacetyline torch. This caused severe destruction of the building and its priceless contents. The sub-contractors argued that the main contractor was not liable and that therefore they were also not liable. The Court of Appeal decided that Clause 6.2 imposed unlimited

---

10 (1990) 48 BLR 108.
11 [1954] 1 WLR 459 (hereafter *James Archdale*); the House of Lords reached the same decision in *Scottish Special Housing Association v. Wimpey* (1986) 34 BLR 1 (hereafter *Scottish Special Housing*) which arose from a contract that contained similar provisions on fire damage to existing structures of the Employer; cf *Dorset County Council v. Southern Felt Roofing Co. Ltd* (1989) 48 BLR 96 which arose from an ad hoc contract.
12 See n. 9; see also: *Dorset v. Southern Felt Roofing Co.*; *London Borough of Barking and Dagenham v. Stamford Asphalt Co. Ltd and Others* (1997) 82 BLR 25 (hereafter *Barking & Dagenham v. Stamford*).

liability on the main contractor for the loss that had occurred and that the failure of the plaintiff to insure was an entirely different matter that did not affect the Contractor's responsibility for the risk. *James Archdale* and *Scottish Special Housing* were distinguished on the grounds that, unlike in the present case, the contracts in those cases expressly stated that the risk of fire damage was solely the employer's.

The scope for this type of controversy in relation to the JCT 98 has been reduced considerably because Clause 20.2 now states expressly that the liability and indemnity of Contractor under that clause, 'where Clause 22C.1 applies, excludes loss or damage to any property required to be insured thereunder caused by a Specified Peril'. The property required to be insured under Clause 22C.1 is existing structures and their contents either owned by the Employer or for which he is responsible. The loss to be insured against is 'the full cost of reinstatement, repair or replacement'. The Contractor is therefore not liable for these costs arising from damage or loss to existing structures and their contents from a Specified Peril. This was the conclusion as to the effect of the same terms in the JCT80 in *Ossory Road (Skermersdale) Ltd v. Balfour Beatty Building Ltd and Others*.[13] In that case, Judge Fox-Andrews QC held that the Contractor under the JCT 80 was not liable to the Employer for the costs of reinstatement of damage to existing structures and their contents by fire caused by the Contractor's own negligence.

The insurance required under Clause 22C.1 covers only material damage. Consequential losses, e.g. cost of disruption and other additional costs incurred in the carrying on of any business by the Employer in the premises, are not required to be insured against. The Contractor's liability for this associated type of damage was raised in *Kruger Tissue (Industrial) Ltd v. Frank Galliers Ltd and Others*[14] which arose from a contract let on the JCT 80. Judge Hicks QC started off by construing Clause 20.2 as imposing liability on the Contractor for loss (including consequential loss) arising from damage to property. He then decided that Clause 22C.1 was an insurance clause and that the Contractor's liability under 20.2 was curtailed only to the extent that the Employer was required to insure against the risk concerned. As the insurance required under Clause 22C.1 was of the material damage variety, there was no contractual obligation on the Employer to insure against his consequential loss arising from fire damage. *Ossory Road* was therefore distinguished as limited to only material damage claims.

7.2.3 Personal injury and property damage from Excepted Risks

The effect of Clause 21.3 is that the Employer is not entitled to indemnification against the risks of personal injury or damage to property caused by any of the Excepted Risks, which are defined in Clause 1.3 as consisting of:

ionising radiations or contamination by radioactivity from any nuclear fuel or from any nuclear waste from combustion of nuclear fuel, radioactive toxic explosive or other hazardous properties of any explosive nuclear assembly or nuclear component thereof, pressure waves caused by aircraft or other aerial devices travelling at sonic or supersonic speeds.

---

13 (1993) CILL 882 (hereafter *Ossory Road*).
14 *Kruger Tissue (Industrial) Ltd v. Frank Galliers Ltd Others* (1998) 57 Con. LR 1.

7.2.4 Risk of damage or loss to Works and Site Materials

Under Article 1 and Clause 2.1 the Contractor undertakes to carry out and complete the Works in accordance with the Contract. Such an undertaking carries with it an obligation to reinstate any damage to work executed or unfixed materials at no additional cost to the Employer unless the Contract is terminated by frustration or, in respect of any particular risk, the Contract itself states expressly that the Contractor is not to have such responsibility. The responsibility of the Contractor to reinstate any damage was asserted in *Charon (Finchley) Ltd v. Singer Sewing Machine Co. Ltd*[15] as a general principle applicable to construction contracts in these terms:

> Indeed, by virtue of the express undertaking to complete (an in many contracts to maintain for a fixed period after completion) the contractor would be liable to carry out his work again free of charge in the event of some accidental damage occurring before completion even in the absence of any express provision for protection of the risk.

It is provided in Clause 21.3 that the Contractor is not to indemnify the Employer in respect of loss or damage to the Works and Site Materials from the Excepted Risks. It is submitted that, subject to frustration at common law, that provision does not in any way exempt the Contractor from the responsibility to reinstate the Works. The effect of Clause 21.3 is probably that the Contractor is not responsible for the cost of dealing with contamination from the Excepted Risks.

7.3 JCT 98 insurance requirements

As explained earlier, the Contractor's obligations to indemnify the Employer or to reinstate any damage to the Works and Site Materials are useless unless he has the necessary funds. It was also explained that insurance is the customary way of ensuring availability of the necessary funds. It is for these and other reasons explained later that the JCT98 provides for insurance against the following:

1. the Contractor's liability under Clause 20.1: liability for death and personal injury (Clause 21.1.1.1);
2. the Contractor's liability under Clause 20.2: liability for injury and damage to real and personal property (Clause 21.1.1.1);
3. injury or damage to property due to collapse, subsidence, vibrations, weakening or removal of support or lowering of groundwater attributable to the carrying out of the Works (Clause 21.2.1);
4. damage to the Works and Site Materials (Clause 22A, or 22B or 22C.2);
5. damage to existing structures and their contents arising from the Specified Perils (applies to contracts involving extensions or refurbishment of existing facilities) (Clause 22C.l);
6. loss or damage to off-site materials or goods from the Specified Perils (Clause 30.3);
7. the Employer's loss of liquidated damages arising from extension of time for 'loss or damage occasioned by any one or more of the Specified Perils' (Clause 22D).

---

15 (1968) 112 SJ 536.

With most of these items of insurance, the party to insure is to obtain a type of cover referred to in the Contract as a 'Joint Names Policy'. Prior to its Amendment 16, the JCT 80 defined this term under Clause 1.3 as ' a policy of insurance which includes the Contractor and the Employer as the insured'. The intended effect of this requirement was to prevent an insurer of those risks from exercising its subrogation rights against the Contractor or the Employer. An insurer's subrogation right includes that of being able to stand in the shoes of the insured and to seek remedies against whoever actually caused the relevant damage or loss. In *Petrofina (UK) and Others v. Magnaload Ltd and Others*[16] it was held that where a policy names more than one party as insured under it, the insurer could not exercise its subrogation rights against a co-insured after meeting a claim from another co-insured. However, the decision in *Barking and Dagenham v. Stamford*[17] cast some doubts as to whether a Joint Names Policy as defined above was effective in limiting the insurer's subrogation rights in the intended way. In that case, it was argued that where the insurance covers damage to property, the insurer may still have subrogation rights against any insured without any insurable interest in the property. Accepting this argument, the trial judge held that if the Employer takes out a Joint Names Policy against damage to the Employer's existing buildings and their contents, the insurer would still be entitled to look to the Contractor for compensation if the Contractor caused the damage because of his lack of insurable interest. The trial judge's decision was upheld on appeal but on different grounds. Although the Court of Appeal expressly declined to consider that argument, the JCT still responded by re-defining the Joint Names Policy as:

> a policy of insurance which includes the Employer and the Contractor as the insured and under which the insurers have no right of recourse against any person named as an insured, or, pursuant to Clause 22.3, recognised as an insured thereunder.

This requires that the policy expressly waives the insurer's rights of subrogation against every party named in the policy as an insured.

It is to be noted that, under Clause 21.3, there is no obligation on the Contractor to insure against any personal injury to or death of any person or any damage to any property (the Works and Site Materials included) by the effect of the Excepted Risks. As insurance is not available in the insurance market at affordable rates, this provision reflects commercial reality.

7.3.1 Insurance against personal injury or death

This category of insurance is to cover the Clause 20.1 liability of the Contractor for death and personal injury. By Clause 21.1.1.2 the Contractor's insurance against this risk must comply with all relevant legislation. For insurance purposes, it is important to distinguish between the Contractor's own direct employees and others. The latter group includes the public at large and other parties working on the site. The minimum cover against injury to the Contractor's employees is as stated in the Employer's Liability (Compulsory

---

16 (1983) 25 BLR 37.
17 (1997) 82 BLR 25. See also *Amec Civil Engineering Ltd v. Cheshire County Council* [1999] BLR 303 in which it was held by Judge Gilliland QC that a Contractor under the 5th edn of the Institution of Civil Engineers Conditions is entitled to payment by the Employer under the Contract for dealing with a risk insured under a joint names policy without any allowance for receipts under the policy.

Insurance) Act 1969. The Contractor's insurance against liability for death and injury of people other than his own employees should comply with the minimum amount of cover stated in the Appendix.

The cover is for any occurrence or series of occurrences of death or injury arising from the same event. If the Contractor's actual liability exceeds the amount of cover he must make up the shortfall from his own resources. It follows therefore that the minimum amount of cover should be decided bearing in mind the extent of the risk and the Contractor's financial capabilities. The Contractor's potential liability under Clause 20.1 may be substantially more than the minimum cover required under the contract. This is because the amounts of damages for personal injuries being awarded by the courts have increased tremendously over the years. For this reason, Clause 21.1.1 is sometimes amended to require unlimited cover. This strategy must be reconsidered because, in recent years, unlimited cover is rarely available in the insurance market.

By law the Contractor must maintain Employer's Liability Insurance for as long as he is an employer. The JCT 98 is silent on the question of the period for which the Contractor is to maintain his insurance against death of and injury to people other than his employees. The common practice is to maintain the cover, at least, up to practical completion.

7.3.2 Insurance against damage to property

The Contractor is to take out and maintain insurance against his liability under Clause 20.2, i.e. liability in respect of damage or injury to real and personal property. The amount of cover is subject to a minimum stated in the Appendix.

7.3.3 Insurance against Employer's Clause 21.2.1 risks

The Contractor is only responsible for damage to property due to negligence, breach of statutory duty, omissions or default on the part of the Contractor or of parties for whom the Contractor is responsible in law. Sometimes, damage can occur without any fault on the part of the Contractor, e.g. the damage in the case of *Gold v. Patman*.[18] Against this type of damage, the Employer may have some protection through the professional liability insurance of his professional team if the damage is due to their professional negligence. Where the damage is due to neither the fault of the Contractor nor professional negligence, the Employer will have to foot the bill. Clause 21.2.1 envisages that a prudent Employer may wish to insure against this type of liability. If that is the case, it should be indicated in the Appendix that insurance under Clause 21.2.1 may be required. The amount of cover per occurrence or series of occurrences arising from the same event must also be stated. Furthermore, the Architect must instruct the Contractor to take out the insurance. The cost of effecting and maintaining this category of insurance is to be added to the Contract Sum.

Clause 21.2.1 insurance covers injury and damage to any property caused by collapse, subsidence, vibrations, weakening or removal support or lowering of groundwater attributable to the carrying out of the works. The clause expressly excludes from the cover damage and injury:

---

18 See n. 4.

- for which the Contractor is responsible under Clause 20.2;
- attributable to defective design (which would normally be covered by the professional indemnity insurances of the designers);
- which are reasonably foreseeable as the inevitable consequence of the carrying out of the Works;
- to existing structures and their contents if Clause 22C applies;
- to the Works and Site Materials;
- covered by any other insurance already arranged by the Employer;
- arising from the consequences of war, invasion, etc;
- arising from or caused by the Excepted Risks;
- caused by or arising from pollution or contamination of buildings, land, water, or the atmosphere;[19]
- that results in the Employer's liability for his breach of contract.

The reason for their exclusion is that either they are covered by insurances required under other clauses of this Contract or other contracts or cover is not available in the insurance market at reasonable rates.

The insurance must be in the joint names of the Contractor and the Employer and placed with insurers approved by the Employer. The Contractor must send the policy and all premium receipts to the Employer through the Architect. If the Contractor fails to take out or maintain the insurance, the Employer can arrange remedial insurance or pay the premiums himself. Apart from seeking damages under the general law, the Contract does not provide the Employer with any sanction against this type of default. In practice, the cover required by Clause 21.2.1 is commonly provided as an extension to the Contractor's public liability policy (explained later). Where this practice is not followed, it is underwritten as a separate policy.

7.3.4 Insurance of Works and Site Materials

There are three alternative clauses for the insurance of the Works and Site Materials: 22A or 22B or 22C. Whichever of these is to apply to the particular project should be completed in the Appendix.[20] Each of these clauses requires the same insurance cover described as 'All Risk Insurance'. However, there are differences between them regarding who is responsible for taking out and maintaining the cover and the administrative procedures to be complied with concerning verification of compliance and management of occurrence of the risk.

Clause 22A is intended for projects involving new work where the intention of the parties is that the Contractor is to be responsible for the insurance. Clause 22B is also intended for new work but where it is the Employer who is responsible for obtaining the cover. At first sight, it appears strange that an Employer should take on such responsibility when it is the Contractor who is responsible for and carries out the work. This arrangement is clearly not suitable for inexperienced employers. However, where the Employer has repetitive programmes of procurement, e.g. developers, local authorities and retail chains,

---

19 The insurance requirement covers pollution or contamination caused by a sudden identifiable, unintended and unexpected incident which occurs completely at a specific moment during the period of the insurance.

20 See Section 1.5.2 dealing with Appendix entry in relation to Clauses 22A/22B/22C.

he may be in a position to obtain cheaper insurance on account of bulk business and the economies of scale. Clause 22C is to be used where the Works consist of refurbishment of, or modifications to, or extension of existing structures. Here the Employer is responsible for the All Risk Insurance of the Works. The rationale for putting this responsibility on the Employer is that, as an owner of the existing buildings, the likelihood is that he will already have building insurance cover. It would therefore be more convenient and cheaper for the Employer to negotiate the insurance required under Clause 22C as an extension to any such cover.

7.3.4.1 All Risk Insurance

'All Risk Insurance' is defined in Clause 22.2 as 'insurance which provides cover against physical loss or damage to work executed and Site Materials'. However, costs of repair, replacement or rectification of the following are expressly excluded:

1. defects in property due to wear and tear, obsolescence, deterioration, rust or mildew;
2. defective work and Site Materials attributable to design, materials or workmanship of poor quality;
3. loss/damage caused by war, invasion, revolution, etc., ordered by any government or public authority;
4. loss/damage from disappearance or shortage revealed only by an inventory and not attributable to any specific event;
5. loss or damage from the Excepted Risks;
6. loss or damage caused by civil commotion (applies only to contracts in Northern Ireland);
7. loss or damage from terrorist activities (applicable only to contracts in Northern Ireland).

From the long list of exceptions, it should be apparent that All Risk Insurance does not in any way cover all risks of damage to the Works and Site Materials and that the term is therefore a misnomer.

7.3.4.2 Insurance under Clause 22A (by Contractor)

By Clause 22A.l the Contractor is responsible for taking out and maintaining this insurance which must be a Joint Names Policy against the risks covered by the Clause 22.2 definition of 'All Risk Insurance'. As explained earlier, the intention behind the requirement for a Joint Names Policy is to prevent the possibility of the insurer pursuing the Employer for damages or a contribution through exercise of its rights of subrogation.

The minimum amount of cover should be the full reinstatement value of the Works plus a percentage for professional fees.[21] This percentage is to be stated in the Appendix. As the reinstatement value of the works will increase over the contract period, ideally the policy must allow a gradual increase of the amount of maximum cover but this is rarely done in practice. The worst possible loss is where the Works are destroyed just before practical completion. At this stage their value may be more than the Contract Sum because of inflation and Variations. Whilst conservatism would call for an appropriate allowance for these increases, it is often argued that, as a complete loss of the Works is uncommon, such

---

21 The full reinstatement value is to be inclusive of VAT if the Employer's status for VAT purposes is exempt or partially exempt.

an approach would be wasteful of premiums. The strategy to adopt would depend upon the risks inherent in the particular project and the parties' preferred ways of dealing with them. However, it has to be borne in mind that where the preferred strategy demands a departure from the requirements of Clause 22A, the clause should be amended accordingly.

The insurance is to be maintained up to and including the date of whichever of these occurs first: (i) the issue of the Certificate of Practical Completion; or (ii) determination of the employment of the Contractor under the Contract. There is therefore a need on the part of the Architect to advise the Employer to ensure that the right insurance is in place by the time the Contractor's insurance comes to an end.

Sub-clause 22A.2 makes elaborate provisions for the enforcement of the obligation of the Contractor to take out and maintain this element of insurance. First, the insurers must be approved of by the Employer. Second, the Contractor must send to the Employer through the Architect, the policy, premium receipts, and any endorsements for safekeeping. Finally, if the Contractor fails to comply with any of the insurance obligations, the Employer may take out any remedial cover. In such an event, the premiums paid or payable are recoverable from the Contractor against payments due or as debt. It is submitted that the Employer may also be entitled to recover other sums as damages for breach of contract, e.g. cost of managerial time expended.

Sub-clause 22A.3.1 recognizes the common practice whereby contractors take out, on an annual basis, a single policy, the so-called Contractor's All Risk Insurance, to cover all their contracts undertaken in the year covered by the insurance. Provided the annual policy meets the Joint Names Policy requirements and covers the Works, the Contractor will not be required to take out a policy specific to the present Contract. Also, although there is no requirement that the annual policy is deposited with the Employer, the Contractor must provide him with evidence that the policy is being maintained, e.g. premium receipts. If the Contractor fails to comply with these requirements the Employer may take out and maintain the relevant insurance with consequences similar to those explained already.

Upon discovery of any occurrence of loss or damage covered by the policy, the Contractor must forthwith give notice in writing to both the Architect and the Employer (Clause 22A.4.1), specifying the nature, extent, and location of the loss or damage. He must also comply with the requirements of the policy. This will usually mean informing the insurer and submitting a claim on the insurer's standard form. It is common practice for the insurer to instruct loss adjusters to visit the Works and assess the amount of damage before paying on the policy. After inspection by the insurer or his agents the Contractor must carry out all the necessary work of repair, restoration, and clearance and disposal of wastes. He is also to proceed with the carrying out of the Works to completion.

The Contractor and Nominated Sub-contractors insured in respect of the damages must instruct the insurer to pay the insurance proceeds to the Employer (Clause 22A.4.4). Without this provision, the insurer would have to make payment in the joint names of all the insured. After deducting the professional fee element, the rest is to be paid to the Contractor by instalments under payment certificates. Eaglestone[22] notes that, in practice, in getting paid for the remedial work, the Contractor deals directly with the insurer and that the Architect and the Employer are hardly ever involved.

The damage is to be ignored when carrying out valuation for certification (Clause 22A.4.2). This means that, in effect, the Contractor is paid twice for the same work. It is

---

22 Eaglestone, F., *Insurance under the JCT Forms*, 2nd edn (Blackwell Science, 1996), at p. 175.

to be noted that for the remedial work the Contractor is paid no more than the insurance proceeds less the professional fees. Any shortfall between actual costs and the residue of the proceeds is therefore to be borne by the Contractor.

7.3.4.3 Insurance of Works under Clause 22B (by Employer)

The provisions in this clause are a mirror image of Clause 22A. However, there are two key differences. First, the positions of the Contractor and the Employer are interchanged regarding the insurance obligations. It is the Employer who must take out the Joint Names Policy. Second, the remedial works are to be valued and paid for as a Variation in accordance with Clause 13. The Employer must therefore meet any shortfall between the insurance proceeds and the total valuation of the remedial work.

7.3.4.4 Insurance of Works, Clause 22C.2 (by Employer)

These provisions are virtually the equivalent of those under Clause 22B. However, there is one major difference. Either party may determine the employment of the Contractor within 28 days of the occurrence of the loss or damage provided it is just and equitable to do so. The procedure for any such determination and the financial settlement are discussed in Chapter 16.

7.3.4.5 Insurance of existing structures under Clause 22C.1 (by Employer)

The cover required here is referred to in some circles as 'Fire and Special Perils' policy. The Employer is to take out and maintain a Joint Names Policy against damage to the existing structures and their contents from the Specified Perils. The amount of cover is the full cost of reinstatement, repair or replacement arising from the damage. The existing structures are to include from the relevant date any relevant part handed over under Clause 18. The insurance is to be maintained up to and including the date of whichever of these occurs first: (i) the issue of the Certificate of Practical Completion; or (ii) determination of the Employment of the Contractor under the Contract whether or not the determination is contested.

When a claim is submitted for compensation for loss or damage arising from the insured risks, the Contractor and any Nominated Sub-contractor covered must authorize the insurer to pay the insurance proceeds to the Employer.

7.3.4.6 Terrorism cover

'Terrorism' is defined under Clause 22.2 as 'the use of violence for political ends and includes any use of violence for the purpose of putting the public or any section of the public in fear'. The definition is so wide that it could well cover violence by religious fundamentalists, animal rights activists, or even a students' union campaigning for a change in the education policy of the Government. It is therefore likely that terrorism cover will continue to be an issue in construction contracts.

The definition of All Risk Insurance in relation to work in England and Wales covers damage to the Works from terrorism. The insurance of the Work and Site Materials under Clauses 22A.1, 22B.1, 22C.2 must therefore cover damage or loss from terrorism. Also, because there is nothing in the Clause 1.3 definition of Special Perils to exclude damage from terrorism, insurance of existing structures and their contents under Clause 22C.1 must also include such damage. However, the effect of the definition of All Risk Insurance is that, where the work is in Northern Ireland, there is no requirement for cover against damage or loss to the Works and Site Materials from terrorism. This position reflects commercial unaffordability, or unavailability, of such cover in the Northern Ireland situation. There is a Government scheme providing equivalent cover.

In the 1990s, increased terrorist activity in England, e.g. the bomb explosions in Canary Wharf in London and in the commercial centre of Manchester, resulted in withdrawal by insurers of full cover in respect of terrorism from their standard commercial policies on account of astronomical rises in levels of claims. The Government stepped in to make full cover available by establishing Pool Reinsurance Company Ltd, the function of which is to take the risk away from insurers. This enables terrorism cover withdrawn from the standard policies to be bought back with premiums graded by the risk associated with the zone of the location of the works. It is these premiums that go to fund the Government's scheme. However, the Government has retained the right to cancel the scheme whenever it proves unworkable, giving rise to the possibility that terrorism cover effected at commencement may cease to be unavailable prior to Practical Completion.

Amendment TC/94, issued in April 1994, was designed to deal with this eventuality. It has not been incorporated into the JCT 98 because it was considered convenient to maintain it as a separate amendment for use only as long as there is a need to do so. Its main provisions are as follows. Either party is to inform the other if the insurer notifies him that terrorism cover will cease or will no longer be available from a specified date. The Employer may elect to continue with the project or determine the Contractor's employment. If the Employer elects to continue, he thereby accepts complete responsibility for damage or loss to executed work or Site Material from terrorism. Any work of restoration and related work arising from such damage or loss is therefore to be paid for as a Variation. Before the date on which the cover is to cease or from which it will be unavailable, the Employer must notify his election in writing to the Contractor.

7.3.4.7 Insurance under Clauses 22A/22B/22C and Sub-contractors

Under Clause 22.3, all Nominated Sub-contractors are to be protected against liability for damage/loss to the Works from the Specified Perils. Two methods to this end are provided: either it is provided in the insurance policy that, in respect of those risks, each Nominated Sub-contractor is insured or it is stated that the insurer's rights of subrogation against the sub-contractor are waived.[23] Where Clause 22C applies, similar protection is also to be provided against damage to existing structures and their contents from the Specified Perils.

With one exception, the same protection is to be provided to domestic sub-contractors: Clause 22.3.2. The exception is that, where Clause 22C applies, there is no requirement to protect them against liability for damage/loss to existing structures and their contents from the Specified Perils. This exception has two serious consequences for the liability of a domestic sub-contractor for damage or loss to the existing structures and their contents from a Specified Peril caused by his negligence. First, as evidenced in *British Telecommunications plc v. James Thomson and Sons (Engineers) Ltd*,[24] the Employer can successfully sue the sub-contractor in tort. In that case, which arose from a contract in the

---

23 The *Barking and Dagenham v. Stamford* litigation suggests that merely stating that the sub-contractor is an insured may not be enough and that it is safer to require the insurer to waive its subrogation rights.
24 *British Telecommunications plc v. James Thomson and Sons (Engineers) Ltd* (1999) BLR 35.

terms of the JCT 80, a domestic sub-contractor, invoking the so-called 'contract structure' defence to liability in tort, argued that, in view of the Employer's obligation to insure against the risk, it would not be fair, just and reasonable to impose on them a duty of care in tort to avoid causing the damage or loss. Rejecting the argument, the House of Lords pointed out that it was instructive that the main contract provided expressly that Nominated sub-contractors, but not domestic sub-contractors, were to be protected against liability. Second, the Employer's insurer would be able to exercise its subrogation rights against the sub-contractor. As noted by their Lordships, it would therefore appear that the exclusion of domestic sub-contractors from the protection of the Employer's policy was intended to reduce the premiums for such cover.

7.3.5 Insurance of off-site materials

Under Clause 30.3, the value of off-site materials that are on the list of materials annexed to the Contract Bills and intended for the Contract must be included in interim certificates. A precondition for such inclusion is that such materials are insured for their full value against damage or loss from the Specified Perils: Clause 30.3.5. The cover is required for the period commencing with transfer of ownership of the materials to the Contractor until their delivery to the site. It is further required that the policy protects the interests of the Employer and the Contractor in respect of damage by the Specified Perils. The Employer's interest is as an owner of the materials where, in accordance with Clause 16, they have been included in interim certificates and paid for by the Employer. To protect this interest, the policy must, at the very least, provide that the Employer is an insured and that he is to be indemnified against loss or damage to the materials from the Specified Perils after they have been so paid for.

7.3.6 Insurance against loss of liquidated damages

Liquidated damages are estimates of the loss likely to be incurred by the Employer if the building is not completed by the date for completion. For example, this could be the cost of getting alternative accommodation for the period of delay. Under Clause 25.4.3, loss or damage caused by any of the Specified Perils is a Relevant Event, i.e. the time for completion is extended. In such an event, the Employer loses his right to recover liquidated damages for the duration of the extension although he stills stands to incur losses from the delay in completion. In situations where the Employer cannot avoid liability for delay and the risk of delay from the Specified Perils is significant, insurance may be his best course of action.

Clause 22D is designed for this type of situation. It applies only if it is indicated in the Appendix that insurance under that clause may be required. The maximum period of delay to be insured against is also to be stated there. On contract award, the Architect must either inform the Contractor that the insurance is no longer required or instruct him to obtain a quotation for an agreed amount of cover. The quotation is to be sent to the Architect, who, after consultation with the Employer, must finally instruct the Contractor whether or not to take out the policy. The expense of taking out and maintaining the policy is to be added to the Contract Sum.

7.4 Enforcement of insurance requirements

Reflecting the importance of having appropriate insurance cover, the Contract, in addition to stating the insurance requirements, addresses how they may be enforced. Generally, a party with the responsibility to arrange the cover must, upon request by the other party, provide evidence of compliance. Furthermore, in all cases of failure of a party to take out and maintain a policy as required by the Contract, the other party may do so at the expense of the defaulting party. In practice, this is often almost impossible to do because the party to arrange in default will not usually have the information required by insurers as part of their underwriting procedures. For example, the Employer is unlikely to have details of the Contractor's safety records.

7.5 Effect of Joint Fire Code

As a result of their grave concerns over high levels of claims on account of fires on construction sites, construction insurers produced a Joint Fire Code of Practice that imposes obligations on employers, contractors, and construction professionals involved on construction projects. The object of the code is to ensure that adequate detection and protection systems are incorporated during the design and planning stages of projects and that work on site is carried out to the highest standards of fire safety. Although the application of the code is not mandatory, whether or not it applies affects the level of premiums or even the availability of cover from some insurers. Under the JCT98, it applies only if it is completed against Clause 22FC.1 in the Appendix that it applies. Where it applies, it should be further completed in the Appendix whether or not the insurer of the Works has specified the Works as a 'Large Project'.[25] Clauses 22FC.2.1 and 22FC.2.2 require the Employer and the Contractor, respectively, to comply with their obligations under the Code. A party suffering loss from a breach of the Code for which the other is responsible under the Contract is entitled to an indemnity from the other.

Clause 22FC.3.1 contemplates a situation where an insurer of the Works, in response to a breach of the Code, specifies Remedial Measures to be carried out by a specified date, referred to as the 'Remedial Measures Completion Date'. It is the Contractor's responsibility, in compliance with related AIs, to carry out the measures by the specified date. Under Clause 22FC.3.2, the Employer may arrange the carrying out of the Remedial Measures at the expense of the Contractor if the Contractor, without reasonable cause, fails to start the remedial work within 7 days of receipt of the insurer's notice or to proceed regularly and diligently[26] with the work. It is important to note that there is no requirement for the Contractor to be reminded in writing of this default before the Employer is entitled to intervene in this way. It is also to be noted that there is no provision for the Contract Sum to be adjusted on account of the costs incurred in carrying out the Remedial Measures. It follows therefore that the Architect must not take them in to account in

---

25 This calls for the Contractor, as part of his tendering procedures, to have contacted insurers and collected this information.
26 For explanation of the meaning of this term see Section 16.5.2.

payment certificates. It is for the parties themselves to recover this by way of set-off or as debt.

7.6 Commercial insurance products

There is some disparity between the terminology of the JCT 98 and the jargon of the insurance market. This disparity calls for awareness of the terms used by the insurance market to describe their standard products intended for the construction industry. It has to be noted that, where the name of a product is similar to the wording of the JCT 98, it will not necessarily provide the cover required by the Contract.

There are three main types of insurance product available to the construction industry to meet the risks discussed in this chapter: (i) Employer's Liability Insurance; (ii) Public Liability Insurance; (iii) Contractors' All Risk Insurance. These have not been designed to meet the insurance requirements of any particular standard form of contract. Also, for the same type of product, terms may vary from insurer to insurer. It is therefore important to examine the products to ensure that, together, they meet the requirements of the Contract.

As already explained, Employer's Liability Insurance protects the Contractor as an employer of people against his legal liability for accidents to or diseases sustained by his employees in the course of their employment. The reason why it is compulsory under the Employer's Liability (Compulsory Insurance) Act 1969 is to ensure that an injured employee, or his dependants, obtains compensation regardless of whether or not the Contractor has the financial resources to meet the employee's claim.

The Contractor's Public Liability Insurance, also called 'Third Party Policy' covers the Contractor's liability for injury to or death of people other than the Contractor's own employees and damage to property other than the construction works. It therefore covers only part of the Contractor liabilities under Clause 20. Sometimes, it is extended to provide the cover required by Clause 21.2.1

The Contractor's All Risk Policy covers loss/damage to the Contractor's construction works and site materials. As already explained, the name is a misnomer as no policy covers all forms and causes of loss or damage. Although the JCT 98 does not require it, for reasons of prudence, this policy is often extended to cover the Contractor's temporary accommodation, plant, tools and equipment.

7.7 Amendments to the insurance provisions

There may be a need to amend the provisions either to remedy perceived shortcomings or to meet the special requirements of the particular project or parties. Where this is the case, care has to be exercised to avoid the amendments being invalidated by Clause 2.2.1 of the Contract.[27]

---

27 See Section 1.5.5 for discussion on the problems associated with this clause.

7.8 Role of the Architect

The Architect is under a duty to check that the insurance arrangements of the Contractor and the sub-contractors are as required under the Contract. For details on this duty see Section 2.4.5.

8

Novation, assignment and domestic sub-contracting

It is common practice for a party to a building contract to involve third parties in its performance either as recipients of some of his benefits or undertakers of his obligations under the contract. The common legal mechanisms whereby such third party involvement may be allowed include novation, assignment, and sub-contracting. This chapter first explains the general principles governing the operation of these mechanisms. Their intended operation under JCT 98 is then examined.[1]

8.1 Novation

This is a process where a contract between A and B is converted into a contract between A and is C. The contract is then said to be novated to C. This can be achieved only with the agreement of all the three parties concerned. There are two common situations in which novation of building contracts occurs. First, where an employer is in serious financial difficulties, the contractor, a financier of the development and the employer may enter into an agreement whereby the financier steps into the shoes of the employer to complete the project by acting as the employer for the outstanding work. The same principle applies to the contracts of engagement with the insolvent employer's consultants. Second, a receiver or liquidator of an insolvent building contractor may, with the agreement of the employer, novate a building contract entered into by the contractor to another contractor who is prepared to take it over at an agreed price.[2]

8.2 Assignment

To understand correctly the concept of assignment as used in relation to building contracts, it essential to be aware of the distinction between the burdens of a contract and its benefits. In a construction contract the benefit to the contractor is to be paid in accordance with its terms whilst his burden is the obligation to carry out and complete the work. The

---

1 Practice Note 9, published by the JCT, provides some guidance on assignment and sub-letting.
2 See Section 16.6 for discussion on novation for the Contractor's insolvency.

employer's benefit is his right to have the work completed whilst his burden is to make payment.

Assignment refers to the legally recognized transfer by a party to contract of his rights or obligations under it to a third party to the contract, the assignee. After the transfer, the party effecting the transmission (the assignor), no longer has any rights or obligations under the contract whilst the assignee can himself sue and be sued under the contract. Because of the doctrine of privity of contract, it is not possible to assign the burdens or benefits of contract at common law except by a novation. This means that, where a contractual party wishes to achieve the effect of assignment, the consent of the two other parties has to be obtained.

However, the benefits of a contract can be validly assigned under statute or in equity without the consent of the other contractual party unless the contract itself expressly forbids such assignment.[3] The rationale for this is that the benefit of a contract, i.e. the right to call upon the other party to perform his obligations, has some economic value and is considered as belonging to a class of personal property called 'choses in action' or 'things in action'. As such, it is capable of being assigned under the Law of Property Act 1925 (LPA 25) which allows assignment of this class of property. For assignment of choses in action to be effective, s. 136 of the Act requires that it:

- is in writing;
- is signed by the assignor;
- is absolute, i.e. the entire right must be assigned and not just a part of it (e.g. a contractor assigning his right to payment due from the employer must assign his entitlement to all the payment);
- is not purported to be by way of charge only (e.g. payment on interim certificates to a stakeholder would not be assignment);
- is accompanied by express notice in writing to the debtor (the party against whom the assignor had the contractual right). The consent of the debtor to the assignment is not necessary unless the contract giving rise to the right contains a specific requirement for his approval.

Where there was an intention in a transaction to assign a contractual right but s. 136 of the LPA 25 has not been satisfied completely, there may be valid assignment at the discretion of the court. This type of assignment is referred to as 'equitable assignment' or 'assignment in equity'.

An example of this type of third party involvement is assignment by a contractor of his right to payment from the employer to a bank that is financing his performance of the contract. On the employer's side, the right to require the builder to make good defects after completion may be assigned to persons occupying the buildings, such as purchasers from the employer or his tenants.

In most situations owners and contractors choose whom to contract with on the basis of mutual expectations of good performance and a cooperative attitude in the event of difficulties. For example, the contractor would be expected to apply interim payment towards the carrying out of the works and to overlook minor breaches of the contract by the employer, e.g. very short delays in payment, in the interests of future good relations.

---

3   This principle was expressly affirmed by the House of Lords in the *Linden Gardens* case, see below, n. 5.

Proper performance of the work and maintenance of good relations with the employer may not be very high in an assignee bank's list of priorities. It is for this reason that many standard forms of contract contain provisions limiting the parties' rights of assignment.

8.2.1 Assignment under JCT 98

Clause 19.1.1 of JCT 98 states that 'neither the Employer nor the Contractor shall, without the written consent of the other, *assign this contract* [our emphasis]'. Whilst statute and equity allow the assignment of the benefits of a contract without the consent of the other contracting party, there is no valid way of doing the same for the burdens of a contract. Thus, it is impossible, in the strict sense, to 'assign a contract', i.e. both burdens and benefits, without the consent of the other party.

The meaning of contractual provisions of the kind in Clause 19.1.1 has therefore been a matter of considerable controversy. It was argued in the 10th edition of *Hudson's Building and Engineering Contracts* that since it is impossible to assign a contract in the strict sense, the intention behind clauses which prohibit 'assignment of a contract' is not to prevent assignment of the benefits of the contract but to prohibit the parties from obtaining vicarious performance of contractual obligations without consent. In *Helstan Securities Ltd v. Hertfordshire County Council*[4] it was held that the effect of a similarly worded prohibition in the contract was to make the purported assignment of the benefits ineffective. The issue finally came before the House of Lords in *Linden Gardens Trust Ltd v. Lenesta Sludge Disposal & Others*.[5]

> Stock Conversions engaged McLaughlin & Harvey as main contractors on a JCT 63 contract to carry work on their premises including removal of blue asbestos. Clause 17 of JCT 63 prohibited assignment in terms similar to those of JCT 98 Clause 19.1.1. Lenesta were in effect nominated sub-contractors for the removal of the asbestos. On 3 July 1985 Stock Conversions issued a writ against Lenesta claiming damages for breach of a direct warranty provided by them. In August 1985 Stock Conversions transferred their leasehold interest in the building to Linden Gardens. In January 1987 Stock Conversions assigned all rights of action under the contracts to Linden Gardens who then became second plaintiffs in the action already commenced. The required consent was not obtained. In March 1989 McLaughlin & Harvey and a third party were brought into the action as defendants. Linden Gardens were seeking to recover the cost of remedial works incurred by Stock Conversion and costs of additional works incurred by themselves after the assignment. The House of Lords held that the effect of Clause 17 was to make the purported assignment by Stock Conversions without consent invalid. Linden Gardens could not therefore enforce the benefit of the contract.

Many construction professionals and even students have been incredulous about the debate surrounding 'assigning a contract'. To them it is simple logic that if one is forbidden from two courses of action one of which is impossible to do anyway, it simply means the possible course is forbidden. It is ironical that a visit to the House of Lords was necessary to confirm what most people in the construction industry have always known. The

---

4   [1978] 3 All ER 262.
5   (1993) 63 BLR 1 (hereafter *Linden Gardens*); applied by the Court of Appeal in *Darlington B. C. v. Wiltshire Northern Ltd* (1994) 69 BLR 59 (hereafter *Darlington*).

layman's approach to the debate is hardly different from Lord Browne-Wilkinson's statement in his leading judgment that:[6]

> Although it is true that the phrase 'assign this contract' is not strictly accurate, lawyers frequently use those words inaccurately to describe an assignment of the benefit of a contract since every lawyer knows that the burden of a contract cannot be assigned.

It should be noted that assignment is subject to the consent of the Employer himself and not the Architect and that there is no requirement for the withholding of this consent to be reasonable. It follows therefore that any refusal of consent by either party is a dispute.

Under Clause 27.2.1.4 the Employer may determine the employment of the Contractor for assignment without consent. Similarly, under Clause 28.2.1.3 the Contractor may determine his own employment if the Employer assigns without consent.

8.2.2 Assignment after Practical Completion

On completion the Employer may wish to transfer some interest in the completed facility, e.g. leasing or selling the building, to other parties. With outright sale the purchaser would not normally have remedies against the Employer because of the *caveat emptor* ('let the buyer beware') rule. In leases containing fully repairing covenants the Employer may even be entitled to insist on the lessees themselves making good the defects. Thus, although the Employer would be entitled to bring proceedings against the Contractor to enforce the terms of their contract, there will usually be virtually no incentive on his part to do so after the sale or lease of the building. It is therefore only prudent for potential purchasers and lessees to insist on the Employer assigning to them, as part of the sale or lease transactions, the right to bring such proceedings. The ability to effect such an assignment would therefore make the property more marketable.

Even where the Employer is ready to institute the proceedings, such a course of action suffers from two principal drawbacks. First, it is far more convenient for the purchasers and lessees to do so themselves. Second, where the Employer transferred the building for its full market, he would be entitled to only nominal damages because he suffers no loss. This flows from the general principle that in an action to recover damages for breach of contract, the claimant can recover only his own losses even if the contract was intended to benefit the third party who suffers the loss.

Clause 19.1.2 gives the Employer the right to make such an assignment. It is important to note that the clause applies only if it is indicated in the Appendix to be the case. The terms of any such assignment must be such that the assignee is bound by any prior agreement reached between the Employer and Contractor regarding the rights assigned. For example, if the Employer reached a compromise with the Contractor regarding errors in the setting out of the works or departures from the specification, the assignee should not be able to go back on the compromise.

The need for Clause 19.1.2 may have been reduced by the House of Lord's decision in *St Martins Corporation Ltd v. Sir Robert McAlpine & Sons Ltd*[7] which was heard together with the *Linden Gardens* case.

---

6   Ibid, at p. 427.
7   (1993) 63 BLR 1 (hereafter *St Martins*).

St Martin's Property Corporation (Corporation) engaged Sir Robert McAlpine & Sons on a development contract which incorporated a version of JCT63 Clause 17 of which was exactly the same as in the *Linden Gardens* case. Corporation later transferred the property and assigned the benefit of the contract with McAlpines to St Martin's Property Investments Ltd (Investments) for its full market value. After practical completion defects were discovered and were made good at great expense to Investments. Corporation and Investments were joint plaintiffs in a claim against McAlpines for damages for breach of contract. The House of Lords held that Investment's claim should fail because the purported assignment underlying their claim was invalid without the consent required under Clause 17. On Corporation's claim, McAlpine argued that they were entitled to only nominal damages because they had suffered no loss from their breach. However, the House of Lords decided that, as an exception to the general principle in McAlpine's argument, Corporation were entitled to recover Investment's loss on their behalf. It was explained that the exception applies to large developments which, to the knowledge of the employer and the contractor, were going to be occupied or purchased by third parties. In such cases an employer may recover for the benefit of third parties their losses arising from the contractor's breach.

The potential impact of this decision is considerable. Most construction contracts between developers and contractors should fall into this exceptional class. The *St Martin's* decision was also followed almost immediately by the Court of Appeal in *Darlington*.

8.3 Domestic sub-contracting

The term 'sub-contracting' or 'sub-letting' is used to refer to the situation whereby a party to a contract obtains vicarious performance of his duties by third party to the contract, whilst still remaining primarily responsible to the other party to the contract for the adequacy of the performance of the third party. This is in contrast to both assignment and novation where the assignor and the novator cease to have rights and/or obligations, as the case may be, under the contract.

Whether or not a party to a contract is entitled to sub-let the performance of any of his duties under it depends on the nature of the duties and any express contractual provisions on the point. As a general rule, under the general common law, where a duty is of a personal nature, the obligation to perform it cannot be discharged by sub-letting. Indeed, sub-letting in such circumstances would be a breach of the contract.

Davies v. Collins[8]: the appellants operated a dyeing and cleaning business to which the respondent US army officer entrusted his uniform for cleaning and minor repairs. The appellants sub-contracted the cleaning to sub-contractors who failed to return the uniform. The Court of Appeal decided that, on the facts of the case, the sub-letting was a breach of contract.

However, if by the nature of the duty, it does not matter who does the work, sub-letting without consent may be allowed.

---

8   [1945] 1 All 247; see also Court of Appeal's application of this principle to construction work in *Southway Group Ltd v. Wolff and Wolff* (1991) 57 BLR 33.

British Waggon Co. v. Lea & Co:[9] Parkgate hired out their railway waggons to Lea & Co. It was a term of the hire contract that Parkgate would keep the waggons in good repair. Subsequently Parkgate assigned the benefits of the contract, i.e. the right to rents, to British Waggon Co. and arranged with them to carry out the repair obligations. After taking over the workmen of Parkgate, British Waggon was ready and willing to perform these obligations. One of the issues in dispute was whether or not Lea were bound to accept performance of the repair duties by a third party. It was held that, where a party contracts to carry out work of so rough a nature that ordinary workmen conversant with the business would be perfectly able to execute the work, the party may be entitled to sub-let it without the consent of the other contracting party.

There can be little doubt that the service rendered by the contractor in a construction contract is of a very personal nature. This is the only conclusion that can be reached considering the elaborate vetting procedures usually followed in the selection of contractors. It follows therefore, that even in the absence of express provisions disallowing sub-letting of any part of the works, such as contained in JCT 98 Clause 19.2, the Contractor may not be entitled to sub-let in any case.

8.3.1 Sub-contracting under JCT 98

There are two types of sub-contractor under JCT 98: Domestic Sub-contractors, and Nominated Sub-contractors. The rest of this chapter is about Domestic Sub-contractors. Nominated Sub-contractors are dealt with in Chapter 9.

8.3.2 Domestic Sub-contractors

Most of the provisions on Domestic Sub-contractors are contained in Clause 19. Clause 19.2.1 defines a Domestic Sub-contractor as a person to whom the Contractor sub-lets any portion of the Works other than a Nominated Sub-contractor whilst a Nominated Sub-contractor is defined under Clause 35.1.

JCT 98 provides for two methods of appointing Domestic Sub-contractors. With the first method, the choice of Sub-contractor is solely the Contractor's subject only to the written consent under Clause 19.2.2 of the Architect to the sub-letting of that portion of the Works. The Architect must not withhold his consent unreasonably. Reasonableness of objection to sub-letting will usually relate to the integrity and the technical, financial and organizational capability and capacity of the sub-contractor concerned. This restriction, in widest sense, applies to sub-letting to 'labour only sub-contractors'(LOSC) even though in practice most contractors engage them without the consent of the Architect. A more prudent course of action is to obtain a blanket consent to engagement of LOSCs.

Sub-letting without consent entitles the Employer to determine the Contractor's employment (Clause 27.2.1.4).

The Conditions do not expressly empower the Architect to vet Domestic Sub-contractors. This is because the Contractor is only required to obtain consent to sub-letting and not to his choice of sub-contractors. In principle, the Contractor can therefore obtain consent of sub-letting before he even considers the question of whom to appoint. However,

---

9   [1880] 5 QB 149.

in practice, the Architect may vet sub-contractors by simply withholding his consent until he is satisfied with the identity of the organization or person to whom the work will be sub-let.

There is anecdotal evidence of architects demanding, as a condition of giving consent to proposed sub-letting, the provision of assignable collateral warranties in favour of employers from the sub-contractors. Such practice is open to attack on the basis that withholding consent on grounds of failure to provide the desired warranties is unreasonable. This risk may be reduced by amending JCT98 to give the Architect such powers. It is to be noted that inclusion of such an amendment in the Contract Bills may fall foul of Clause 2.2.1.[10]

The second method of appointing a Domestic Sub-contractor is provided for by Clause 19.3. It requires the annexation of lists of sub-contractors to work packages defined in the Contract Bills. Although each work package is priced by the Contractor, it is to be carried out by the Contractor's choice from the list annexed to the particular work package. The list must contain a minimum of three sub-contractors but, subject to agreement between them, the Employer or the Contractor may add other names to the list before the final choice is made. If there are withdrawals from the list resulting in less than three available sub-contractors before the sub-contract is awarded, there are two alternative courses of action:

- The list is expanded to the minimum of three by the Employer or the Contractor subject to their agreement (not to be withheld unreasonably). It is to be noted that here the Architect cannot himself add to the list.
- The work is given to the Contractor who may sub-contract it under Clause 19.2.2 if he so wishes.

For the provisions for specified sub-contractors to work smoothly they must be ready and willing to submit tenders for the sub-contract work packages before the Contractor returns the tender for the main works. Without such sub-contract tenders, the Contractor would be in the untenable position of pricing work without any knowledge of how much he will have to pay the sub-contractor selected at the end of the day. Failing such tenders a prudent contractor would make allowance for this uncertainty in his tender. There is a further problem for the Contractor in that if there are withdrawals from the list, the Contractor bears the costs and schedule consequences of any delays in reaching agreement as to how to proceed.

It is suggested that the Contractor may be able to get round these problems by pricing the work packages on the basis of *bona fide* quotations from competent sub-contractors irrespective of whether or not they are specified in the Contract Bills. He can then seek the agreement of the Employer or the Architect to the inclusion of those sub-contractors not already listed in the Contract Bills. In the event of consent being refused, the likelihood is that the Contractor will succeed in adjudication or arbitration if the reason for seeking their inclusion is failure of listed sub-contractors to submit quotations or withdrawal from the lists.

---

10 See Sections 1.4.3, 1.5.5 and 6.14.2 for discussion of the clause 2.2.1 problem and its implications for amending the Contract.

8.3.3 Domestic Sub-contract conditions

JCT 98 does not require the use of any particular standard form of contract for domestic sub-contracts. However, it specifies certain terms which must be included in all such sub-contracts irrespective of whether or not the sub-contractor is appointed from a list annexed to the Contract Bills. If the Contractor fails to incorporate these terms he would be in breach of contract and thus entitling the Employer to damages.

8.3.3.1 Access for the Architect

Under Clause 11 the Architect is entitled to all access at reasonable times to the Works and workshops and other locations where work is being done for the Contract. The Contractor must ensure that the requirements of Clause 11 are stepped down into sub-contracts. Without such provisions, the Architect would have no rights of access to workshops and the like of domestic sub-contractors.

8.3.3.2 Effect of determination of the Contractor's employment

The sub-contract must provide that the employment of the sub-contractor under the sub-contract is automatically determined upon the determination of the employment of the Contractor under the main contract (Clause 19.4.1). This requirement is calculated to avoid any problems with removing the Domestic Sub-contractors from the site after such determination. It is to be noted that if the Contractor complies with Clause 19.4.1 to the letter, the employment of the Domestic Sub-contractor may not be terminated if the Employer, rather determining under the contract, terminates the contract under the general law for a fundamental or a repudiatory breach.

8.3.3.3 Unfixed materials

Clause 19.4.2 requires that, in respect of unfixed materials and goods brought onto the site by a Domestic Sub-contractor for the purpose of carrying out the sub-contract works, the sub-contract must provide that:

- such materials and goods must not be removed from the site for any purpose other than the carrying out of the Works without the written consent of the Contractor who must not consent unless he obtains the written consent of the Architect as required under Clause 16.1;
- where such materials and goods are included in any Interim Certificate they become the property of the Employer after the amount properly due to the Contractor is discharged by the Employer to the Contractor and the Sub-contractor must not dispute this transfer of title;
- such materials and goods become the property of the Contractor after the Sub-contractor is paid for them by the Contractor.

Under Clauses 30.2.1.2 and 30.2.1.3 the Architect must include the value of unfixed materials, including materials belonging to sub-contractors, in Interim Certificates. Payment for unfixed materials presents certain risks to the Employer, which are explained in Section 15.4. Certain provisions in JCT 98 are designed to minimize these risks. Basically, they are calculated to ensure that materials paid for by the Employer are actually used for the Works. The purpose of 19.4.2 is to step these provisions down into sub-contracts so as to be binding on sub-contractors.

However, it has to be borne in mind that even complete compliance with Clause 19.4 would be ineffective to transfer title to goods and materials to the Contractor or Employer unless the sub-contractor concerned acquired title in them in the first place. For example, where a supply contract reserves titles to the supplier until materials are paid for in full, title in the materials cannot pass to the Contractor or Employer until this condition is complied with. The maxim *nemo dat quod non habet* (Lat. 'you cannot give away what you do not own') catches the essence of the problem.

8.3.3.4 Interest on overdue payment

Clause 19.4.3 requires the sub-contract to provide that the sub-contractor shall be entitled to simple interest if the Contractor fails to meet any payment due under the sub-contract by its final date for payment. The rate of interest is to be 5 per cent over the Base Rate of the Bank of England that was current when the payment became overdue. The sub-contract must further provide that the right to interest is without prejudice to the sub-contractor's statutory right to suspend performance of the sub-contract for non-payment or the contractual right to determine his own employment. This is designed to step down into sub-contracts the corresponding right of the Contractor under Clause 30.1.1.1.

8.3.3.5 Fluctuations under Clauses 38 or 39

Finally, where fluctuations under Clauses 38 or 39 apply to the Contract, the terms of the same fluctuation clause must be incorporated into all domestic sub-contracts (Clauses 38.3 and 39.4).

8.3.3.6 Incorporation of the terms into sub-contracts

Unless the terms required to be incorporated are actually incorporated into a sub-contract they cannot be enforced by or against the sub-contractor concerned because of lack of privity of contract between the Employer and the sub-contractor. In such an event, the Employer's only remedy is to recover any damages from the Contractor for failure to comply with the requirement. However, this remedy is likely to be only academic because, in the circumstances in which the Employer is likely to need it, the Contractor would normally be insolvent. One way of avoiding this problem is for the Architect to make his consent to sub-contracting conditional upon an appropriate sub-contract being entered into and to require the Contractor to terminate the sub-contract if it does not comply.

DOM/1, the standard form specially drafted jointly by representatives of main contractors and sub-contractors for use for domestic sub-contracts under JCT 98, complies with the requirements of Clause 19.4. The easiest way of complying with Clause 19.4 is therefore to use DOM/1. However, it is common knowledge that it is usually heavily amended by main contractors before use. However, as explained above such compliance does not guarantee the Employer complete protection against the risk of *de facto* owners of goods and materials claiming them after they have been paid for through Interim Certificates.

Where an *ad hoc* (special purpose) form is used, it is not uncommon for contractors to seek to comply with the clause by stipulations such as 'the sub-contractors shall carry out the sub-contract works in accordance with the terms of the main contract' or 'the sub-contractor shall be deemed to have notice of all the provisions of the main contract'. Whether such stipulations are effective to incorporate the relevant terms of the main contract into the sub-contract is a matter of the interpretation of the particular sub-contract.

The following two cases illustrate that different outcomes are possible depending upon the interpretation of the contracts in the context of all their surrounding circumstances.

Chandler Bros Ltd v. Boswell:[11] the sub-contract terms were generally similar to those of the main contract. Although the engineer's power under the main contract to instruct the main contractor to remove a sub-contractor was not a term of the sub-contract, one of its recitals required the sub-contractor to carry out the sub-contract works in accordance with the terms of the main contract. The Court of Appeal refused to imply into the sub-contract a term that the contractor is entitled to remove the sub-contractor if so instructed by the Engineer. The fact that the sub-contract, in addition to the general reference to the main contract, repeated provisions on most other matters but omitted to do the same for the term in dispute suggested that it was never intended to be part of the sub-contract.

Sauter Automation Ltd v. Goodman (Mechanical Services) Ltd[12] this concerned ownership of materials and goods brought onto site by a sub-contractor. The sub-contractor's quotation contained a *Romalpa* clause. The Contractor's order in response to the quotation stated the terms of conditions of the main contract (the then current version of the GC/Works/1) a term of which stated that every sub-contract was to contain a term vesting in the Contractor ownership of materials and goods brought onto the site. It was held that the terms of the contractor's order governed the sub-contract and not those of the sub-contractor's quotation and that the reference to the main contract conditions was effective to vest ownership in the Contractor.

8.3.4 Sub-contracting and the chain of liability

Sub-contracting requires the use of at least two contracts. First, there is the contract between the employer and the contractor: the main contract. Second, there is a separate contract between the contractor and the sub-contractor in question: the sub-contract. A particular problem posed by sub-contracting is that very often the sub-contractor undertakes some special duties such as design that are not part of the main contract. It follows from the doctrine of privity of contract that within the framework of the two contracts, the employer has no remedy if the sub-contractor defaults in the performance of these special duties. However, in cases where the sub-contractor is recommended to the contractor by the employer on the strength of promises made to the employer by the sub-contractor, the law may treat the promise as a separate contract between the employer and the Sub-contractor. This type of contract existing alongside a main contract is referred to as a 'collateral warranty'. A classic case which illustrates this principle is that of *Shanklin Pier v. Detel Products Ltd.*[13] A supplier of paint made a statement that his paint would last between 7 and 10 years to the plaintiff employer. On the strength of this statement the employer specified the defendant's paint to his contractor. The paint did not last anywhere

---

11 [1936] 3 All ER 179; in the *Dawber Williamson* litigation the attempt to incorporate these provisions into the sub-contract by a stipulation in the sub-contract that the sub-contractor was deemed to know of the main contract failed to achieve the intended effect.
12 (1986) 34 BLR 84.
13 [1951] 2 All ER 471.

near even a year. It was held that the statement constituted a warranty and the employer was therefore entitled to damages for its breach.

It is now common practice for employers to demand collateral warranties from sub-contractors. In such warranties duties which are not part of the main contract are sometimes imposed on sub-contractors. This is often the case where the sub-contractor provides free design services. In addition, there will be greater opportunities for third parties to contracts to acquire enforceable rights under such contracts when the Contracts (Rights of Third Parties) Bill currently going through Parliament becomes law very shortly.

9

Nominated Sub-contractors

Nominated sub-contracting is peculiar to the UK and certain of its former colonies although its use has declined considerably over the years. The essence of this scheme is that the employer, or more usually his architect/engineer acting on his behalf, selects a sub-contractor who is then appointed by the main contractor. This essence is best captured by Lord Reid's description of the system of nomination in *(T. A.) Bickerton & Son v. North West Metropolitan Regional Hospital Board*[1] as:

> an ingenious method of achieving two objects which at first sight might seem incompatible. The Employer wants to choose who is to do the prime cost work and settle the terms on which it is to be done, and at the same time to avoid the hazards and difficulties which might arise if he entered into a contract with the person whom he has chosen to do the work.

This chapter explains some of the reasons why employers may wish to select the sub-contractor and how the scheme works in contracts incorporating the JCT 98.

9.1 Reasons for nomination

There are a variety of advantages to an employer of this arrangement. They include the following:

1. Some specialist work may have long delivery times. Examples of such work include electrical and mechanical installations and structural steelwork. To meet the employer's programme, it is often necessary to place orders for such work long before the main contractor can be appointed.
2. Many specialist sub-contractors undertake some design which is frequently outside the training, expertise and experience of the traditional design team. As evaluation of bids for such services often involves some trade-off between quality of design (as regards life-cycle costs) and sub-contract price, it is often in the employer's long-term financial interest to control this trade-off.

---

1 [1970] 1 WLR 607, at 611G (hereafter *Bickerton*).

3. In many cases the appropriate design expertise is available only from organizations that provide not only design but also installation. Without nomination, the design risk would have to be allocated to the contractor. Risk allocation theory suggests that a general contractor, because of his lack of control over and understanding of the risk, would price the risk more highly than would a specialist sub-contractor with that type of design expertise.

4. The employer may, for various reasons, e.g. time, cost, quality, and good business relations, prefer a particular sub-contractor to carry out certain parts of his project. This is especially the case with sub-contractors with proprietary systems.

5. Tendering for specialist works with a design element often takes much longer than tendering for the run-of-the mill building trades. The reason for this is that the design element takes up additional time. Where the time between the contract award and construction is limited, a main contractor may not be able to obtain competitive tenders for these works. For such works, tenders can be sought before design is complete, thereby keeping the employer's costs down.

6. As a result of specialization to cope with the increasing complexity of today's projects, specialist sub-contractors are usually able to execute certain parts of projects at a lower cost than general contractors. If nomination were not used, the general contractor would normally sub-contract such parts and take the advantage of the lower cost. However, this benefit goes to the employer where nomination is used.

An option available to an employer is to enter into separate contracts with these specialists but this course of action entails the employer losing the advantage of single point responsibility. It is also a recipe for each separate contractor making claims against the employer founded upon alleged acts and omissions of the other contractors. For example, under Clause 25.4.8.1 of the JCT 98, the Contractor is in principle entitled to extension of time if he is delayed by 'the execution of work not forming part of this Contract by the Employer himself or by persons employed or otherwise engaged by the Employer as referred to in Clause 29 or failure to excecute such work'. Under Clause 26.2.4.1 the Contractor is also entitled to recover direct loss and/or expense arising from the same cause. Under Clause 28.1.3.3, the Contractor is also entitled to determine his own employment if, by reason of the execution or non-execution of the works of a direct contractor, the carrying out of the Works has been suspended for a continuous period exceeding the maximum period of delay indicated in the Appendix. Furthermore, matters relating to public liability insurance become very complicated for, under Clause 20, as between the Contractor and the Employer, the Employer carries the risk of loss or damage caused by his other direct contractors.

9.2 Circumstances permitting nomination

Clause 35.1 defines a Nominated Sub-contractor and describes circumstances under which nomination is allowed. These are:

1. a prime cost sum is entered in the Contract Bills;
2. a sub-contractor is named in the Contract Bills to carry out specified work;
3. a prime cost sum is created by an AI as to the expenditure of a provisional sum;

4. an AI as to the expenditure of a provisional sum names a sub-contractor to carry out the work;
5. a prime cost sum is created by an AI which is in effect a Variation resulting in additional work of a kind for which the Architect is already entitled to nominate a sub-contractor;
6. a sub-contractor is named to carry out work by an AI which is in effect a Variation resulting in additional work for which the Architect is already entitled to nominate a sub-contractor;
7. the Contractor and the Architect, on the behalf of the Employer, agree to use a prime cost sum for part of the contract;
8. the Contractor and the Architect, on behalf of the Employer, agree to the use of a specified sub-contractor to carry out part of the work.

As a general rule applicable to all construction contracts, where an employer, or an architect/engineer on his behalf, reserves to himself the final selection of a sub-contractor to carry out certain work, the contractor has neither the right nor the obligation, in the absence of a separate agreement between the employer and the contractor, to carry out such work himself: *Bickerton*. Part of this principle is stated expressly in Clause 19.5.2.

However, the Conditions anticipate that the Contractor may be able to carry out work reserved for performance by Nominated Sub-contractors. Under Clause 35.2.1, where such work is listed in the Appendix the Contractor may tender for it if the Architect is prepared to consider his tender. Also, if an AI as the expenditure of a provisional sum in the Contract Bills or in a Nominated Sub-contract creates a prime cost sum then the work covered by that prime cost sum is deemed to have been included in the Contract Bills and listed in the Appendix. The Contractor may therefore tender for such work subject to the Architect's willingness to consider his tender. It is stated expressly that the Employer is not thereby obliged to accept the lowest, or any, tender. This is unnecessary because this is the common law position in any case.

If the Contractor's tender is accepted, the contract involved is not to be treated as a Nominated Sub-contract. Rather, it is to be considered a separate contract. Variations to the work are valued in accordance with the principles under Clause 13. However, the equivalent documents in the tender accepted are used in place of the Contract Bills and the Contract Drawings under the Main Contract Conditions.

9.3 Contract structure

Appointment of a Nominated Sub-contractor requires use of the following documents.

Tender NSC/T:	The Standard Form of Nominated Sub-contract Tender, 1998 Edition;
Agreement NSC/A	The Standard Form of Articles of Nominated Sub-contract Agreement between a Contractor and a Nominated Sub-contractor, 1998 Edition;
Conditions NSC/C:	The Standard Conditions of Nominated Sub-contract, 1998 Edition, incorporated by reference into Agreement NSC/A;
Agreement NSC/W:	The Standard Form of Employer/Nominated Sub-contractor Agreement, 1998 Edition;
Nomination NSC/N:	The JCT Standard Form of Nomination Instruction for a Sub-contractor.

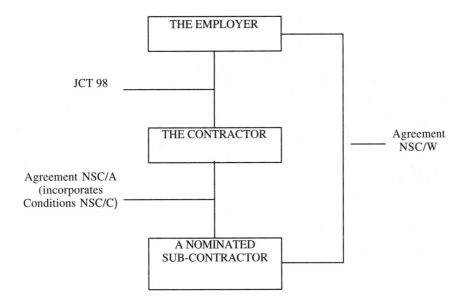

Fig. 9.1 Contractual Arrangements in Nominated Sub-contracting

9.3.1 Tender NSC/T

This is a standard form consisting of three parts:

Part 1: The Architect's Invitation to Tender to a Sub-contractor
This part contains an abstract of the particulars of the Main Contract and the proposed Nominated Sub-contract, which are to be completed by the Architect and included in the Sub-contract tender documents.

Part 2: Tender by a Sub-contractor
This part is a standard form for the Sub-contractor's tender to carry out the Sub-contract Works. Matters to be completed by the Architect or the tendering sub-contractor include the Sub-contract tender sum, earliest dates of commencement and completion, attendances to be provided by the Main Contractor free of charge, and the period for which the tender is to remain open.

Part 3: Particular Conditions
This part contains matters concerning the Sub-contract that must be agreed between the Contractor and the Sub-contractor. This agreement can be reached only after the Contractor has received the Architect's Nomination Instruction. They include details of the programme for carrying out the Sub-contract and for the submission of the Sub-contractor's drawings, special attendances both proposed and agreed, and changes to information given in NSC/T Part 1.

9.3.2 Agreement NSC/W

There are two main sources of risk to the Employer regarding the use of Nominated Sub-contractors. The first source consists of certain duties performed by Nominated Sub-contractors but which are not part of the Contractor's responsibilities under Main Contracts, e.g. design of the Sub-contract works. Such duties represent a break in the chain of liability. The second source of risk is that certain defaults of a Nominated Sub-contractor make the Main Contractor entitled to extra payment and/or extension of time, e.g. delayed completion of the Sub-contract Works. As a form of protection to the Employer against these risks. Agreement NSC/W imposes the following undertakings on the Sub-contractor:

- to attempt to agree the particulars in Part 3 of Tender NSC/T with the Contractor;
- to comply with the obligations of designers under the CDM Regulations to the extent that he is to carry out any design of the Sub-contract Works;
- to exercise reasonable skill and care in any design of the Sub-contract Works or selection of goods and materials for them for which he is responsible;
- to supply the Architect with information so as not to delay the Main Contract Works and thus lead to the Contractor being entitled to extension of time or loss and/or expense;
- not to commit any default which will cause the Architect to instruct the Contractor to determine the Sub-contractor's employment;
- to indemnify the Employer against extra costs due to re-nomination arising from the Sub-contractor's failure to complete the Sub-contract Works;
- to indemnify the Employer against remedial work being required after the Sub-contract final accounts.

If, as a result of a breach of these promises, the Employer suffers a loss, he therefore has a direct recourse in contract against the guilty Sub-contractor. The Employer, on his part, undertakes to:

- pay for pre-contract works or services executed or rendered by the Sub-contractor at the request of the Architect (where the nomination is abortive the payment is to cover only reasonable expenses and is not to include profit);
- make direct payments in accordance with Clause 35.13 of the Main Contract.[2]

The Agreement also contains dispute resolution procedures that dovetail those applicable to the Main Contract.

9.3.3 Agreement NSC/A and Conditions NSC/C

This constitutes the formal contract between the Contractor and the Sub-contractor after its completion and execution by both parties. In a similar arrangement as in the Main Contract Articles of Agreement, this agreement consists of recitals, articles, an attestation and a list of documents incorporated into the Sub-contract. Similarly, it incorporates the terms of Conditions NSC/C, which, in most respects, mirrors the JCT 98 Conditions with

---

2   See Section 9.6

the Contractor and the Nominated Sub-contractor being in a relationship similar to that between the Employer and the Contractor.

9.4 Nomination procedure

The main steps involved are as follows (some of the steps are not mentioned in the Conditions but must be inferred):

1. The Architect draws up a list of prospective sub-contract tenderers.
2. The Architect carries out a preliminary enquiry as to whether those on the list are able to tender for the sub-contract.
3. The Architect completes NSC/T Part 1 and the first page of Agreement NSC/W.
4. The Architect dispatches tender documents to each tendering Sub-contractor. The tender documents consist of:
    - a copy each of the completed NSC/T and NSC/W,
    - drawings/specification/bills of quantities/the Planning Supervisor's Health and Safety Plan/the Principal Constractor's Health and Safety Plan as applicable;[3]
    - the Appendix to the Main Contract completed to the extent possible at that stage.
5. Each tendering sub-contractor:
    - prepares his tender;
    - completes Part 2 of NSC/T to reflect his tender and signs all copies;
    - executes NSC/W and returns it to the Architect.
6. Upon receipt of complete sets of Tender NS/Ts and Agreement NS/Ws from tenderers, the Architect:
    - assesses the tenders and selects a sub-contractor;
    - arranges for the Employer to execute Agreement NSC/W and to sign NSC/T Part 2 of the successful tender as approved or signs it as approved on behalf of the Employer[4]
    - sends a certified copy of the executed NSC/W to the sub-contractor.
7. The Architect issues Nomination NSC/N to the Contractor. This is in effect an instruction to the Contractor to engage the sub-contractor. The instruction should be accompanied with copies of:
    - the completed NSC/T Parts 1 and 2,
    - the numbered tender documents,
    - the executed Agreement NSC/W,
    - the Principal Contractor's Construction Health and Safety Plan.
8. The Architect sends a copy of his nomination instruction to the sub-contractor together with copies of:
    - the executed Agreement NSC/W,
    - the completed Appendix to the Main Contract.

---

3   These are the numbered tender documents which, after execution of the Sub-contract, become the Numbered Documents.
4   The Sub-Contractor and the Employer are bound by Agreement NSC/W from this point even if at the end of the day the Nominated sub-contract is not entered into. For example, the Employer must pay for any pre-contract work beyond preparation of the tender. Ideally the extent of such work must be agreed beforehand.

If this is the first time the sub-contractor is informed of the identity of the Main Contractor he can withdraw his tender without having to give any reasons. This will often be the case where the sub-contractor, for reasons already explained earlier, is selected before the award of the Main Contract. If the sub-contractor withdraws the Architect is to re-nominate or issue a Variation instruction omitting the work involved.

9. On receipt of the nomination instruction, the Contractor must notify the Architect in writing of any objections to the sub-contractor at the earliest practicable moment and in any case, within 7 working days from the receipt.[5]

10. If the Contractor has no objections and the sub-contractor has not withdrawn, the Contractor is to:

- negotiate the matters in NSC/T Part 3 with the sub-contractor,
- complete it and have it signed by or on behalf of himself and the sub-contractor,
- enter into Agreement NSC/A with the sub-contractor.

9.5 Some nomination problems

Failure to complete nomination within a reasonable time would entitle the Contractor to extension of time and recovery of any consequent direct loss and/or expense for not having received in due time necessary instructions (Clauses 25.4.6 and 26.2.1). These entitlements are subject to Contractor's compliance with the procedures and requirements of Clauses 25 and 26. Problems that may delay the issue of the nomination instruction include:

- objections of the Contractor to the nomination;
- delay in reaching agreement between the Contractor and the proposed sub-contractor;
- failure of the Contractor and the sub-contractor to enter into the sub-contract.

9.5.1 Objections to a nomination

Under Clause 35.5.1 the Contractor has a right to raise reasonable objections to any nomination. There are no hard and fast rules as to whether or not an objection is reasonable. However, failure of the sub-contractor to satisfy the following criteria will often render an objection reasonable if raised on those issues:

- compatibility of the sub-contractor's programme with the Contractor's master programme;[6]
- the financial and organizational capability of the sub-contractor;
- capacity, e.g. a capable sub-contractor may well be overloaded already with work;
- financial stability;
- previous experience;
- unwillingness of the sub-contractor to enter into a suitable Sub-contract.

---

5 See Section 9.5.1.
6 *Fairclough Building Ltd v. Rhuddlan Borough Council* (1985) 30 BLR 26 (hereafter *Fairclough v. Rhuddlan*).

Where the nomination is to replace a previously Nominated Sub-contractor, the Contractor's objection would be reasonable if the ground for it is that the proposed replacement sub-contractor will not accept an obligation to carry out the remainder of the sub-contract to a programme that allows the Contractor to complete the Main Contract Works on time.

> *Percy Bilton v. Greater London Council;*[7] Bilton entered into a contract to construct dwellings for the GLC. The form of contract was substantially the same as the then current version of the JCT 63. Lowdells were Nominated Sub-contractors for mechanical services whose employment was determined when they went into liquidation, causing delays for which there was no entitlement to extension of time under the contract. Bilton argued that failure of the Sub-contractor to complete was the Employer's fault or breach of the Main Contract and that, as there was no provision for extending time, time for completion had become at large. The Court of Appeal decided that failure of a Nominated Sub-contractor was neither the Employer's fault nor his breach of contract. It was however pointed out that the Contractor had some protection against this problem which he could have used. This was that the Contractor had a right to refuse the re-nomination unless the new sub-contractor undertook to carry out the remainder of the Sub-contract Works to a schedule compatible with the Contractor's obligation to complete by the Main Contract's completion date or there is a separate agreement with the Employer on a later completion date for the Main Contract.[8]

The Contractor is to make his objection in writing and at the earliest practicable moment. This is subject to a maximum of 7 working days from receipt of the nomination instruction. Considering that the Contractor may have to conduct some investigation into the proposed nominee and negotiate the terms of the sub-contract, the timetable for raising objection is very demanding and the Contractor must therefore act very fast. In raising objections to a nomination, the Contractor must be careful not to leave himself open to charges of defamation.

If the Architect does not accept the Contractor's grounds as reasonable, the Contractor can contest the validity of the nomination instruction in adjudication on a question of whether or not the issue of an instruction is empowered by the Conditions. It has been suggested that, in many cases, the Contractor's right to object to a nomination may be only academic. The unrealistic time limits aside, it is argued that all too frequently the Contractor hardly possesses sufficient knowledge of the nominee or of the nature of the specialist work to support any objections.

9.5.2 Delays and failure to reach agreement

A common problem in the nomination process is that the proposed sub-contractor and the Contractor are unable to agree on the matters in NSC/T Part 3, e.g. schedule details. By Clause 35.8, the Contractor is to use his best endeavours to reach agreement with the sub-contractor. If, despite his endeavours, he is unable within 10 working days from receipt of the nomination instruction to enter into the Nominated Sub-contract, the Contractor is to

---

7   (1981) 17 BLR (CA); (1982) 20 BLR 1 (HL) (hereafter *Percy Bilton*).
8   These *dicta* were applied by the Court of Appeal in *Fairclough v. Rhuddlan* which arose from similar facts.

give written notice of his inability to the Architect. In the notice the Contractor must either indicate when he expects to enter into the sub-contract or give reasons for the non-compliance with the Nomination Instruction. Presumably, the reasons are to be given where there is such a gulf between positions of the Contractor and the nominee that it is not possible to put forward a date regarding the execution of the sub-contract. A footnote to the clause gives the following examples of possible reasons for non-compliance:

- any discrepancy in, or divergence between, the numbered tender documents;
- any divergence between the numbered tender documents and the Contract Drawings, Contract Bills, AIs, or any further drawing issued by the Architect;
- any reason given to the Contractor by the sub-contractor for not being prepared to take the steps necessary to complete the nomination procedures;
- difference between the Main Contract particulars sent with the sub-contract tender documents and those submitted with the copy of the nomination instruction;
- changes in the following matters which occur between the time of submission of the Sub-contract tender and time of nomination:
  - (i)   the particulars of the Main Contract,
  - (ii)   obligations and restrictions to be imposed by the Employer,
  - (iii) the order of carrying out of the Works,
  - (iv) the type and location of access to the site.

On receipt of the notice, the Architect has several courses of action depending upon the contents of the notice. Where the notice indicates that agreement is very likely, the Architect can fix a later date by which the Nominated Sub-contract must be executed (Clause 35.9.1). It is doubtful whether the Contractor is entitled to extension of time for this category of delay in executing the sub-contract. If the Architect does not consider the delay justified by the reasons given by the Contractor, the Architect is to inform the Contractor in writing of his opinion and require him to get on with the execution of the sub-contract. However, where the delay is justified, the Architect may take any of the following actions depending upon the circumstances (Clause 35.9.2):

- issue further instructions to get over the impasse;
- withdraw the nomination and re-nominate another sub-contractor;
- withdraw the nomination and issue a Variation instruction omitting the work involved.

A copy of the AI conveying any of these courses of action is to be sent to the sub-contractor.

It may be that the reason for the delay is that the sub-contractor will not accept to undertake towards the Contractor obligations that the Contractor undertakes towards the Employer under the Main Contract. If the Architect still insists on the engagement of that sub-contractor, the provisions of Clause 35.22 are to be noted: the Contractor's liability to the Employer is reduced correspondingly.

9.6 Payment to Nominated Sub-contractors

Clause 30.2 of the Main Contract Conditions requires the Architect to include in Interim Certificates amounts in respect of Nominated Sub-contract Works determined in accordance with Conditions NSC/C. Under Clause 30.1.1.1, the final date for payment pursuant to an Interim Certificate is 14 days from its issue. This obligation is subject to rights of set-off in favour of the Employer allowed in situations specified in the Contract.[9]

In respect of each Interim Certificate, the Architect is to indicate the amounts included in it as interim or final payment to Nominated Sub-contractors (Clause 35.13.1.1). The Architect must also inform each Nominated Sub-contractor of the amount due him in the Certificate (Clause 35.13.1.2). The RICS form 'Statement of Retention and of Nominated Sub-contractor Values' is normally completed for this purpose.

Clause 35.13.2 further requires the Contractor to pay each Nominated Sub-contractor pursuant to an Interim Certificate by the final date for payment under Conditions NSC/C. Clause 4.16.1.1 of NSC/C states this as 17 days from the issue of the relevant Interim Certificate. It is therefore envisaged that the Employer would have paid the Contractor before he is obliged to pay his Nominated Sub-contractors. However, payment by the Employer is not a precondition for payment to a Nominated Sub-contractor. The Contractor is obliged to pay on the Interim Certificate within the 17 days regardless of whether or not he has meanwhile been paid by the Employer.

There are three sanctions against the Contractor's failure to pay Nominated Sub-contractors. First, under Clause 4.16.4 of NSC/C, the Nominated Sub-contractor is entitled to simple interest on the amount not paid on time at a rate of 5 per cent above the Base Rate of the Bank of England current when the payment became overdue for the period of non-payment. Second, exercise of this entitlement is without prejudice to the Sub-contractor's right, under Clause 4.21, to suspend the carrying out of the Sub-contract Works on account of such non-payment.[10] The third, and most useful sanction from the Nominated Sub-contractor's standpoint, is the right to direct payment from the Employer provided for in Clause 35.13.3. The direct payment mechanism works as follows. Clause 35.13.2 requires the Architect, before issuing any Interim Certificate or the Final Certificate, to demand reasonable proof, e.g. receipts and bank statements, from the Contractor that monies included in previous Interim Certificates have been discharged to the Sub-contractors concerned. Obviously, this step is not necessary with the first Interim Certificate. If the Contractor fails to provide proof of payment, by JCT 98 Clause 35.13.5.1, the Architect is to issue a certificate to that effect specifying the amount still to be discharged to the Sub-contractor concerned. A copy of the certificate is to be issued to the Nominated Sub-contractor concerned. Sometimes, the sole reason for any inability on the part of the Contractor to provide the proof of payment is that the Sub-contractor concerned failed to provide the necessary documentation, e.g. receipts. The Architect should normally contact the Sub-contractor concerned to verify the position. If from his investigations, he reaches the conclusion that the Sub-contractor's failure to provide the necessary evidence of payment is the cause of the problem, he is to treat the Contractor as having provided the proof and not to issue the certificate of non-payment (Clause 35.13.4).

Upon receipt of a certificate of non-payment the Employer is obliged under both the Main Contract and Agreement NSC/W to reduce the amounts of future payments otherwise due to the Contractor by the amount of his non-payment (Clause 35.13.5.2). The Employer can then directly pass on such monies to the Sub-contractors concerned. The operation of the direct payment mechanism is subject to a number of qualifications:

---

9   See Sections 15.1.5 and 15.16 for details on set-off by the Employer against assessment to the Contractor.
10 See Section 15.5.1 for comment on the effect of the Late Payment of Commercial Debts (Interest) Act 1998.

- The Employer is not obliged to pay a Nominated Sub-contractor amounts in excess of the amount available for reduction (Clause 35.13.5.2).
- The Employer must first deduct any amounts due him under the Main Contract, e.g. liquidated damages and other set-offs allowed under it (Clause 35.13.2).[11]
- The set-off and direct payment to the Nominated Sub-contractor is to be made at the same time that the Employer pays the Contractor the balance; if there is no balance due, it is to be made within 14 days of issue of the Interim Certificate (Clause 35.13.5.3.1).
- Where the amount in an Interim Certificate is in respect of the Contractor's Retention the amount of set-off is not to exceed the Contractor's own retention (i.e. the Employer must not set-off against retentions of other Nominated Sub-contractors) (Clause 35.13.5.3.2).
- Where the Employer has to set-off in favour of two or more Nominated Sub-contractors and the amount available to set-off against is insufficient for all of them, the Sub-contractors are to be paid *pro rata* to their respective amounts of non-payment. However, the Employer may adopt any other method of apportionment that is fair and reasonable in the circumstances (Clause 35.13.5.3.3).
- The Employer must not operate the direct payment mechanism if at the time when payment would otherwise be due either a petition has been presented to a court to wind up the Contractor or a resolution has been properly passed to do so for reasons other than amalgamation or reconstruction (Clause 35.13.5.3.4).

Under Agreement NSC/W, the Employer undertakes to pay the Sub-contractor for work and services performed or provided prior to the nomination. Clause 35.13.6 provides that, after the execution of the Nominated Sub-contract, any such payment for work or services included in the Sub-contract Sum is to be deducted from Interim Certificates under the Main Contract which contain amounts in favour of the Nominated Sub-contractor concerned (Clause 35.13.6). The amount of any such deduction should not exceed the sum stated by the Architect as the Sub-contractor's.

When, in the opinion of the Architect, a Nominated Sub-contractor has not only achieved practical completion of the Nominated Sub-contract Works but has also complied sufficiently with his obligations under Clause 5E.5 of Conditions NSC/C in respect of CDM, the Architect is required to issue a Certificate of Practical Completion of the Nominated Sub-contract Works. At any time thereafter, the Architect may include in an Interim Certificate the final accounts of the Sub-contractor concerned. Upon expiry of 12 months from the date of the Certificate of Practical Completion of the Sub-contract, it is mandatory to include the final accounts in Interim Certificates. However, this duty is subject to compliance by the Sub-contractor with two obligations: (i) completion of making good defects which are his responsibility under the Sub-contract; and (ii) submission of all documents necessary for the preparation of the final accounts to the Architect through the Contractor. It is also stated in Clause 30.7 that an Interim Certificate containing the final accounts of all Nominated Sub-contractors must be issued at least 28 days before the issue of the Final Certificate. The aim of this provision is to allow the Architect to take all the steps which are conditions precedent to the Employer being entitled to make direct payments against the Final Certificate.

---

11 See Sections 15.1.5 and 15.1.6 for details on set-off by the Employer against payment to the Contractor.

9.7 Contractor's set-off under the Sub-contract

Clause 35.13.2 provides that the amount certified in favour of a Nominated Sub-contractor shall be paid by the Contractor to the Nominated Sub-contractor in accordance with Conditions NSC/C. The stipulation that the payment must be in accordance with NSC/C implies that the Contractor is entitled to withhold or deduct sums from the amount stated by the Architect as payable where such withholding of deduction is authorized under Conditions NSC/C. As between the Contractor and a Nominated Sub-contractor, Clause 4.16.1.2 of Conditions NSC/C expressly recognizes that the Contractor may be entitled to withhold or deduct amounts from sums stated by the Architect as payable to a Nominated Sub-contractor. Sums for which there is express authority to withhold or deduct are:

- the Contractor's loss or damage caused by failure of the Sub-contractor to complete his work in accordance with the contractual programme of the Sub-contract (NSC/C Clause 2.9);
- the costs incurred by the Contractor to give effect to his directions issued to the Sub-contractor but with which he has failed to comply (NSC/C Clause 3.10);
- deductions made by the Employer under the Main Contract caused by any act, omission, or default of the Sub-contractor (NSC/C Clause 4.16.2.2);
- costs of disruption on the Main Contract resulting from any act, omission or default of the Sub-contractor (NSC/C Clause 4.40);
- expenses incurred by the Contractor in effecting insurances that the Sub-contractor has failed take out (NSC/C Clause 6.8);
- the Contractor's direct loss and/or expense caused by determination of the employment of the Sub-contractor for his breach (NSC/C Clause 7.5).

The procedure and timetable for set-off are summarized in Fig. 9.2. Within 5 days of the date of issue of an Interim Certificate the Contractor must give written notice to each Nominated Sub-contractor to whom money is stated by the Architect to be payable (NSC/C Clause 4.16.1.1). In respect of the amount stated as payable, referred to as 'Amount A', the notice must explain what it relates to and how it was calculated. There is a similar requirement in respect of any other amount due (referred to as 'Amount B') to the Sub-contractor under the Sub-contract.[12] The Employer has a corresponding duty to the Contractor in respect of the amount stated in Interim Certificates.[13] The Contractor needs the Employer's statement to extract the relevant particulars relating to Nominated Sub-contractors. It is therefore unfortunate that the 5 days within which the Contractor must do this runs from the date of issue of the Relevant Interim Certificate rather than the date of receipt of the Employer's statement.

Where the Contractor intends to withhold or deduct from the amount stated by the Architect as payable to a Nominated Sub-contractor, the Contractor must serve notice of his intention not later than 5 days before the final date for payment (Clause 4.16.1.2 of Conditions NSC/C). The notice must specify the amount to be withheld or deducted and the grounds for the withholding or deduction. There is a similar requirement for notice of intention to withhold or deduct from Amount B.

---

12 For example, the Sub-contractor's claims for loss and/or expense under Clause 4.39 of Conditions NSC/C.
13 See Section 15.1.4.

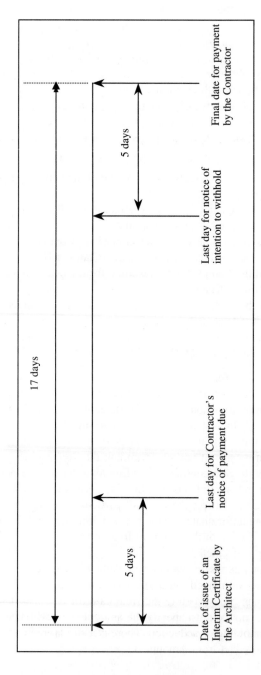

Fig. 9.2 Payment to Nominated Sub-contractors

9.8 Contractor's responsibility for a Nominated Sub-contractor's work

The JCT 98 expressly states the extent of the Contractor's responsibility for the work of a Nominated Sub-contractor. The provisions accord with principles developed under the general law.[14] The obligation of the Contractor under Clause 2.1 to carry out and complete the Works covers all work regardless of whether or not parts are to be executed by Nominated Sub-contractors. This is amplified in Clause 19.5.1 which provides that, unless otherwise stated in the Conditions, the Contractor is responsible to the Employer for the carrying out and completion of Nominated Sub-contract Works as if they were the Contractor's own direct work. Under the Main Contract, the Contractor is not responsible to the Employer for any of the following:

- design of any part of Nominated Sub-contract Works by a Nominated Sub-contractor (Clause 35.21.1);
- selection of kinds of materials and goods for Nominated Sub-contract Works by a Nominated Sub-contractor (Clause 35.21.2);[15]
- compliance with performance specifications or requirement where they are imposed on the Sub-contractor under Agreement NSC/W (Clause 35.21.3);
- a Nominated Sub-contractor's failure to supply information that the Architect needs to supply to the Contractor in accordance with JCT 98 Clauses 5.4.1 and 5.4.2 (Clause 35.21.4);
- liability excluded by the Sub-contractor under the Nominated Sub-contract.

These exclusions represent a major risk to the Employer. The main function of Agreement NSC/W is to protect the Employer against this risk. As explained in Section 9.3.2, it creates a contractual relationship between the Employer and the Nominated Sub-contractor in which the latter undertakes to indemnify the Employer against his losses arising from those matters.

9.9 Determination of employment of a Nominated Sub-contractor

Determination of the Nominated Sub-contractor's employment may be by the Contractor, the Sub-contractor himself or as an automatic consequence of the determination of the Contractor's employment under the Main Contract.

9.9.1 Determination by the Contractor

There are three main grounds upon which the Contractor can determine the employment of a Nominated Sub-contractor: (i) that the Sub-contractor has committed any of 'the

---

14 *Young & Marten Ltd v. Mcmanus Childs Ltd* (1969) 9 BLR 77 (hereafter *Young & Marten*), *Gloucestershire C. C. v. Richardson* (1968) 2 All ER 1181.
15 The Contractor would still be liable to the Employer for poor workmanship and defects attributable to materials and goods being of poor quality: *Young & Marten*.

specified defaults' listed in Clause 7.1.1 of the Nominated Sub-contract (NSC/C); (ii) insolvency; (iii) corruption in relation to this Sub-contract or any sub-contract under any other contract between the same Contractor and the Employer. The specified defaults are equivalent to the grounds upon which the Employer may determine the employment of the Contractor under Clause 27.2.1. By Clause 35.25 the Contractor must not determine on any of these grounds without a written instruction from the Architect to do so. The determination procedure specified in Conditions NSC/C is summarized in Figure 9.3. The provisions for determination of the Sub-contractor's employment for insolvency are also essentially step down versions of similar provisions under Clause 27.4 of the Main Contract. Where determination for insolvency is optional, the Contractor must not determine without the written consent of the Architect (Clause 35.25). Where the Employer requires the Contractor to determine the Sub-contractor's employment because of corruption, the Contractor must comply. The meaning of corruption in this context is explained in Section 16.7 of this book.

It is a common requirement in all the determination procedures that the Architect is kept informed of all relevant developments. He is required to be served with copies of all relevant communication between the Contractor and the Sub-contractor. There are two main reasons for this. First, the Architect can prepare to re-nominate where necessary so as not hold up the Contractor's progress. Secondly, the Contractor can be stopped from determining the Sub-contractor's employment without justification.

9.9.2 Determination by a Nominated Sub-contractor

Under Clause 7.7.1 of NSC/C, a Nominated Sub-contractor has a right to determine his own employment on either of two grounds: (i) that the Contractor has wholly or substantially suspended the Works before completion without reasonable cause; (ii) that the Contractor has failed without reasonable cause to proceed with the Works, thereby seriously affecting the progress of the Nominated Sub-contract Works. This right is expressed to be without prejudice to other rights of the Sub-contractor. This has the effect that the Sub-contractor is also entitled to determine his own employment under the common law.[16]

The procedure for determination by the Sub-contractor under the Sub-contract mirrors that applicable where it is the Contractor who is determining his employment under the Main Contract. However, the Sub-contractor is not required to obtain the instruction of the Architect before determining. Neither is there a duty to inform him after the event. He is only required to give to the Contractor written notice of the default in question. If the Contractor continues the default for 14 days upon receipt of the notice, the Sub-contractor acquires a right to determine which must be exercised within 10 days from the expiry of the 14 days.

The Contractor bears the cost and schedule consequences of any valid determination by the Sub-contractor. It follows therefore that the Contractor must challenge any attempt at determination without justifiable cause. The Contractor can then treat the attempt as repudiation by the Sub-contractor and determine his employment if he stops working. Although none of the provisions expressly state that the Architect is to act as an umpire of

---

16 See Sections 16.1 and 16.2 for explanation of the distinction between determination under the contract and under the common law.

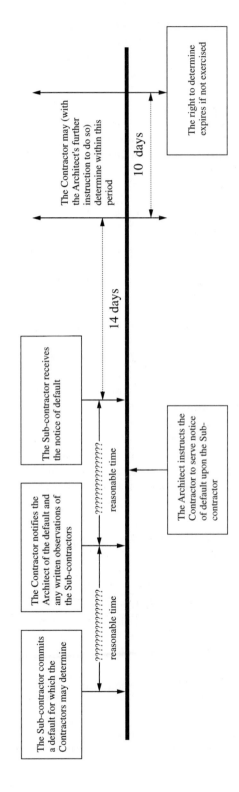

Fig. 9.3 Procedure for Determining the Employment of a Nominated Sub-contractor

whether or not any determination by the Sub-contractor is valid, in practice, this is the most likely position. The reason for this assertion is that, in deciding whether or not to instruct the Contractor to determine, the Architect would first have to consider the validity of the determination by the Sub-contractor.

9.9.3 Automatic determination

There is to be automatic determination of the employment of every Nominated Sub-contractor upon the determination of the employment of the Contractor under the Main Contract (Clauses 7.10 and 7.11 of NSC/C).

9.10 Re-nomination

As already discussed, where a proposed nomination does not result in the formation of a sub-contract the Architect may nominate another sub-contractor. However, re-nomination is mandatory in any of the following circumstances unless the relevant work is omitted by a variation:

1. the Contractor, on orders from the Employer to do so, determines the employment of the Sub-contractor for corruption (Re-nomination under Clause 35.24.7.4);
2. determination for insolvency of the Sub-contractor (Re-nomination under Clause 35.24.7.3);
3. the Contractor lawfully determines the employment of the Sub-contractor under Clause 7.1 of Conditions NSC/C (Re-nomination under Clause 35.24.6.3);
4. the Sub-contractor validly determines his own employment under the Sub-contract (Re-nomination under Clause 35.24.8.1);
5. work of re-execution (i.e. having already been properly executed by the Sub-contractor) arising from the exercise by the Architect of his powers under Clauses 7, 8.4, 17.2 and 17.3 and the Sub-contractor is not obliged to, and refuses to carry out the work (Re-nomination under Clause 35.24.8.2).

As a general principle, where re-nomination arises from failure of any Nominated Sub-contractor to complete the Sub-contractor Works, the replacement Sub-contractor is to be responsible for not only the outstanding work but also any remedial works necessary.[17] Also, the re-nomination clauses expressly state' this requirement.

By Clause 35.24.10, all re-nominations are to be made within a reasonable time after the obligation to do so has arisen.[18] It is not clear when the obligation to re-nominate arises. Consider the situation where the employment of a Nominated Sub-contractor is determined with the consent of the Architect. Does the obligation arise when the determination took place or when the Contractor requested a re-nomination?

The cost implications of a re-nomination depend upon the circumstances giving rise to it. For the circumstances under items (1), (2) and (3) the amount payable to the new Sub-

---

17 *Fairclough v. Rhuddlan*; see also *dicta* of Sir David Cairns in *Percy Bilton Ltd v. GLC* (1981) 17 BLR 1.
18 This term was first introduced to reflect the House of Lord's decision in *Percy Bilton*, that such a term would be implied into a contract that does not contain it.

contractor is to be added to the Contract Sum and included in Interim Certificates (JCT 98 Clause 35.24.9). This means that the Employer incurs the extra bill in respect of any remedial work and increased pricing of the remainder of the Sub-contract Works by the replacement Sub-contractor. In theory, the Employer is entitled to recover this extra from the Sub-contractor under NSC/W if this is possible. However, in the end, if the original Sub-contractor does not pay up, it is the Contractor who bears the costs of making good defects in the work carried out by the original Sub-contractor. This is because, under Clause 30.6.2.1, the value of such defective work is to be subtracted from the Contract Sum when preparing the Final Certificate. This is very much in line with the basic principle that the Contractor is responsible to the Employer for the workmanship and materials of Nominated Sub-contractors.

With re-nomination arising under items (4) and (5) the amount properly payable to the new Sub-contractor is to be added to the Contract Sum and included in Interim Certificates (Clause 35.24.9). However, any extra payments to the new Sub-contractor over those which should have been made to the original Sub-contractor are recoverable from the Contractor as set-offs or debt. Presumably, the Architect would be required to advise the Employer of the amount of the extra payment.

9.11 Delays caused by a Nominated Sub-contractor

This subject is covered in Section 11.3.8.

9.12 Extension of time for Nominated Sub-contractors

The period within which the Sub-contractor is to complete the Nominated Sub-contract Works is to be specified in the relevant sub-contract documents. However, under Clause 2.2.1 of the Conditions NSC/C the Sub-contractor is entitled to extension of time on grounds, referred to here as the Sub-contract Relevant Events that mirror the Relevant Events under the Main Contract. The only difference is that the Sub-contract time may be extended if the Sub-contractor is delayed by any 'act, omission, or default of the Contractor or of parties for whom the Contractor is responsible' although it is not referred to in the Sub-contract as a Relevant Event (Clause 2.3.1 of Conditions NSC/C).

In claiming extension, the procedures that the Sub-contractor has to follow are similar to those with which the Contractor has to comply under the Main Contract:

- he must give to the Contractor written notice that delay has actually occurred or is likely to occur;
- he must provide to the Contractor details of the cause of the delay: material circumstances, whether or not the cause of the delay is a Sub-contract Relevant Event, particulars of the effects, estimates of expected delays in completing the Nominated Sub-contract Works;
- revised notices, particulars and estimates as reasonably necessary or required by the Contractor.

The Contractor must inform the Architect of any notice of delay given by the Sub-contractor together with any comments of the Contractor regarding the delay. He is also to forward to the Architect any details submitted by the Sub-contractor. The Contractor is to join the Sub-contractor in applying to the Architect for extension of time. If the Architect reaches the opinion that the cause of the delay is one or more of the Sub-contract Relevant Events or that it is due to 'an act, omission, default' of the Contractor or his agent, the Architect is to estimate a fair extension of time and give consent to that extension. The Architect has 12 weeks from receipt of the notice and all necessary details to make and inform the Contractor of his decision. However, if the time left to the due date of completion of the Sub-contract Works is less than 12 weeks, the decision must be made by the date of completion. The consent is actually given to the Contractor who must then grant that extension of time to the Sub-contractor. If the Architect reaches the view that no extension of time is due, the Contractor and the Sub-contractor must be informed accordingly.

Clause 35.14.1 of the Main Contract conditions expressly forbids the Contractor from granting extensions of time otherwise than in accordance with the Nominated Sub-contract. This implies that the Contractor must always get the consent of the Architect to any extension. In situations where the cause of the delay is not a Relevant Event under the Main Contract, the Contractor may be tempted not to grant extension of time to Nominated Sub-contractors even where the Sub-contractors are properly entitled to extension, e.g. disruption caused by the Contractor himself, or suspension of the Sub-contract Works for non-payment. This is primarily because, under the Main Contract, the Contractor may himself be entitled to extension of time if the Nominated Sub-contractor fails to complete by the due date for completion. By refusing to grant extension, the Contractor puts the Sub-contractor in this position, thereby gaining some benefit. The role of the Architect in overseeing the operation of the extension of time provisions of the Nominated Sub-contract is therefore intended as an impartial judge. However, it is often pointed out that in many cases the Architect has neither a sufficient understanding of the nature of the Nominated Sub-contract Works nor the underlying correspondence between the Contractor and the Sub-contractor to act fairly.

If delays on the Sub-contract Works also cause, or are likely to cause, delay on the Main Contract Works, the Contractor must also comply with the requirement of Clause 25.3.1 of the JCT 98: notice of actual or likely delay, indication of whether the cause of delay is a Relevant Event, estimates of delay to completion and provision of all other necessary information. If the cause of delay is a Relevant Event, the Contractor would himself be entitled to extension of time. However, the extension granted to the Contractor does not have to be of the same duration as the Sub-contract extension. The reason for this is that the effects of the same event on the completion dates can be different.

9.13 Delayed completion of the Nominated Sub-contract Works

The JCT 98 provides for this in Clause 35.15. If, on the due date for completion of the Sub-contractor Works, completion is not achieved the Contractor must notify the Architect, who must issue a certificate to that effect within 2 months of the notification if he is satisfied that all extensions of time to which the Sub-contractor is entitled have been granted. The issue of this certificate is one of the conditions precedent to the Contractor's right to recover liquidated and ascertained damages for delayed completion of the Sub-contract Works.

10

Nominated Suppliers

Generally, the source from which a construction contractor purchases materials or goods for contract works is entirely a matter for him to decide. However, for reasons similar to those for the use of Nominated Sub-contractors, many construction contracts contain provisions whereby the contract administrator selects a supplier of certain materials or goods for the contractor to use. Suppliers engaged in this manner are referred to as 'nominated suppliers'. They do not carry out any work or fixing of materials on site. The aim in this chapter is to explain the JCT 98 provisions on Nominated Suppliers. For simplicity, the word 'materials' is used hereafter to cover materials and/or goods.

10.1 Definition of a nominated supplier

Clause 36.1.1 defines a 'Nominated Supplier' as a supplier nominated by the Architect in any of the following ways to supply materials:

1. A prime cost sum is included in the Contract Bills in respect of the materials and a supplier of the materials is named in either the Contract Bills or an AI.
2. An AI as to the expenditure of a provisional sum included in Contract Bills makes the supply of identified materials the subject of a prime cost sum. The supplier is either named in the same or subsequent AI.
3. An AI as to the expenditure of a provisional sum specifies materials which are available from a sole supplier. In such a case, the Architect is bound to make the supply of those materials the subject of a prime cost sum. The sole supplier is then deemed to be a Nominated Supplier.
4. A variation issued or sanctioned by the Architect in writing requires the use of materials available from a sole supplier. Such materials are to be made the subject of a prime cost sum and the supplier treated as a Nominated Supplier.

It is important to note that a Nominated Supplier is created automatically in any of the above situations even if that was never the intention of the person who prepared the Bill of Quantities priced by the Contractor or of the Architect in issuing the relevant instruction. Advantage should be taken of Clause 36.1.2 where there is an intention to compel the Contractor to use a particular supplier without creating a Nominated Supplier. It provides that as long as the item is not covered by a prime cost sum in the Contract Bills, specifications may include materials available from a sole source or named suppliers in the Contract Bills without creating a Nominated Supplier. It follows therefore, that the Contractor must price the risk associated with the supply of such materials. In this respect, there is a distinction between nomination of sub-contractors and that of suppliers in that, under Clause 35.1, a sub-contractor named in the Contract Bills is to be treated as a Nominated Sub-contractor.

10.2 Contractor's responsibility for the materials of a Nominated Supplier

As explained in Section 2.3.1.1, a main contractor is responsible to the employer for the quality of materials of nominated suppliers unless the particular circumstances of the main contract indicate that, regarding those matters, no reliance had been placed in the skills of the main contractor.[1] However, where the circumstances of a main contract compel the contractor to accept materials from a particular supplier, it cannot be said that the employer relied upon the contractor on the matters of quality. The contractor would not therefore be liable for defects in the materials.[2]

10.3 Terms of supply contracts

Whilst the JCT 98 prescribes the form of contract to be used in the engagement of Nominated Sub-contractors, there is no similar constraint with regard to Nominated Suppliers. However, it specifies certain terms that must be incorporated into all supply contracts with Nominated Suppliers. Some of these terms are designed to pass down to the Nominated Supplier the like liability of the Contractor under the main contract. Clause 36.4 protects the Contractor from having foisted upon him a Nominated Supplier who will not accept any of those terms by providing that, unless the Architect and the Contractor agree otherwise, the Architect must not nominate such Supplier. The specified terms concern:

- quality
- defects
- delivery programme
- payment and discounts
- determination of employment of Contractor
- ownership of materials delivered to site
- arbitration
- other terms.

Many commentators point out that most suppliers would find the provisions, particularly those regarding ownership of materials and joint arbitration, totally unacceptable.[3] It is

---

1   *Young & Marten Ltd v. McManus Childs Ltd* (1969) 9 BLR 77.
2   *Gloucestershire C. C. v. Richardson* (1968) 2 All ER 1181.

common practice for many suppliers to send out quotations incorporating their own standard terms of business. This means that the Contractor must pay particular attention to the problem referred to as the 'battle of the forms' if he is to avoid unwittingly entering into the contract of sale on the supplier's own terms. The problem was highlighted in *Butler Machine Tool Co. v. Ex-Cell-O Corporation (England) Ltd*:[4]

> The plaintiff (seller) offered to deliver machinery to defendant on the plaintiff's standard terms, which included a price variation clause. The defendant (buyer) replied with an order, which was on its own standard terms, which contained no variation of price clause. The order had a 'tear-off' acknowledgement slip, which was signed and returned. The defendant then failed to take delivery early, which was one of the conditions entitling the plaintiff to an increase of price. The Court of Appeal held that defendant's order was a counter-offer because the terms differed in substance from the seller's terms. The plaintiff's acknowledgement was deemed an acceptance of the counter-offer. The plaintiff's variation clause did not therefore form part of the contract.

10.3.1 Quality (Clause 36.4.1)

The supply contract must provide that: (i) the Nominated Supplier undertakes that the materials supplied shall be of the quality and standard specified; (ii) to the extent that matters of quality and standards of materials are left to the opinion of the Architect, they are to be to his reasonable satisfaction. Those terms are intended to pass down to the Supplier the Contractor's responsibility under Clause 2.1 for quality and standards.

10.3.2 Defects (Clause 36.4.2)

The Nominated Supplier is responsible for defects in materials which appear before the expiry of the Defects Liability Period. The Supplier must therefore make good the defects or replace the defective materials depending upon the nature of the defects. Furthermore, the Nominated Supplier must indemnify the Contractor against any costs reasonably incurred as a direct consequence of such defects. This responsibility is qualified in two ways. First, after the materials have been fixed, the Supplier is not responsible for defects that should have been revealed by reasonable examination by the Contractor. The onus is therefore on the Contractor, before using any material, to ensure that there are no reasonably discoverable defects. Second, the Supplier is not responsible for defects due to workmanship, or poor storage, misuse or neglect by parties other than the Supplier.

Under the general law, a supplier would be liable under the supply contract for the normal limitation period in contract, i.e. 6 years for simple contracts or 12 years for deeds. It is possible, with very careful drafting, to produce a clause complying with Clause 36.4.2 without excluding liability after expiry of the Defects Liability Period.

---

3 For example, Parris, J., *Arbitration: Principles and Practice* (Granada, St Albans, 1983), at p. 46.
4 [1979] 1 All ER 965; see also *Sauter Automation Ltd v. Goodman (Mechanical Services) Ltd* (1986) 34 BLR 81; *Chichester Joinery Ltd v. John Mowlem & Co. plc* (1987) 42 BLR 100.

10.3.3 Delivery programme (Clause 36.4.3)

The Supplier must comply with any delivery programme agreed with the Contractor. Such agreement may allow revision of the programme on account of any of the following:

- *force majeure*;
- civil commotion, local combination of workmen, strikes, or lock-out;
- variations and AIs as to the expenditure of provisional sums;
- failure to supply the Architect in due time, information requested by the Nominated Supplier;
- exceptionally adverse weather conditions.

Three key points have to be borne in mind regarding the revision of the delivery programme. First, inclusion of terms allowing such revision is optional only. Second, whilst the Contractor cannot grant extensions of time to Nominated Sub-contractors without the Architect's consent, there is no requirement for the Architect to approve the revision of the delivery programme. Third, the ground 'civil commotion, local combination of workmen, strike or lock-out' is not limited to those 'affecting any trades employed upon the Works or any trades engaged in the preparation, manufacture or transportation of any of the goods required for the Works', a material qualification to a similar ground for extension of time under the main contract (see the Relevant Event under Clause 25.4.4). The Contractor would therefore be best advised to place a similar limitation in the supply contract.

Where there is no agreement on a delivery programme, the Supplier must comply with any reasonable direction of the Contractor. However, except perhaps for materials which are definitely not critical to progress, the Contractor would be ill advised not to have an agreed programme of delivery. The reason for this is that, in the absence of such a programme, it would be very difficult to determine whether or not there has been 'delay on the part of' of a Nominated Supplier for purpose of extension of time for the Relevant Event under Clause 25.4.7.

10.3.4 Payment to Nominated Suppliers (Clause 36.4.4)

The Contractor must discharge payment in respect of materials within 30 days of the end of the month during which they were delivered (Clause 36.4.6). The Nominated Supplier must allow the Contractor a discount of 5 per cent (1/19 of the net amount paid or payable) if the Contractor complies with this obligation (Clause 36.4.4).

It is be noted that there is no provision for a Nominated Supplier to be paid directly by the Employer in the event of the Contractor failing to discharge payment to the Supplier.

10.3.5 Determination of employment of the Contractor (Clause 36.4.5)

The Nominated Supplier's obligation to make delivery of materials terminates upon determination of the employment of the Contractor under the Contractor. The only exception is that, where at the time of determination the materials had been paid for in full, the Supplier would still have to deliver the materials.

10.3.6 Ownership of materials (Clause 36.4.7)

Ownership in materials is to pass to the Contractor upon delivery by the Nominated Suppliers. This means any retention of title clauses in the contract of sale would be a contravention of the JCT 98. The purpose of this provision is to address problems explained in Sections 8.3.3 and 15.4.

10.3.7 Arbitration (Clause 36.4.8)

The effect of a term complying with Clause 36.4.8 is that any arbitration between the Contractor and a Nominated Supplier should be in accordance with the JCT 98 edition of CIMAR current at the Base Date. Also, by virtue of Clause 41B.4, both the Nominated Supplier and the Contractor have consented to appeals to the court either on a question of law arising during arbitration proceedings or a question of law arising from an award.

The primary intention under Clause 36.4.8 is to allow joinder of multiple parties to arbitration proceedings. For reasons given in Section 17.4.3.4, although such a term would help, it does not guarantee the possibility of consolidations of such proceedings.

10.3.8 Other terms (Clause 36.4.9)

A supply contract may incorporate terms additional to those specified. However, it must be a specified term that the other specified terms are to prevail over any other term in the contract. This is probably to frustrate any attempt to avoid Clause 36.4 by incorporating the specified terms, thereby complying with the letter of the clause, but defeating its purpose by means of other terms to the contrary.

10.3.9 Implied terms

Until towards the end of the nineteenth century, the principle of *caveat emptor* (Latin for 'Buyer beware') applied to contracts for the sale of goods. This meant that it was for the seller to satisfy himself as to the quality of the goods and their suitability for his purpose before making the purchase. After the sale, the seller was not answerable for quality matters not expressly provided for in the contract. As a result of the industrial revolution and the complex sale systems that sprang up to distribute its products from the factory to the ultimate consumer, the courts gradually implied terms into sale contracts to provide minimum protection to buyers. Parliament codified these terms into the Sale of Goods Act 1893. Various amendments were made to the 1893 Act in subsequent legislation.[5] To avoid the complexity created by scattering the law across different Acts, the law was consolidated in the Sale of Goods Act 1979 (SGA 1979). SGA 1979 was itself amended by the Sale and Supply of Goods Act 1994.

The sale of goods legislation covers both a 'contract of sale' and an 'agreement to sell'. A contract of sale is defined in s. 2(1) of SGA 1979 as a 'contract by which the seller transfers or agrees to transfer the property in goods to the buyer for a money consideration, called the price'. For our purposes 'property in goods' is the same as ownership of them.

---

5   Misrepresentation Act 1967; Consumer Credit Act 1974; Unfair Contract Terms Act 1977.

An 'agreement to sell' is defined in s. 2(5) as a contract of sale where 'the transfer of the property in the goods is to take place at a future time or subject to some condition later to be fulfilled'. A contract of sale gives the buyer not only rights to damages for breach of the contract but also proprietary remedies. For example, if property was transferred with the sale, the buyer can enforce his rights as an owner against the seller and the world at large. By the same token, the buyer carries the risk of damage to the goods even where they remained in the possession of the seller. With an agreement to sell, the buyer is entitled to only damages if the terms of the agreement are broken whilst the risk of damage remain with the seller. Most contracts between the Contractor and suppliers of materials and goods are subject to the sale of goods legislation. SGA 1979 implies to contracts for the sale of goods terms concerning title to, and quality of, the goods.

Some of the implied terms are conditions whilst others are warranties. For breach of a condition, the buyer may terminate the contract and claim damages for the breach. He may also affirm the contract and claim damages.[6] Damages would include refund of payment already made and cost of improvements made to the goods by the buyer. The implied term on charges and encumbrances is only a warranty. This means that, although the buyer can recover damages, he is not entitled to terminate the contract. There are additional statutory provisions on consequences of a breach of a condition. s. 15A provides:

> (1) where in the case of a contract of sale–
>
> > (a) the buyer would, apart from this subsection, have the right to reject goods by reason of a breach on the part of the seller of a term implied by section 13, 14, or 15, but
> > (b) the breach is so slight that it would be unreasonable for him to reject them.
>
> then, if the buyer does not deal as consumer, the breach is not to be treated as a breach of condition but may be treated as a breach of warranty.
>
> (2) This section applies unless a contrary intention appears in, or is to be implied from, the agreement.
>
> (3) It is for the seller to show that a breach fell within subsection 1(b) above.

10.3.9.1 Title

Under s. 12 of SGA 1979:

> (1) In a contract of sale, . . ., there is an implied [condition] on the part of the seller that in the case of sale, he has a right to sell the goods, and in the case of an agreement to sell, he will have a right to sell the goods at the time when the property is to pass.
>
> (2) In a contract of sale, . . ., there is also an implied [warranty] that–
>
> > (a) the goods are free, and will remain free until the time when the property is to pass, from any charge or encumbrance not disclosed or known to the buyer before the contract is made, and
> > (b) the buyer will enjoy quiet possession of the goods except so far as it may be disturbed by the owner or other person entitled to the benefit of any charge or encumbrance so disclosed or known.

---

6   See section 16.1.1 for explanation of the distinction between a condition and a warranty.

It is to be noted that the implied term as to the seller's right to sell is a condition.

10.3.9.2 Sale by description
S. 13 provides:

(1) Where there is a contract for the sale of goods by description, there is an implied [condition] that goods will correspond with the description.

. . .

(2) If the sale is by sample as well as by description it is not sufficient that the bulk of the goods correspond with the description.

An example of a sale by description is where the buyer places an order for the goods in reliance upon their description in a catalogue or the description is otherwise included as a term of the contract, such as where there is a purchase on the buyer's terms which include a detailed specification. In *Arcos Ltd v. Ronaasen & Son*,[7] in which the dimensions of timber supplied were not as specified, it was stated that 'if the written contract specifies conditions of weight, measurement and the like, those conditions must be complied with. A ton does not mean about a ton, or a yard about a yard'. An extreme application of this condition was in *Re Moore & Co. and Landauer & Co. Ltd*[8] in which the contract required tinned fruit to be in cases of 30 tins per case. The Court of Appeal held that the buyer was entitled to reject a consignment of fruit some cases of which contained 24 tins although the total number of tins was as required in the contract.

10.3.9.3 Satisfactory quality
S. 14(2) states:

(2) Where the seller sell goods in the course of a business, there is an implied [condition] that the goods supplied under the contract are of satisfactory quality.

(2A) For the purposes of this Act, goods are of satisfactory quality if they meet the standard that a reasonable person would regard as satisfactory, taking account of any description of the goods, the price (if relevant) and all the other relevant circumstances.

. . .

(2C) The term implied by subsection (2) above does not extend to any matter making the quality of goods unsatisfactory–

(a) which is specifically drawn to the buyer's attention before the contract is made,
(b) where the buyer examines the goods before the contract is made, which that examination ought to reveal, or
(c) in the case of a contract for sale by sample, which would have been apparent on a reasonable examination of the sample.

---

7 [1933] AC 470.
8 [1921] 2 KB 519.

Most of the relevant case law pre-dates the SGA 1979 and refers to an implied condition that the goods would be of 'merchantable quality'. S. 14(6), which has now been replaced, stated that goods are of merchantable quality 'if they are as fit for purpose or purposes for which goods of that kind are commonly bought as it is reasonable to expect having regard to any description applied to them, the price (if relevant) and all the other relevant circumstances.' Although the concept of 'merchantable quality' has now been replaced by that of 'satisfactory quality', the application of the former in the case law provides some insights into the likely application of the latter. For example, in *Wren v. Holt*[9] it was held that where a person bought and drank a glass of beer in a public house and fell ill as a result of it, the condition implied at common law as to merchantable quality was broken because the normal purpose of beer is human consumption in reasonable quantities without any risk to the life of the drinker. It is important to note from the statutory definition that goods may still be of merchantable quality even if they are not fit for the particular purpose to which the buyer puts them. In *Rotherham Metropolitan Borough Council v. Frank Haslam Milan & Co. Ltd and M. J. Gleeson (Northern) Ltd*[10] which arose from a JCT 63-based contract, part of the Contract Bills stated:

> Granular hardcore shall be well graded or uncrushed gravel, stone, rock fill, crushed concrete or slag or natural sand or a combination of any of these. It shall not contain organic material, material susceptible to spontaneous combustion, material in frozen condition, clays or more than 0.2% of sulphate ions as determined by BS 1377.

Steel slag used in cellars expanded, resulting in heaving of the ground floor and cracking of reinforced concrete slabs. One of the issues in the litigation was whether the steel slag was of merchantable quality. This was a contract for work and materials and SGA 1979 did not therefore apply. Only the implied condition of merchantability at common law therefore applied. However, there was no difference between the statutory and common law obligations because the common law definition of 'merchantable quality' was adopted in the definition in the superseded s. 14(6) of SGA 1979. It was found that the slag was perfectly suitable for use as hardcore in road construction and any other locations where it would be unconfined. The Court of Appeal decided that as the slag was fit for some of the purposes within the description under which it was saleable without price reduction, it was of merchantable quality. The fact that it was not fit for use in confined conditions did not mean lack of merchantable quality. However, a new s. 2B provides that, in appropriate cases, 'fitness for *all* the purposes for which goods of the kind in question are commonly supplied' is an aspect of quality. It is therefore arguable that satisfactory quality imposes a higher standard in such cases.

10.3.9.4 Fitness for purpose
S. 14(3) provides:

> Where the seller sells goods in the course of a business and the buyer, expressly or by implication, makes known–
>
>> (a) to the seller, or

---

9 [1903] 1 KB 610.
10 (1996) 78 BLR 1.

(b) where the purchase price or part of it is payable by instalments and the goods were previously sold by a credit-broker to the seller, to that credit-broker,

any particular purpose for which the goods are being bought, there is an implied [condition] that the goods supplied under the contract are reasonably fit for that purpose, whether or not that is a purpose for which such goods are commonly supplied, except where the circumstances show that the buyer does not rely, or that it is unreasonable for him to rely, on the skill or judgment of the seller or credit-broker.

The ingredients for the implication of this condition are to be noted: (i) the supplier sells the goods in the course of a business; (ii) the contractor, expressly or by implication, made known to the supplier or his agent the particular purpose of the goods. Where the goods are for a unique purpose, e.g. a hot water-bottle,[11] the purpose of the purchase is made known by implication. However, where the goods may be used for several different purposes, the contractor must have stated his particular purpose. Where these two ingredients exist, there is a presumption that the contractor relied upon the skill or judgement of the supplier as to the suitability of the goods for that purpose. However, this presumption may be rebutted, e.g. evidence that the supplier disclaimed special knowledge of the relevant matters or that the contractor is more expert on them. Ordering an article by its brand or trade name may also be evidence of absence of reliance upon the skill of the seller.

10.3.9.5 Sale by sample
Under s. 15, where there is a sale by sample, the sale will be subject to the implied conditions that: (i) the bulk will correspond with the sample in quality; (ii) the buyer will have reasonable opportunity to compare the bulk with the sample regarding quality; (iii) the goods will be free from defects making their quality unsatisfactory, which would not be apparent on reasonable examination of the sample. S. 15 is most relevant to new materials which are often bought on the strength of samples from salespeople. Depending on the contents of any marketing leaflets, the other sections already explained may also apply.

10.3.9.6 Exclusion of terms
S. 6(2) of the Unfair Contract Terms Act 1977 (UCTA) provides that 'As against a person dealing as consumer liability for breach of the obligation arising from- (a) sections 13, 14, and 15 of the 1979 Act . . . cannot be excluded or restricted by reference to any contract term.' Against a person dealing otherwise than as a consumer the liability referred to may be excluded but only in so far as the term satisfies the requirement of reasonableness.[12] S. 12(1) of UCTA defines a consumer in these terms:

(1) A party to a contract 'deals as consumer' in relation to another party if-

(a) he neither makes the contract in the course of a business nor holds himself out as doing so: and
(b) the other party does make the contract in the course of a business; and
(c) in the case of a contract governed by the law of sale of goods . . . the goods passing under or in pursuance of the contract are of a type ordinarily supplied for private use or consumption.

---

11 *Priest v. Last* [1903] 2 KB 148.
12 S. 6(3) of UCTA.

In the vast majority of cases the Contractor under a JCT 98 Contract will be buying goods and materials for the Works in the course of a business. With such supply contracts, exclusions of the terms are subject to the test of reasonableness. In assessing reasonableness, the court is required to have regard to the following guidelines stated in Schedule 2 to UCTA:

    (*a*) the strength of the bargaining positions of the parties relative to each other, taking into account (among other things) alternative means by which the customer's requirements could have been met;

    (*b*) whether the customer received an inducement to agree to the term, or in accepting it had an opportunity of entering into a similar contract with other persons, but without having to accept a similar term;

    (*c*) whether the customer knew or ought reasonably to have known of the existence and extent of the term (having regard, among other things, to any custom of the trade and any previous course of dealing between the parties);

    (*d*) where the term excludes or restricts any relevant liability if some condition is not complied with, whether it was reasonable at the time of the contract to expect that compliance with that condition would be practicable;

    (*e*) whether the goods were manufactured, processed or adapted to the special order of the customer.

10.4 Exclusion of Nominated Supplier's liability

The JCT 98 recognizes that some suppliers have sufficient bargaining power, e.g. sole suppliers, to impose exclusion clauses in supply contracts, e.g. that the supplier does not accept liability for costs of disruption and other consequential losses arising from late delivery, failure to deliver, or discovery of defects. By Clause 36.5.2, the Contractor is entitled to reject any purported Nominated Supplier who imposes exclusion clauses unless the Architect approves the exclusion clauses in writing. It is to be noted that such approval has a sting in its tail for the Employer! The Contractor's liability to the Employer is similarly excluded (Clause 36.5.1). Prudence therefore dictates that the Employer be consulted before approval is given.

10.5 Nominated Suppliers and claims under the Main Contract

There are three main ways in which the Contractor may become entitled to extension of time or recovery of direct loss and/or expense as a result of an event arising from the nomination process or the Nominated Supplier's performance of the contract of sale:

- compliance with Architect's instructions;
- the Contractor not having received in due time necessary instructions in relation to the materials in question;
- 'delay on the part of' a Nominated Supplier.

10.5.1 Architect's Instructions

Under Clause 25.4.5, if compliance with certain types of AIs results in delay, the Contractor may be entitled to extension of time subject to compliance with the requirements of Clause 25.2. Some of these instructions may concern, or arise from contracts with, Nominated Suppliers, e.g. Variations affecting the materials of a Nominated Supplier. Some of these instructions may also give rise to entitlement under Clause 26 to recovery of the Contractor's direct loss and/or expense, e.g.

- an instruction under Clause 8.3 to open or test work, materials, goods and the verdict is that the work, materials, or goods are in accordance with the Contract;
- an instruction under Clause 23.2 to postponing the works or any part of it;
- an instruction under Clause 13.2 requiring a Variation;
- an instructions under Clause 13.3 on the expenditure of provisional sums.

10.5.2 Non-receipt/late receipt of Instructions

Where the supply intended for a Nominated Supplier is covered by a prime cost or provisional sum, the Architect must issue a nomination instruction before the Contractor is obliged to engage the Supplier concerned. If the Contractor is delayed as a result of the Architect's failure to issue the instruction in due time, the Contractor would be entitled to extension of time and his direct loss and/or expense subject to compliance with the stipulations of Clauses 25 and 26.

10.5.3 Delay on the part of a Nominated Supplier

The discussion of the Relevant Event 'delay on the part of' a Nominated Sub-contractor also applies to Nominated Suppliers. However, there is one major difference. With Nominated Sub-contractors, the Architect performs the role of certifier of non-completion of the Nominated Sub-contract Works. Thus, in effect, it is the Architect who decides whether there has been 'delay on the part of' a Nominated Sub-contractor. This role therefore protects the Employer and the Sub-contractor against unwarranted claims by the Contractor that there has been such a delay. Unfortunately, there is no such role with respect to Nominated Suppliers. It follows therefore that meticulous attention has to be paid to scheduling and monitoring of the supply of materials by Nominated Suppliers.

10.6 Re-nomination

Withdrawal of a Nominated Supplier, whether through business failure or repudiation, without completing the supply contract is always a possibility. Unfortunately, the JCT 98 is silent on this eventuality. Two possibilities are canvassed in *Keating on Building Contracts*. First, on the assumption that the principle in *Bickerton v. North West Metropolitan Regional Hospital Board*[13] applies to Nominated Suppliers, the Architect

---

13 [1970] 1 WLR 607.

would have to re-nominate a supplier. Second, it may be argued that, because the JCT 98 expressly provides for the re-nomination of Nominated Sub-contractors, the effect of failure to do the same for Nominated Suppliers is that re-nomination is not required. The same argument may be advanced in respect of the Architect's power to issue a Variation omitting the materials in question. If this argument were valid, then there would be frustration if a sole supplier were involved. It would appear, therefore, that the Architect may re-nominate or issue appropriate Variation instructions.

10.7 Standard documents for Nominated Suppliers

The JCT has published two standard documents for use in connection with Nominated Suppliers: (i) Standard Form of Tender by a Nominated Supplier (Tender TNS/1); (ii) Standard Form Warranty by a Nominated Supplier (Agreement TNS/2). Their use, which is optional, is explained in Practice Note 15.

10.7.1 Tender TNS/1

This is a standard form on which tenders for supply contracts are invited by the Employer, or more usually the Architect, from suppliers. The Contractor is then instructed to accept the successful tender.

The form consists of three parts: (i) Form of Tender, Schedule 1 and Schedule 2. The Form of Tender is the supplier's formal offer to perform the supply contract and must be completed by the tenderers. It is important to note that one of the undertakings made in this offer is that none of the conditions of the supply contract to be entered into with the Contractor will conflict with the requirements of the main contract. Schedule 1 details certain matters that are relevant to the supply, e.g. description of the materials, access to the Works, and particulars of the main contract. Schedule 2 contains a reprint of Clauses 36.3 to 36.5 of the JCT 98, which are the parts most likely to be in conflict with suppliers' terms of business of suppliers. This Schedule therefore ties in very well with the undertaking in the Form of Tender to avoid any such conflict.

10.7.2 Warranty TNS/2

This form serves a function similar to that of Warranty NSC/W in the case of a Nominated Sub-contractor, i.e. to protect the Employer against any break in the chain of liability and to indemnify the Employer against his liability to the Contractor and any other loss caused by the defaults of the Supplier.

10.8 Contractor's right to object to a nomination

Whilst the Contractor is given a general right to raise reasonable objections to the nomination of a Sub-contractor, the right to raise objections to the nomination of a supplier is limited to two situations:

- where the supplier will not accept the terms stipulated in Clause 36.4;
- where the contract of sale contains exclusion or limitation clause not approved by the Architect in writing.

It is therefore possible that the Contractor could be compelled to deal with unsuitable suppliers, e.g. suppliers without necessary financial resources, or suppliers whose delivery programmes do not mesh with the Contractor's own master programme.

10.9 Payment to the Contractor

The amount properly chargeable to the Employer by the Contractor in respect of the materials of a Nominated Supplier is the total amount paid or payable by the Contractor in respect of the materials. Clauses 36.3.1 and 36.3.2 expressly state that this amount is to include:

- cash discounts (the benefit of any other discount has to be passed to the Employer);
- taxes other than VAT and statutory duties payable in respect of the materials which cannot be recovered under any other clause of the Contract;
- net cost of packing, packaging, transporting and delivering the materials less any credit for return of packaging to the Supplier;
- price adjustment other than discounts;
- expenses which, in the opinion of the Architect, have been properly incurred by the Contractor in obtaining the materials from the Supplier which cannot be reimbursed under any other clause of the Contract.

To avoid the risk of the Employer disputing the validity of any amount being charged to him, it is advisable for the Contractor to get the Employer to agree the payment terms, particularly terms regarding price variation, applicable to contracts with Nominated Suppliers.

11

Delays, extension of time and liquidated damages

In most construction contracts, the contractor is expected to complete the contract works by a specified date. This date may be revised under the provisions of the contract in defined situations. A contractual provision which allows such revision is referred to as an 'extension of time' clause. Failure of the contractor to complete by the due date of completion would normally result in liability for damages for breach of contract. A common practice is to include a liquidated damage clause that states the amount payable in case of delayed completion. This chapter explains, first, the legal principles governing delays, extension of time, and liquidated damages relating to contracts in general. The specific provisions of the JCT 98 on these issues are then discussed.

11.1 Concept and application of liquidated damages

According to principles of the law of contract, to succeed in a claim for damages for breach of contract, a claimant must prove to the satisfaction of the court that:

1. the defendant's breach of contract caused losses in the amount claimed;
2. the loss was not too remote at the time of formation of the contract;
3. he took all reasonable steps to mitigate his loss.

It is then the duty of the court to decide the amount in respect of which the claimant has furnished the required proof and to make an award in that amount. Damages so assessed by the courts are referred to as 'unliquidated' damages. This often involves time-consuming and costly litigation. To avoid this difficulty, it has become common practice, particularly with construction contracts, to state expressly in the contract itself the amount that will be payable in the event of its breach. The amounts so stated are referred to as 'liquidated damages' whilst clauses in which they are stated are called 'liquidated damage clauses'.

11.1.1 Concept of liquidated damages

The basic essence of liquidated damages is that they represent a genuine pre-assessment of the likely loss that will flow from the breach of contract in question. This follows from the general principle that the aim of damages is to place the innocent party in the position he would have occupied had the contract been performed without a breach. Both parties then enter into the contract in full awareness of their monetary rights and liabilities in the event of a breach. When the breach occurs, the claimant can then recover his loss from the defendant without time-consuming litigation.

Construction workers often refer to liquidated damages as 'penalties'. This is a misnomer because penalty clauses are void at law and therefore unenforceable through the courts. The claimant will then have to go through the trouble of proving unliquidated damages. When a claim for liquidated damages is made, the defendant can challenge it on the grounds that the liquidated damage clause is in fact a penalty clause and must be ignored. In such an event, the courts apply well-established principles for distinguishing liquidated damages from penalties. These principles were stated by the House of Lord in the famous case of *Dunlop Pneumatic Tyres Co. v. New Garage and Motor Co.*[1] The facts of that were as follows. Dunlop sold tyres to New Garage under a contract that contained terms restricting New Garage as to the prices at which they could retail the tyres. There was an undertaking to pay Dunlop £5 per tyre sold in contravention of the terms. The House of Lords had to decide whether the undertaking was a genuine liquidated damage clause or a penalty. In deciding the case, Lord Dunedin put forward the following propositions which have now received universal acceptance as guiding principles for distinguishing between penalties and liquidated damages:

1. Whether the parties call a payment 'damages' or a 'penalty' is not conclusive. The court must determine whichever it is in truth.
2. The essence of a penalty is a payment of money stipulated *in terrorem* (Lat. to frighten) of the offending party, i.e. its purpose is to strike terror on mere contemplation of the breach; the essence of liquidated damages is a genuine covenanted pre-estimate of likely damage.
3. The question of whether a sum stipulated is a penalty or liquidated damages is one of construction to be decided upon the terms and inherent circumstances of each particular contract judged as at the time of making it and not as at the time of its breach. However, in *Philips Hong Kong v. The Attorney General of Hong Kong*[2] the Privy Council suggested what actually happened at the time of breach may provide evidence as to what could reasonably have been expected to be the likely loss at the time of contract formation.
4. It will be held to be penalty if the sum stipulated is extravagant and unconscionable in comparison with the greatest loss that could conceivably be proved to have flowed from the breach. In the *Philips Hong Kong* case the Privy Council also provided some guidance on the application of this principle. In that case it was held that, to prove that a provision is penal, it will not normally be sufficient to identify a hypothetical situation where the application of the provision could result in a larger sum than the actual loss being recovered.
5. If the breach consists only of the non-payment of money and the amount stipulated is greater than the sum which ought to have been paid, it will be held to be a penalty.
6. Where a single sum is payable on the occurrence of one or more of several events, some of which may cause serious loss and others but trifling loss, there is a

---

1   [1915] AC 79.
2   (1993) 63 BLR 41 (hereafter *Philips Hong Kong*).

presumption that it is a penalty. However, a minimum figure for liquidated damages in contracts with variable liquidated damages is not necessarily penal.[3] It follows from this principle that the practice of stating liquidated damages as an amount 'per week or part thereof' of delay is open to challenge that the amount is a penalty. In the situations contemplated by such a provision, it is therefore better practice to convert the amount into damages per day of delay.

7.   It is no obstacle to the sum stipulated being a genuine pre-estimate of loss that it is impossible to make a precise estimation of the loss consequent upon the breach. This principle is particularly helpful in situations where losses are not easily quantifiable, e.g. churches and public sector projects.

11.1.2 Concept and sectional completion obligation

In construction contracts requiring completion of sections at different specified times, purported liquidated damage clauses have been construed as penalties by the application of the sixth principle cited above. Delay in completing only one section would not be as serious as delayed completion on all the sections. This means that if the liquidated damages are expressed as a single sum for the whole contract, it would be construed as penalty on the grounds that the same sum is payable for both breaches even though they would have different consequences. The same problem arises with some contracts involving the construction of a number of standard units, e.g. housing, because delayed completion of only one unit would not be as damaging as delayed completion of every unit. The facts of *Bramall & Ogden Ltd v. Sheffield City Council*[4] illustrate the application of this principle to contracts involving sectional completion or multiple units:

> This case arose from a contract in the terms of the JCT 63 for the construction of 123 dwellings. Clause 22 provided for the deduction of liquidated damages for failure to complete the entire contract by the completion date. The Appendix stated the liquidated damages payable as £20 per week for each uncompleted house. The contract did not provide for sectional completion. This meant that if any number of dwellings remained uncompleted after the completion date the Employer would have been entitled under the terms of the contract to deduct liquidated damages for the whole contract, i.e. in respect of all the houses (123 x £20/week). Judge Hawsher QC decided that Clause 22 was a penalty because the contract did not allow for the liquidated damages recoverable to be varied with the number of houses uncompleted. It was therefore unenforceable. Interestingly, the Employer's claim for liquidated damages in respect of only the uncompleted houses was also rejected on the grounds that the contract did not provide to that effect. The reasoning of the judge was that the liquidated damage clause was *prima facie* invalid and having been struck down by the invalidity, there was no effective liquidated damage clause. No liquidated damages of any kind were therefore owed.

This case provides a good lesson on the importance of completing contract documents with due care. It also highlights the fact that liquidated damage clauses are construed

---

3   See n. 2.
4   (1983) 29 BLR 73; see also *Stanor Electric Ltd v. R Mansell Ltd* (1988) CILL 399.

strictly *contra proferentem* in case of ambiguity.[5] There is now a *Sectional Completion Supplement* for use with the JCT 98.[6]

11.1.3 Application of liquidated damages

Once the breach covered by a liquidated damage clause occurs, and provided the clause is neither invalidated by the Unfair Contract Terms Act 1977 nor considered a penalty, the only damages payable are the amounts stated in the clause. If the innocent party suffers no actual loss or even benefits from the breach, the damages would still be payable. This outcome follows from the principle that liquidated damages payable are assessed as at the time of the making of the contract rather than at the time of its breach.

> *Clydebank Engineering and Shipbuilding Co. Ltd v. Don Jose Ramos Yzquierdo y Castaneda*:[7] Clydebank contracted with the Spanish Government to supply them with four torpedo boats. Liquidated damages were set at £500 per week of delay. Delivery of the boats was delayed and the Spanish Government claimed liquidated damages for the period of the delay. Clydebank argued that, by delaying, they had done the Spanish Government a favour because soon after the due delivery date, the American fleet had sunk the greater part of the Spanish fleet. The House of Lords rejected this argument and upheld the claim for damages.

Where liquidated damages are payable, they represent the total remedy available to the innocent party. Whether his actual losses exceed the amount of the liquidated damages is irrelevant.

> *Cellulose Acetate Silk Co. Ltd v. Widnes Foundry Ltd*:[8] the contract involved the construction and delivery of a chemical plant. Liquidated damages were set in the contract at £20 per week of delay. Delay of 30 weeks occurred. The purchasers stood to suffer actual losses of substantially more than £600 to which they were entitled under the contract. The House of Lords rejected the purchasers' claim for their actual losses in the sum of £5850.

> *Temloc Ltd v. Errill Properties Ltd*:[9] this case is an extreme illustration of this principle and provides yet another lesson on the importance of completing contract documents with great care. Against the Appendix entry for liquidated damages '£NIL' was entered in a JCT 80 contract for a shopping development. Completion was delayed and the Employer, a developer, was sued by his prospective tenants. The Employer's claim to recover his liability from the Contractor as general damages failed in the Court of Appeal. This decision has been severely criticized. In *Baese Property Ltd. v. R. A. Building Property Ltd*,[10] which concerned substantially the same facts, the Supreme Court of New South Wales refused to follow the decision in *Temloc*. However, the Scottish case of *John Maxwell & Sons (Builders) v. Simpson*[11] was decided along the lines of *Temloc*.

---

5   See also *Peak v. McKinney*; see below, n. 17; *Temloc*, see below, n. 9.
6   See Section 1.5.5.
7   [1905] AC 6; see also *BFI Group of Companies Ltd v. DCB Integration Systems Ltd.* (1987) CILL 348.
8   [1933] AC 20.
9   (1987) 39 BLR 30 (hereafter *Temloc*).
10 (1989) 52 BLR 130.
11 (1990) SCLR 92.

11.1.4 Fixing liquidated damages in construction contracts

In most commercial situations, it is not very difficult to estimate with sufficient accuracy the loss that the employer stands to suffer in the event of delayed completion. One approach would be to determine the profits that the employer would have made in the period of delay. A second approach involves assessing the cost of alternative accommodation plus an amount for business disruption. However, in ecclesiastical and some public sector buildings, these approaches are not always suitable. For such projects, a common approach is founded upon the premise that, at about the time of the delay, a large proportion of the contract sum, typically 80–90 per cent, would have been paid to the contractor. The employer would therefore be out of pocket by this amount without the benefit of using his building. It is then argued that the interest that the employer would have earned by putting that sum in a bank represents an acceptable measure of his loss. This approach is favoured by the Society of Chief Surveyors in Local Government. In *J. F. Finnegan v. Community Housing Association*[12] an Official Referee upheld a liquidated damages clause that required application of the same approach to 85 per cent of the estimated contract price to arrive at the liquidated damages payable. In the Australian case of *Multiplex Construction Pty Ltd v. Abgarus Pty Ltd*[13] the court accepted a similar approach to fixing liquidated damages. That approach involved determination of the damages by a formula that calculated the interest that would have been charged by the commercial banks on the actual payments made to the contractor.

11.2 Effect of delay: general principles

The reality of construction projects is that a variety of events can prevent the contractor from completing the works by the agreed date. Some of these events may be due to the acts or omissions of the employer, or of people for whom the employer is responsible in law. Failure of the employer to give possession of the site of the works to the contractor is an example of an omission for which the employer would be directly responsible. An example of an omission of other parties for which the employer would be responsible is failure of his architect to give appropriate instructions in time. For the sake of simplicity both types of events are henceforth referred to as events for which the employer is responsible.

On the subject of delays, extension of time, and liquidated damages, four general principles have to be borne in mind. Although these principles apply to all contracts that require performance within defined periods and entail liquidated damages, they are explained mainly in the context of construction.

The first principle provides an answer to the fundamental question of whether or not the contractor is under a strict obligation, irrespective of delaying influences, to achieve completion by the specified date. The position of the law is that the contractor is so obliged unless he is prevented by factors for which the employer is partly or solely responsible. Where the project is affected by such factors, the contractor is no longer obliged to

---

12 (1993) 95 BLR 103.
13 (1992) 33 NSWLR 504.

complete by the specified date but within a reasonable time. In legal parlance, the date for completion is 'at large'.

> *Dodd v. Churton*:[14] this case concerned a contract which had variation and liquidated damage clauses. However, there was no provision for the revision of the agreed date for completion. Extra work was ordered and the builder was delayed as a result. Allowing a fortnight for the extras, the employer purported to set-off for liquidated damages for that part of the delay beyond the fortnight. The court held that the liquidated damage clause no longer applied. The important issue involved here was that, as the contract did not contain a provision for extending the time for completion, no new date could be fixed under the contract. The date of completion therefore became at large.

However, there is an exception where the contractor expressly and unequivocally agreed to complete by the original completion date even if the employer commits acts of prevention. In such a case, the contractor could be held to his promise.[15] However, such one-sided provisions tend to be construed so *contra proferentem* the employer that if the actual cause of delay is not clearly within the acts of prevention contemplated in the contract, the contractor would be released from the obligation to complete by the due completion date.[16]

The second principle is that where the delay is partly or wholly attributable to an event for which the employer is responsible, the liquidated damage clause may be kept alive only by the existence in the contract of an extension of time clause which covers that cause of delay:

> *Peak Construction (Liverpool) Ltd v. McKinney Foundations Ltd*:[17] Peak was the main contractor of Liverpool Corporation on a housing contract. The contract contained both liquidated damage and extension of time clauses. The extension of time clause empowered the architect to grant extensions of time for, *inter alia*, extras and additions, *force majeure* and 'other unavoidable circumstances'. After the defendant, nominated sub-contractors for piling, finished their work and left the site, some of the piles were found to be defective. Everybody concerned agreed that work should be suspended pending investigation of the piles by a consulting engineer. The Corporation delayed not only in appointing the engineers but also in authorizing the report produced. The Court of Appeal held that, as between the employer and the main contractor, time for completion had become at large because part of the delay was caused by the employer. It is interesting to note that the court did not accept the argument that delays in making the appointment and authorizing the report constituted 'other unavoidable circumstances' for which time could be extended under the contract.

Peak v. McKinney illustrates the tendency of the courts to construe extension of time clauses very narrowly. Unless an extension of time clause is specific as to the cause of the delay it may not be effective.

---

14 [1897] 1 QB 562; see also *Holme v. Guppy* (1838) 2 M & W 387.
15 *Jones v. St John's College Oxford* (1870) LR QB 115; *Percy Bilton v. Greater London Council* (1982) 20 BLR 1.
16 *Roberts v. Bury Commissioners* (1870) LR 5 CP 310.
17 (1970) 1 BLR 114 (hereafter *Peak v. McKinney*); see also *Rapid Building Group Ltd v. Ealing Family Housing Association Ltd* (1984) 29 BLR 5 and *Percy Bilton v. Greater London Council* (1982) 20 BLR 1.

A corollary of the second principle is that, where the contractor is delayed by an event for which the employer is responsible and that event is not covered by the extension of time clause, completion time becomes at large. It has often been thought that a catch-all provision which allows the architect to extend time in any circumstance is the most effective way of ensuring that a liquidated damage clause is always alive. This generalist approach does not always work because the courts are reluctant to allow the architect to extend time for employer-caused delays unless the extension of time clause is very specific on the cause of delay. For example;

> *Wells v. Army & Navy Co-operative Society Ltd*:[18] a building contract provided for extension of time for a list of specific causes of delay and 'other causes beyond the contractor's control'. The Court of Appeal ruled that the purported catch-all phrase could not include breaches of contract or other types of interference for which the employer was responsible and that, consequently, liquidated damages were not recoverable.

The third principle is that, even with the existence of an extension of time clause covering the cause of a delay, time for completion will become at large if the extension of time to which the contractor is entitled is not granted in accordance with the contract.

> *Miller v. London County Council*:[19] an extension of time clause stated: 'it shall be lawful for the engineer, if he shall think fit, to grant from time to time, and at any time or times by writing under his hand, such extension of time for completion of the work and that either prospectively or retrospectively, and to assign such other time or times for completion as to him may seem fit'. Another clause provided for liquidated damages for delays. Eight months after completion of the works, the engineer issued a certificate granting extension of time and certifying the amount due the employer as liquidated damages. It was held that the wording of the extension of time clause did not empower the engineer to extend time after the completion of the works. The use of the word 'retrospectively' only allowed him to wait until the cause of the delay had ceased and then, within a reasonable time thereafter, to grant extension of time. The extension granted was therefore not in accordance with the contract, with the further consequences that time for completion was at large and liquidated damages irrecoverable.

The fourth principle is that where time for completion becomes at large, any liquidated damage clause becomes ineffective and the employer is only entitled to unliquidated damages if the contractor fails to complete within a reasonable time.[20] Apart from the difficulty of proving his damages, there is the additional disadvantage that the employer may not be entitled to set-off such damages against interim certificates if the contractor objects.

It is a source of common debate whether the failed liquidated damages constitute an upper limit to the amount the employer can recover as unliquidated damages. In other words, is the employer hoist with his own petard? In *Esley v. J. G. Collins Insurance*

---

18 (1902) 86 LT 764; see also *Peak v. McKinney* in which the catch-all phrase 'or other unavoidable circumstances' failed.
19 (1934) 151 LT 425.
20 *Rapid Building Group Ltd v. Ealing Family Housing Association Ltd* (1984) 29 BLR 5 (hereafter *Rapid Building v. Ealing*); *Peak v. McKinney*, see, n. 17.

Agencies Ltd[21] the Supreme Court of Canada held that where liquidated damages are struck down for being penal, recoverable unliquidated damages cannot exceed the sum found to be penal. However, this point is still moot in English law so far as construction contracts are concerned. In *Cellulose Acetate Silk Co. v. Widnes Foundry Co.*[22] the House of Lords left the question open. The Court of Appeal did the same in *Rapid Building v. Ealing.*[23]

It follows from the discussions so far that an extension of time clause is for the benefit of both parties. The advantage to the contractor is that his liability to pay liquidated damages is restricted to situations where the employer is not responsible for the delay. On the employer's side, it prevents the date of completion becoming at large when the contractor suffers delay for which the employer is responsible. The widespread belief in the construction industry that extension of time clauses are solely for the benefit of the contractor is therefore very much mistaken.

11.3 Grounds for extension of time under JCT 98

Clause 25.4 lists the causes of delay, collectively referred to as the 'Relevant Events', for which the time for completion may be extended. Examination of the list makes two facts very clear. First, they include not only events attributable to the Employer but also those outside his control, for example *force majeure*. Second, the list is by no means exhaustive of possible events for which the Employer would be responsible. It follows from the principles already discussed that if an event for which the Employer is responsible, but which is outside the list, delays the Contractor, the Architect would have no jurisdiction to grant extension of time. Time for completion would therefore become at large in spite of the extension of time clause.

The Relevant Events are listed under Clause 25.4 as:

1. *force majeure* (Clause 25.4.1);
2. exceptionally adverse weather conditions (Clause 25.4.2);
3. loss/damage due to the Specified Perils (Clause 25.4.3);
4. civil commotion, strikes, or lockouts, etc. (Clause 25.4.4);
5. compliance with an Architect's Instruction (Clause 25.4.5);
6. failure of Architect to comply with the Information Release Schedule (Clause 25.4.6.1);
7. failure of the Architect to provide further drawings and details that have become necessary (Clause 25.4.6.2);
8. delays on the part of Nominated Sub-contractors and Nominated Suppliers (Clause 25.4.7);
9. delay due to execution or non-execution of work directly by Employer or his other contractors (Clause 25.4.8.1);
10. delays caused by failure by the Employer to supply materials and goods or the manner in which he supplied them (Clause 25.4.8.2);
11. statutory intervention (Clause 25.4.9);

---

21 (1978) 83 DLR (3d) 1.
22 See, n. 8.
23 *See* n. 20.

12. unforeseen shortages of labour, and materials (Clause 25.4.10);
13. delays caused by local authorities and statutory undertakers (Clause 25.4.1 1);
14. failure to give ingress to, or egress from, the Site (Clause 25.4.12);
15. deferment of giving possession of Site (Clause 25.4.13);
16. execution of work covered by an Approximate Quantity in the Contract Bills (Clause 25.4.14);
17. change in Statutory Requirements (Clause 25.4.15);
18. terrorism (Clause 25.4.16);
19. compliance with obligations in respect of CDM regulations (Clause 25.4.17);
20. suspension of performance for non-payment by the Employer (Clause 25.4.18).

In principle, if the Contractor is delayed by any of these events, he would be entitled to extension of time. However, it must be pointed out that mere occurrence of any of these events does not automatically result in a right to extension of time. The Contractor must actually be delayed by the event. In this regard, it is the Contractor's actual progress and not his planned progress which must be considered. It follows therefore that, if the Contractor is ahead of schedule at the time of the delaying event, he will not be entitled to extension of time although he may be entitled to recover his direct loss and/or expense for disruption. However, it is not uncommon for contractors to claim extension of time in those circumstances as protection against future delays for which extension of time may not be available.

It sometimes happens that an event, although a Relevant Event, would not have affected the works had the Contractor not been subject to an earlier delay caused by his own fault. It would appear that the Contractor would still be entitled to extension of time.

> *Walter Lawrence & Son v. Commercial Union Properties Ltd*:[24] a Contractor under a JCT63 contract fell behind with his programme. The Architect refused to grant extension of time for adverse weather conditions on the grounds that had the Contractor followed the programme he would not have been affected by the weather conditions. Rejecting this argument, the court held that the Contractor was entitled to extension of time.

The right to extension of time is further qualified in three ways. First, under Clause 25.3.4.1 the Contractor must always use his best endeavours to prevent delay in the progress of the works irrespective of the cause of potential delay. He is also to continue to use his best endeavours to reduce any delay that still occurs despite such endeavours. What constitutes 'best endeavours' is not without some controversy. One school of thought has it that the Contractor is not expected to incur extra costs to make good the delay but only to take practical steps to reduce it (Clause 25.3.4). An example of a practical step is re-deployment of resources onto another activity if the activity they were originally planned for cannot proceed because of the delaying event. However, it is sometimes also argued that 'best endeavours' means that the Contractor is to use any means whatsoever provided it is within his power and that it matters not that the Contractor will incur additional costs. Second, by Clause 25.3.4.2 the Contractor is to take all reasonable measures necessary to proceed with the works. Such measures must be to the reasonable satisfaction of the Architect. Third, as will become apparent in the discussion of each Event, there are additional restrictions which apply to particular Relevant Events.

---

24 (1984) 4 Con. LR 37.

11.3.1 *Force majeure*

In general, this term refers to Acts of God or man-made events which are beyond the control of the parties, e.g. war, inundation, epidemics and strikes.[25] However, as used in the JCT 98, it must have a restricted meaning as several of the events normally classified under this term, e.g. strikes and lightning, are dealt with separately.

11.3.2 Exceptionally adverse weather conditions

To succeed with a claim on this ground, the Contractor must produce evidence that the conditions complained of are exceptional for that time of year and location. Weather records covering a reasonable period as well as site diaries will normally be demanded by Architects.

11.3.3 Loss/damage from the Specified Perils

The Specified Perils are defined in Clause 1.3 as consisting of:

fire, lightning, explosion, storm, tempest, flood, bursting or overflowing of water tanks, apparatus or pipes, earthquake, aircraft and other aerial devices or articles dropped therefrom, riot and civil commotion, but excluding Excepted Risks.

The Excepted Risks are also defined in the same clause as consisting of:

ionising radiations or contamination by radioactivity from any nuclear fuel or from any nuclear waste from the combustion of nuclear fuel, radioactive toxic explosive or other hazardous properties of any explosive nuclear assembly or nuclear component thereof, pressure waves caused by aircraft or other aerial devices travelling at sonic or supersonic speeds.

The Contractor is obliged to reinstate the Works if they are damaged by any of the Specified Perils. The reason for making such damage a Relevant Event is presumably to give the Contractor time to carry out the reinstatement. These are neutral events, i.e. neither party has any control over them. There is therefore some equity in the sharing of this risk in that the Employer shoulders the consequences of delay whilst the Contractor bears any additional cost from disruption to regular progress.

A question often posed is whether or not the Contractor would still be entitled to extension of time if he is himself the cause of the Specified Peril, e.g. where an employee of the Contractor negligently starts a fire. Case law suggests that the answer is a matter of construction of the particular Specified Peril. In *Computer and Systems Engineering plc v. John Lelliott (Ilford) Ltd and Another*[26] a pipe in a sprinkler system was sheared off through the negligence of a sub-contractor. This caused discharge of 1600 gallons of water that caused damage to the Employer's property. The litigation was whether the damage was caused by 'flood' under an equivalent definition of the Specified Perils. The Court of

---

25 For judicial comments of the meaning of the term at common law see *Oakley v. Portsmouth & Ryde Steam Packet* (1856) 11 Exch. 618; *Matsoukis v. Priestman & Co.* [1915] 1 KB 681; *Lebeaupin v. Crispin* [1920] 2 KB 714.
26 (1990) 54 BLR 1.

Appeal held that the word 'flood' in the definition suggests rapid accumulation of larger volumes of water from an external source and that it did not cover discharge from the sprinkler system. The Court also confined 'bursting of pipes' to damage from internal stresses. This decision therefore suggests that damage to the Works from a Specified Peril caused by the Contractor's negligence is not a Relevant Event.

In addition, the tenor of Clause 25 is that the Contractor is not to benefit from his own default. Besides, the Architect is to grant extension of time if it is fair and reasonable to do so: Clause 25.3.1. The Architect may come to a view that the Contractor's negligence is a breach of his duty under Clause 25.3.4 to use his best endeavours to prevent delay and that, therefore, it is not fair and reasonable to grant extension of time. To avoid any doubt, Clause 25.4.3 is often amended to limit expressly the right to extension of time to only those Specified Perils not caused by the fault of the Contractor or of his agents. Some employers delete it altogether.

For a discussion of the case for taking insurance against the Employer's loss of liquidated damages where completion is delayed as a result of damage to the Works from the Specified Perils, see Section 7.3.6.

11.3.4 Industrial actions

The types of industrial action for which the Contractor is entitled to extension of time include:

> civil commotion, local combination of workmen, strike or lock-out affecting any of the trades employed upon the Works or any of the trades engaged in the preparation, manufacture or transportation of any of the goods and materials required for the Works.

Strikes are the most commonly encountered of these problems although they have declined in recent years. The clause does not discriminate between official and unofficial strikes. However, it is believed that 'working to rule' does not belong to this Relevant Event. Neither does picketing or civil commotion by political agitators, e.g. peace demonstrators, because they are not trades employed upon the Works or engaged in connection with the manufacture or transportation of materials for the Works. Also, it would appear that the clause does not cover strikes by the employees of an statutory undertaker engaged directly by the Employer to carry out work not forming part of the contract.

> *Boskalis Westminster Construction Ltd v. Liverpool City Council:*[27] Boskalis entered into a contract with Liverpool City Council for the construction of dwellings. The Contract incorporated the JCT 63 (1977 revision), Clause 23(d) of which provided for industrial action as a Relevant Event in identical wording. The question put to the court from an arbitrator's award was whether delays caused by the strike actions of the employees of a statutory undertaker employed by the employer to carry out work not forming part of the contract was covered by the clause. The question was answered in the negative.

Considering that JCT 98 Clause 25.4.4 is exactly the same as the clause in question, the same position would apply.

---

27 (1983) 24 BLR 83.

11.3.5 Compliance With Architect's Instructions

The instructions covered are those relating to:

- Clause 2.3: discrepancies within or between documents listed in Clause 2.3;
- Clause 2.4.1: discrepancies between Contractor's Statement in respect of Performance Specified Work and an AI issued after its receipt by the Architect;
- Clauses 8.3 and 8.4.4: opening up work for inspection or testing which shows that the work is in accordance with the Contract;
- Clause 13.2: variations under Clause 13.2 that do not involve acceptance of a 13A Quotation Variation;
- clause 13.3: expenditure of provisional sums (instructions on the expenditure of provisional sums for defined work or Performance Specified Work are excluded);
- Clause 13A.4.1: instruction that a variation for which the Contractor has provided a 13A Quotation should be carried out but valued under Clause 13.4.1;
- Clause 23.2: postponement of any part of the Works;
- Clause 34: antiquities;
- Clause 35: Nominated Sub-contractors;[28]
- Clause 36: Nominated Suppliers.[29]

11.3.6 Failure to comply with the Information Release Schedule (IRS)

Under Clause 5.4.1, the Architect must ensure that the Contractor is supplied with two copies of any item of information referred to in the Information Release Schedule by the date stated in the Schedule for its supply. This obligation is not absolute because it is qualified by 'Except to the extent that the Architect is prevented by the act of default of the Contractor or of any person for whom the Contractor is responsible'.[30] Under Clause 25.4.6.1, it is a Relevant Event if the Architect fails to comply with this obligation. However, failure of the Architect to comply with the Schedule does not in itself necessarily give rise to entitlement to extension of time. Delay which the Contractor has used his best endeavours to avoid or minimize must be caused. In addition, Clause 5.4.1 also provides that the time for release of any item of information may be varied by agreement between the Employer and the Contractor, which is not to be withheld unreasonably. In practice, well in advance of when the information is due, the Architect would have to bring it to the attention of the Employer that he might be unable to supply the information on time. It is then for the Employer to attempt to agree a variation of the Schedule with the Contractor. Withholding agreement unreasonably is a breach of contract that may be taken into consideration in assessing the Contractor's entitlement to extension of time for failure to comply with the original Schedule. The extent to which the proposed change will affect the Contractor's plans for the Works and the timing of the proposal are clearly factors to be taken into account in deciding whether any withholding is unreasonable.

---

28 See Chapter 9.
29 See Chapter 10.
30 Persons for whom the Contractor is responsible include the Contractor's employees, suppliers, domestic sub-contractors and Nominated Sub-contractors and Suppliers.

This Relevant Event is likely to accelerate a new development whereby Employers, rather than resist claims for lack of information from the Architect, look to Architects and other designers for compensation for any loss arising from successful claims.[31]

11.3.7 Failure to provide further drawings and details

Under Clause 5.4.2, the Architect is obliged to provide the Contractor with further drawings and details reasonably necessary to explain or amplify the Contract Drawings. He must also issue instructions necessary for the performance of the Contractor's obligations. Instructions on the expenditure of provisional sums are mentioned specifically. Where the drawing, details or instructions were included in the Information Release Schedule, the timing of release is governed by Clause 5.4.1 Where they are not included, they are to be provided when, in the opinion of the Architect, it is reasonably necessary for the Contractor to receive them. However, the Architect must consider the following in reaching his opinion:

- actual progress
- whether or not the works are ahead of schedule
- the Completion Date.

On the one hand therefore, he must provide the drawings, details and instructions so as not to hold up actual progress. On the other hand, if the Contractor is ahead of programme anyway, the Architect has some flexibility because he is to have regard to the due completion date when deciding when to provide the information. This focus on the contractual completion date is in line with the position already reached at common law under a previous edition of the contract that did not contain Clause 5.4.2. In *Glenlion Construction Ltd v. The Guinness Trust*,[32] which arose from a contract in the terms of the JCT 63, it was decided that the Architect did not have to go out of his way to supply the Contractor with information at times that would allow the Contractor to complete earlier than the contractual completion date.

If the Contractor is likely to be delayed by lack of information not included in the IRS, his role in all this is to advise the Architect of the need for the information. However, he is required to advise only if he is aware that the Architect is ignorant of this need. He does not even have to do so in writing. This represents a considerable watering down of the Contractor's obligation under the pre-Amendment 18 version of the JCT 80 to apply specifically for the information 'on a date which having regard to the Completion Date was neither unreasonably distant from or unreasonably close to the date on which it was necessary for him to receive the same'. The onus is now on the Architect to find out when the Contractor needs what information. This requires close examination of the Contractor's programme and actual progress. It might be prudent on the Architect to serve the Contractor periodically and in writing with a list of information requirements he is aware of. This throws the onus back on the Contractor to bring up other requirements that might have escaped the Architect's attention.

---

31 See *Wessex Regional Health Authority v. HLM Design Ltd* (1995) 71 BLR 32.
32 (1987) 30 BLR 89.

11.3.8 Nominated contractors and suppliers

There are two distinct situations in which a Nominated Sub-contractor may cause delays to the Main Contract Works: (i) where the Sub-contractor causes delay but still completes the Sub-contract Works; (ii) where the Sub-contractor causes delay and withdraws or is withdrawn from the contract without completing the Sub-contract Works.

Where the sub-contractor completes
Where the Sub-contractor manages to complete, there are two possibilities. The first is that, although he has completed, he failed to complete by the due completion date applicable to the Nominated Sub-contract Works and, thereby, caused delay. This cause of delay is a Relevant Event under Clause 25.4.7: 'delay on the part of Nominated Sub-contractors or Nominated Suppliers'. It is a common complaint that it is not fair for the Employer to pay the price for a Nominated Sub-contractor failing to complete on time because that arrangement removes the Contractor's incentive to do everything reasonably within his power to prevent this delay. Furthermore, it is often argued that there is a real possibility that the Contractor could even gain by his own default: causing the delay to the Sub-contractor in the first place. However, there is the counter-argument that the comments in Section 11.3.3, concerning the effect of the Contractor's duty under Clause 25.3.4 to avoid or minimize delay and the requirement under Clause 25.3.1 that extension time must be fair and reasonable, are equally applicable in this situation.

The second possibility is that the Nominated Sub-contractor nevertheless manages to complete on time after having caused delays to the completion of the Main Contract Works. Figure 11.1 shows how this could happen in practice. Activities 1–3, 3–6, and 6–10 constitute the work package of a Nominated Sub-contractor. Activities 1–3, 3–6, and 6–7, 7–12 are on the Contractor's critical path. By definition, this means that if the Sub-contractor takes more than 12 weeks to complete 1–3 and 3–6 the Contractor will suffer delays unless he recovers the schedule slippage by 'crushing' some subsequent activities. However, the Sub-contractor may still be able to complete the Sub-contract Works within the Sub-contract period by 'crushing' activity 6-10.

A decision of the House of Lord suggests that this type of delay is not within the Relevant Event under Clause 25.4.7:

> *Westminster Corporation v. J Jarvis*[33]: Jarvis entered into contract with Westminster incorporating the JCT 63. Peter Lind were Nominated Sub-contractors for piling, which they completed on time. However, they had to return and carry out remedial work when defects were discovered. In the course of this work, they caused delays to Jarvis. The House of Lords held that the delay did not come within 'delay on the part of Nominated', which they construed as meaning failure to complete within the sub-contract period .

The decision that Nominated Sub-contract Works are complete is therefore a source of risk to the Contractor. The risk is that if the Nominated Sub-contractor has to return and make good any defects and the Contractor is delayed as a result of this latter activity, the Contractor would not be entitled to extension of time. Fortunately, the Contractor does have some say in deciding whether or not Nominated Sub-contract Works have reached

---

33 (1970) 7 BLR 64 (hereafter *Westminster v. Jarvis*).

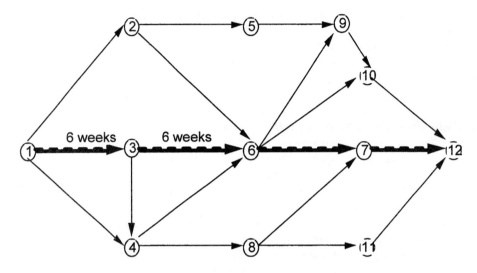

Fig. 11.1 Delays caused by a Nominated Sub-contractor

practical completion. Under the Nominated Sub-contract, the Sub-contractor is to give written notification to the Contractor if the Sub-contractor is of the opinion that the Nominated Sub-contract Works have reached practical completion. The Contractor is then to pass the notification to the Architect with his observations on the position of the works (Clause 2.10 of Conditions NSC/C). However, the final decision as to whether or not the Nominated Sub-contract Works have reached practical completion is the Architect's. When he reaches the view that the Sub-contractor has achieved practical completion and complied sufficiently with his obligations under Clause 5E.8 of NSC/C in respect of the CDM Regulations, he is to issue a Certificate of Practical Completion in respect of the Sub-contract Works with copies to the Contractor and the Sub-contractor (JCT 98 Clause 35.16). The Contractor can contest the validity of the Certificate if there are unfinished parts of the sub-contract by invoking the appropriate dispute resolution mechanism.

With the type of delay, the Contractor's best course of action would be to pursue the Sub-contractor for the acceleration costs or his liability for liquidated damages arising from the Sub-contractor's disruption. As explained in Section 3.5.3 in relation to the Contractor's liability for defects during the Defects Liability Period, the Sub-contractor would liable for such consequential damages.

Where the Sub-contractor withdraws

Withdrawal is used in this context to include determination by the Contractor or the Sub-contractor for whatever reason. Delay caused by a Nominated Sub-contractor before withdrawal would not be a Relevant Event unless the due date of completion of the Nominated Sub-contract Works has expired. This follows from the narrow meaning given to 'delay on the part of Nominated Sub-contractor' in *Westminster v. Jarvis*.

There are three categories of delay that may be caused subsequent to withdrawal:

1. There is the reasonable and inevitable delay caused by the need to find and appoint a substitute sub-contractor. In *Percy Bilton Ltd v. Greater London Council*[34] the House of Lords held that this type of delay was not a Relevant Event under a contract which was substantially the same as the JCT 63. It was also decided that failure of a Nominated Sub-contractor was not a fault of the Employer and that failure of the contract to provide for extension on account of this type of delay did not have the effect of rendering the completion date of the contract at large.

2. There is the further delay beyond that considered reasonable in the circumstances to appoint a substitute sub-contractor. It was suggested in *Percy Bilton* that, although such delay did not come within 'delay on the part of a Nominated Sub-contractor', the Contractor may be entitled to extension of time under what is now Clause 25.4.6 of the JCT 98 on the grounds that he did not receive in due time necessary re-nomination instructions.

3. The fact that the new Nominated Sub-contract contains a completion date later than the original completion of the Sub-contract may cause delays to the Contractor. This delay does not fall under any of the Relevant Events. However, if the Contractor were to object to a proposed nominee on the grounds that the new Sub-contract completion date will delay the completion of the Contract the objection would be reasonable.[35] The Contractor can therefore use this to hold up nomination until there is an agreement to extend the time for completion. Alternatively, on the authority of *Harrison v. Leeds City Council*,[36] the Relevant Event under Clause 25.4.5 (delays caused by compliance with an Architect's instruction under Clause 23.3 to postpone the carrying out of any part of the Works) may apply. In that case, the Court of Appeal held that a nomination instruction to the Contractor to engage a sub-contractor whose programme was inconsistent with the Contractor's programme amounted to an instruction to postpone the carrying out of the affected parts of the Works.

11.3.9 Work not forming part of the contract

It is not uncommon for the Contractor to be responsible for only a part of a larger project, with the remainder to be carried out by the Employer himself, his employees or other contractors directly engaged by the Employer as he is entitled to do under Clause 29. Where this is the case and the Contractor is delayed by the manner of execution or non-execution of the remainder of the project, this Relevant Event would apply.

Some statutory undertakers enter into commercial contracts for the execution of works which are distinct from those they are obliged to carry out by statute. In *Henry Boot v. Central Lancashire New Town Development Corporation Ltd*,[37] which arose from a JCT 63 contract, it was held that work carried out by statutory undertakers under commercial contracts constituted work not forming part of the contract. It may be concluded from this case that, under JCT 98, delays arising from the execution or non-execution of such work would be covered by Clause 25.4.8.1 rather than Clause 25.4.1 1.

---

34 (1982) 20 BLR 1 (hereafter *Percy Bilton*).
35 *Obiter* comments of Sir David Cairns in *Percy Bilton* followed in *Fairclough Building Ltd v. Rhuddlan Borough Council* (1985) 30 BLR 26.
36 (1980) 14 BLR 118.
37 (1980) 15 BLR 1.

11.3.10 Employer's supply or failure to supply goods and materials

This is quite clear. In any case, it is really just an extension of the principle in the previous Relevant Event.

11.3.11 Statutory intervention

For this type of governmental action to constitute a Relevant Event, it must directly affect the carrying out of the works in one or more of the following ways:

- restricting the availability or use of essential labour;
- preventing the Contractor from securing materials, goods, or fuel necessary for the carrying out of the works;
- delaying the Contractor in securing any materials, goods, fuel or energy necessary for carrying out the works.

Examples of this type of event are the imposition of a shorter working week and rationing of fuel. These events may also be covered by *force majeure*. The word 'directly' is operative. It follows therefore that if the effect of the governmental intervention is through a chain of events the Contractor would not be entitled to extension of time.

11.3.12 Shortages of labour, goods and materials

Mere shortages of these resources are not sufficient grounds. To establish a case for entitlement on this ground, the Contractor must be able to show that the shortage was not reasonably foreseeable at the time of tender and that the circumstances responsible for the shortages are beyond his control.

11.3.13 Local authorities and statutory undertakers

This clause relates to the carrying out of, or failure to carry out, work pursuant to their statutory obligations. As already discussed, work arising from commercial contracts entered into with the Employer are not covered here but under Clause 25.4.8.

11.3.14 Ingress and egress

This cause of delay is not the same as failure to give possession of site although it may have the same effect in that the Contractor is unable to take possession of the site. The wording should be carefully noted. For the Relevant Event under Clause 25.4.12 to apply:

- the ingress or egress requirement must have been either defined in the Contract Bills/Drawings or agreed between the Contractor and the Architect;
- the land, buildings and the like must be adjoining or connected with the site and in the possession and control of the Employer.[38]

---

38 *LRE Engineering Services Ltd v. Otto Simon Carves* (1981) 24 BLR 127.

A further restriction is that the Contractor must comply with all appropriate requirements for notices concerning the ingress and egress.

11.3.15 Deferment of possession of site

Where it is stated in the Appendix that Clause 23.1.2 will apply, the Employer has a right under that clause to defer giving the Contractor possession of site for any period not exceeding the stated maximum period of deferment (default maximum is 6 weeks). If the Employer exercises this right and the Contractor is delayed as a result, there is an entitlement to extension of time under this sub-clause.

It is to be noted that deferment of possession beyond the maximum period stipulated in the Appendix would not be part of this Relevant Event. Time for completion would therefore become at large.

11.3.16 Work with Approximate Quantity

Though not expressly stated that the quantity in the Contract Bills must have been underestimated, the rational behind this sub-clause must be that the Contractor could not have allowed for the extra work in his programme.

11.3.17 Change of statutory requirements

The Contractor is deemed to have taken into account in his Statement on Performance Specified Work the need to comply with legislation in existence as at the Base Date. This Relevant Event covers unavoidable delays arising from a need to alter or modify the Statement to comply with legislation introduced after the Base Date. An example is new Building Regulations with which the design of the work is in contravention.

11.3.18 Terrorism

The Contractor would have to demonstrate that: (i) there were threats or acts of terrorism; and (ii) the act of relevant authority that caused the delay was to deal with them. The following would qualify:

- disturbance to suppliers from such acts or threats;
- evacuation of the area covering the site;
- evacuation of areas in which work destined for the site was being carried out.

11.3.19 Compliance/non-compliance with Clause 6A.1

Clause 6A.1 imposes on the Employer an obligation to ensure that the Planning Supervisor and the Principal Contractor perform their duties under the CDM Regulations. The ambit of this Relevant Event is potentially very wide because it covers both compliance and non-compliance with this obligation. Any act by the Planning Supervisor or the Principal Contractor, where this is different from the Contractor, in performance of their duties under the CDM Regulations that causes delay gives rise to an entitlement to extension of time. The only exception is in Clause 6A.3, which provides that the Contractor is not entitled to extension of time for compliance with reasonable requirements of the Principal Contractor in the interests of the CDM Regulations. However, the use of the phrase 'reasonable requirements' is likely to invite disputes as to whether or not a requirement is reasonable.

11.3.20 Suspension for non-payment

Under Clause 30.1.4, the Contractor is entitled to suspend performance for failure of the Employer to meet in full any payment due under the Contract by its final date. This Relevant Event covers delays caused by valid exercise of this right. The validity of suspension of performance is discussed in Section 15.5.2.

11.4 Administrative procedures

The discussion so far has been about the circumstances entitling the Contractor in principle to extension of time. In addition to defining these circumstances, the JCT 98 lays down detailed procedures to be followed in the application for, and the granting of, extension of time.

11.4.1 Obligations of the Contractor

1. The Contractor is to provide the Architect with two copies of his master programme as soon as possible after the execution of the Contract. However, the Contractor is not contractually bound to carry out the work strictly in accordance with the Submitted programme: *Glenlion Construction Ltd v. The Guinness Trust.*[39]

2. Whenever he amends or revises the programme to reflect any extension of time granted by the Architect or a confirmed acceptance of 13A Quotation,[40] he must supply the Architect with two copies of the amendments or revisions within 14 days of the award of the extension or date of issue of the acceptance of the Quotation (Clause 5.3.1.2).

3. The Contractor is to give written notice not only when it becomes reasonably apparent that progress is being delayed but also when progress is likely to be delayed (Clause 25.2). In addition, the Contractor must comply with the notice requirements even if he does not intend to claim extension of time. The notice must:

- give details of all the material circumstances surrounding the delay;
- state the cause of the delay;
- identify which of the causes the Contractor believes to be Relevant Events.

If the material circumstances stated in the notice refer to the work of a Nominated Sub-contractor, a copy of the notice must be sent to the Nominated Sub-contractor concerned.

4. For each of the Relevant Events identified, the Contractor must in the notice, or in writing as soon as possible thereafter, provide:

39 (1987) 39 BLR 89.
40 See Section 6.13 for explanation of a Clause 13A Quotation.

- particulars of the expected effects of the delay;
- an estimate of any delay in completion arising from the delay.

The estimate has to be made for each Relevant Event as if it is the sole cause of the delay. Copies of the estimate of delay are to be given to Nominated Sub-contractors to whom copies of the notice of delay had been sent.

5. The Contractor must, once a notice of delay has been given, continue to monitor events and give such further notices as he thinks necessary, or as may be requested by the Architect.

6. He also has to keep up to date the particulars and estimates previously given. Architects often complain that Clause 25.2 is the cause of a lot of unnecessary paper work.

11.4.2 Duties of the Architect

Clause 25.3.1 defines a timetable within which the Architect must act on applications and notices for extension of time. If he fails to grant extension of time where he should have done so, time for completion becomes at large, with the dire consequences for the Employer in terms of recovery of liquidated damages.[41] If the Employer suffers any loss because of negligent operation of the extension of time provisions, he would be entitled to recover damages from the Architect for breach of his contract of engagement.

If the Architect is of the opinion that any cause of delay notified is a Relevant Event and that it will cause delay in completion, he must then give a written decision granting extension of time by fixing a new Completion Date. The new date is to be fair and reasonable. Where it is not fair and reasonable to grant extension of time, the Contractor is to be so informed in writing. The timetable within which the Architect must reach his decision and act accordingly is the same regardless of the extension of time entitlement: not later than 12 weeks from receipt of the Contractor's notice or of reasonably sufficient particulars and estimates. Failure on the part of the Contractor to supply reasonably sufficient information for the Architect to make a fair judgement postpones the day from which the period of 12 weeks runs. The Contractor must therefore respond very quickly to the Architect's requests for further particulars and estimates. Where the time from the Contractor's notice to the Completion Date is less than 12 weeks the Architect must inform the Contractor not later than the Completion Date.

The Architect must state in any grant of extension the Relevant Events he has taken into account but he does not have to apportion extensions to specific Events. Where variations issued since the last revision of the Completion Date entailed omission of work, the Architect may fix a Completion Date earlier than that last fixed if it is fair and reasonable, having regard to the omission. However, the time stated in a confirmed acceptance of a 13A Quotation as required to complete the work in respect of the Variations is not to be altered.

All Nominated Sub-contractors are to be notified in writing of each revision of the Completion Date whether through fixing under Clause 25.3.1 or acceptance of a 13A

---

41 See Section 11.2.

Quotation. It is not very clear why Sub-contractors whose works are not affected should be informed. Perhaps it is to avoid the administrative overhead of having to investigate which Sub-contractors are affected. If the times within which a Sub-contractor must carry out the sub-contract Works have been affected by an acceptance of a 13A Quotation, the notice to the particular Sub-contractor must state the changes.

11.4.3 Importance of the notices of delay

The nature of the decision-making required of the Architect highlights the importance of timely notices of delays. The aims of the stringent notice requirements are:

- to give the Architect the opportunity to take all reasonable steps available to him to minimize the effect of the delay, e.g. he may issue appropriate variation orders;
- to alert the Architect to watch out for the reasonableness of the Contractor's endeavours to prevent or minimize delays in completing the works;
- to alert the Architect to the effects of the delay as they occur;
- to allow the Architect to advise the Employer of likely delays so that the latter can re-arrange his affairs accordingly.

The implications for extension of time if the Contractor fails to comply with the requirements for notices are discussed in Section 2.3.2. Apart from considerations of extension of time, such failure would constitute a breach of contract for which damages can be claimed from the Contractor.

11.4.4 Review of extensions of time

In most cases, the exact effect of the delaying event cannot be known at the time the Architect is expected to grant extension of time. This may be because either the event is at that time a continuing one or its effects lie in the future. For these reasons, the extensions of time granted before Practical Completion are generally only provisional. The aim of extension of time at this stage is to provide the Contractor with a rough but realistic Completion Date towards which to work. Towards the end of the contract, the Architect should be in a better position to appreciate the actual effect of all the delaying events for which extension of time is grantable. Under Clause 25.3.3, all extensions of time are therefore subject to review towards the completion of construction of the Works. At this review, the Architect must consider all Relevant Events even if not notified by the Contractor. There are two points at which reviews can be carried out: (i) an optional review after the Completion Date if that date occurs before Practical Completion of the Works; (ii) a mandatory and final review within 12 weeks of Practical Completion.
    The final review may lead to any of three possible outcomes.

1. He may fix a later Completion Date than that previously fixed if he had underestimated the delays and such later Date is fair and reasonable. It would appear that he must not fix a later date if the applicable Completion Date was stated in an accepted 13A Quotation and no Relevant Event occurred subsequently. The rationale for this is probably that the Contractor should have considered all previous delays in the 13A Quotation (Clause 25.3.3. 1).
2. He may fix an earlier Completion Date than that previously fixed under Clause 25 or

stated in a confirmed acceptance of a 13A Quotation if it is fair and reasonable to do so having regard to subsequent instructions which omitted some of the Works (Clause 25.3.3.2).

3. The Architect can confirm a Completion Date previously fixed or stated in a confirmed acceptance of a 13A Quotation (Clause 25.3.3.3).

However, under no circumstances should a new Completion Date be fixed to occur at an earlier date than the Date for Completion stated in the Appendix (Clause 25.3.6). In addition, times stated in a confirmed acceptance of a 13A Quotation as required to complete the Works in respect of the relevant Variation must not be altered. Within the same 12 weeks the Contractor must be informed in writing of the result of the review, irrespective of the decision of the Architect. All Nominated Sub-contractors are to be notified in writing if a review results in the fixing of a new Completion Date.

11.5 Deduction of liquidated damages under JCT 98

Under Clause 24.2.1, the Employer is entitled to recover liquidated damages from the Contractor if he fails to complete by the Completion Date. Liquidated damages may be set off against payment due under Interim Certificates and the Final Certificate: Clause 24.2.1.1. If there is no payment due the Contractor under the Contract, the liquidated damages can be recovered as debt in the usual way. However, the issue of a Certificate of Non-completion by the Architect is a condition precedent to the recovery of liquidated damages. In addition, not later than 5 days before the final date for payment pursuant to the relevant Certificate, the Employer must inform the Contractor in writing of his intention to recover liquidated damages by way of set-offs against payment due or as debt. Although notices issued pursuant to Clauses 30.1.4 and 30.8.3 to comply with preconditions of set-offs against payment certificates would meet this requirement if appropriately worded,[42] it would be prudent to serve specific notice of the Employer's intention to recover liquidated damages upon receipt of the Certificate of Non-completion.

Clauses 30.1.1.4 and 30.8.3 impose, as a precondition for valid set-off against payment due under Interim Certificates and the Final Certificate respectively, a requirement that notice of the Employer's intention to do so should be given not later than 5 days before the final date for payment. This means in that in the case of Interim Certificates, the Employer has 9 days from the date issue of the Certificate within which to serve the required notice. With the Final Certificate, it is 23 days from its issue. If there are no further sums due the Contractor under the Final Certificate, the Employer must still, within the same period, require the Contractor to pay him the liquidated damages.

The total amount of liquidated damages recoverable from the Contractor at any point in time is determined by multiplying the rate of liquidated damages inserted in the Appendix (e.g. £ x/week or day of delay) by the total period of delay. If the Employer decides to accept reduced liquidated damages (probably because the actual consequences of the delay are less financially damaging than was anticipated at the time of entering into the contract), he must certify the reduced rate in writing. The purpose of this stipulation is probably to avoid challenges to attempted set-offs by the Employer on the argument that, because the

---

42 See Section 15.1.4 for details on the requirement for these notices.

amount involved is less than the liquidated damages due, it is for something else and therefore not authorized under the contract.

A Contractor may challenge recovery of liquidated damages on grounds that the document actually received from the Employer was not the notice required under the contract. Although *Jarvis Brent Ltd v. Rowlinson Construction Ltd*[43] would probably be decided differently under the JCT 98, it nevertheless provides a good illustration of this type of problem.

> Jarvis carried out work as main contractors for Rowlinson on the terms of the JCT 80. After a Certificate of Non-Completion was issued, the Employer sent a letter enclosing a cheque to Jarvis. A letter sent to Rowlinson by the Quantity Surveyor was also enclosed. The cheque was in the amount due under an Interim Certificate reduced by an amount stated by the Quantity Surveyor in his letter to be recoverable as liquidated damages. Upon the issue of subsequent Interim Certificates, Rowlinson simply sent cheques which reflected deductions for liquidated damages. The plaintiff argued that a written requirement for liquidated damages was a condition precedent to deduction of liquidated damages and that the letter accompanied by the cheque did not constitute such a requirement. It was held that a written requirement was not a condition precedent. The judge emphasized that what was essential was that the Contractor should not be left in any doubt that the Employer was deducting liquidated damages from payment otherwise due under the Certificates. In this instance, the judge found that the Contractor was in no such doubt. It was further decided that, in any event, the letter and the reduced cheques constituted requirements in writing for the purposes of Clause 24.2. 1.

On the question of validity of notices under Clause 24.2.1, Judge Carr took a different approach in *J. F. Finnegan Ltd v. Community Housing Association Ltd.*[44] In that case, an Employer under a JCT 80 contract accompanied a cheque to the contractor with written remittance advice that stated only the number of the certificate to which the cheque related, the amount deducted for liquidated damages and the difference between the value of the certificate and the amount of liquidated damages. This difference was also the amount in the cheque. Judge Carr held that the remittance advice did not meet the requirements of the notice required by Clause 24.2.1. He explained:

> In my judgment the 'requirement in writing' in Clause 24.2.1 should indicate at least the basic details which are being relied upon to justify the deduction including the period of overrun and the figure for deduction which is claimed. These details then become a matter of record and not left to the memories of men several years later as to why a specific figure was claimed for deduction. Such detail would not only concentrate the mind of the employer when making the deduction but also will allow the contractors to know precisely why that deduction is being made and give him an opportunity to challenge it, if he so desires.

Prudence demands that the approach of Judge Carr is preferred. However, the Court of Appeal rejected that the wording of the clause demands such detail even if it is

---

43 (1990) 6 Const. LJ 292.
44 (1993) 65 BLR 103. Upheld by Court of Appeal (1995) 77 BLR 22.

commercially desirable.[45] According to Gibson LJ, a notice stating clearly the sum deducted or to be deducted and that the deduction is for the whole or part of liquidated and ascertained damages would be sufficient.

11.6 Refund of liquidated damages

Clause 24.2.2 provides that if a Certificate of Non-Completion is invalidated by a subsequent extension of time or a confirmed acceptance of a 13A Quotation, the Employer is to repay the liquidated damages recovered on the strength of the superseded Certificate. Many contractors argue that any such refund must carry interest over the period during which the money was in the Employer's hands. In *Department of the Environment for Northern Ireland v. Farrans (Construction) Ltd.*[46] it was decided that interest was payable by the Employer. That litigation arose from a contract let on the JCT 63, which contained provisions similar to Clause 24.2.2 of the JCT 80. This decision was founded on the assumption that deduction of liquidated damages which become refundable as a consequence of subsequent extension of time constitutes a breach of contract for which interest was recoverable as damages under the principle in *Wadsworth v. Lydall*.[47] The correctness of this assumption has been doubted. As explained in Section 2.4, the Employer does not warrant that in assessing quantum, be it payment or extension of time, the Architect will arrive at the figure the Contractor is properly entitled to under the contract. It is a breach of contract only where the Architect fails to operate the extension of time provisions at all. In any case, it is not applicable to the JCT 98 because Clause 24 now clearly anticipates the possibility that extensions of time granted after deduction of liquidated damages may necessitate some refunds. This also follows from the general approach that extension of time are only provisional until after the final review under Clause 25.3.3. As Clause 24.2.2 provides for refund without any mention of interest, it is submitted that the Contractor is not entitled to any interests on refunds.

11.7 Resisting liability for liquidated damages

Three grounds for resisting claims for liquidated damages can be deduced from the discussion so far:

- that the liquidated damages are penal;
- that the liquidated damage clause is inconsistent with the rest of the contract: *Bramall & Ogden v. Sheffield City Council*,[48]
- that time for completion has become at large: *Peak v. McKinney Foundations*.[49]

As already explained, where the resistance is successful, the innocent party is only entitled to unliquidated damages with all the difficulty of proving them. There is yet a fourth

---

45 *J. F. Finnegan Ltd v. Community Housing Association Ltd* (1995) 77 BLR 22.
46 (1981) 19 BLR 1.
47 *Wadsworth v. Lydall* [1981] 2 All ER 401.
48 See, n. 4
49 *Peak v. McKinney Foundations,* see n. 17.

ground of resistance: that the liquidated damage clause is not properly part of the contract. For example:

M. J. Gleeson (Contractors) Ltd v. London Borough of Hillingdon:[50] the litigation arose from a contract for the construction of 300 dwellings. The contract was let in the terms of the then current version of the JCT 63, which did not provide for sectional completion. Under the terms of the printed conditions, the Contractor was only required to complete the whole of the works by a stated completion date or other date properly fixed by the Architect. The Sectional Completion Supplement did not exist at the time. However, the Contract Bills purported to incorporate an annexed programme which entailed sectional completion. As already explained, the liquidated damage clause could be enforced only if the contract allowed for sectional completion. It was held that the programme was not effectively incorporated into the contract because Clause 12(I) (the equivalent of Clause 2.2.1 of the current JCT 80) provided that nothing in the Contract Bills could override the printed terms of the JCT 63.

11.8 Concurrent delays

Delays may be categorized under the following types:

Delays caused by the Contractor: these include delays caused by parties for whom the Contractor is responsible in law. This type is also often referred to as 'culpable delay'. Generally, under most contracts, the JCT 98 included, the Contractor is neither entitled to extension of time nor recovery of loss and/or expense on account of this type of delay.

Delays caused by neutral events: neutral events are those for which neither the Employer nor the Contractor is responsible, e.g. exceptionally adverse weather. Most contracts just allow the Contractor more time to complete but with no corresponding entitlement to recover any loss and/or expense caused.

Delays caused by the Employer: these include delays caused by parties for whom the Employer is responsible in law. In most standard forms, the Contractor is entitled to extension of time and recovery of loss and/or expense caused by this type of delay.

Delays of the last two types are sometimes referred to as 'excusable delay' in the sense that, by being entitled to extension of time, the Contractor is excused liability for liquidated damages which would otherwise have been payable.

The term 'concurrent delays' is used here to describe the situation where a number of delaying events overlap, e.g. non-receipt of information from the Architect and adverse weather conditions. This situation often gives rise to disputes concerning the extent to which each event was responsible for the delay. Although the Architect is not required under Clause 25 to assess extension of time in respect of each Relevant Event, he would have do so if some of the causes of delay are also grounds for recovery of direct loss and/or

---

50 *M. J. Gleeson (Contractors) Ltd v. London Borough of Hillingdon* (1970) 215 EG 165.

expense under Clause 26. There are three approaches commonly adopted to deal with concurrent delays:[51]

1. the first-in-line approach
2. the dominant cause approach
3. the apportionment approach

The suitability of any of these methods depends upon the circumstances of the delay. Even then, none of them can produce precise results in any given situation. Resolution of these problems therefore depends upon the negotiation skills of the parties involved. The first approach assumes that the first event is the cause of the whole delay. On the one hand, this means that if the event is a ground for extension of time the contractor gets the extension even if his subsequent actions compounded the delays. On the other hand, if his own delay was compounded by causes for which the Employer was responsible, the Contractor is not entitled to extension of time.

The dominant approach attributes the entire delay to the dominant event. There are two problems associated with this approach. First, it breaks down if the events have equal influence. Second, the implications for recovery of direct loss and/or expense under Clause 26 could be unfair to one of the parties. For example, where the dominant cause of delay is also a ground for recovery then the Contractor may be entitled to recovery in respect of the whole delay though some of the contributory causes of the delay may not be grounds for recovery. The main problem here is that the Architect may have problems attributing delay to one cause for the purposes of extension of time under Clause 25 and to others when it comes to assessing entitlement to loss and/or expense. For these reasons, the use of the approach was disapproved of in *H. Fairweather & Co. Ltd v. London Borough of Wandsworth*.[52]

The apportionment method attempts to distribute the total delay to the various contributing causes. It would appear that under the JCT 98, where one of the delaying events is not a Relevant Event, the Architect must follow this approach. Clauses 25.2.2.2 and 25.3.1.2 support this view. Under Clause 25.2.2.2, the Contractor is expected to provide the Architect with estimates of the delay caused by each Relevant Event. As Clause 25.3.1.2 requires the Architect to give an extension of time which is 'fair and reasonable', the Architect must make allowances for the Contractor's contribution to the delay.

11.9 Delays within culpable delay

There has been some debate regarding the right of architects to extend time where a contractor already in culpable delay is delayed further by an event which is expressly a ground for extension of time under the contract. Assuming that there is a right to extend time, there is the further question regarding the method of assessing the amount of extension. To illustrate these problems, consider the situation where, after the award of all extensions of time due, a final completion date is fixed for the end of the 100th week. It is now the 120th week and the works are not yet complete. Exceptionally adverse weather

---

51 Egglestone, B., *Liquidated Damages and Extension of Time in Construction Contracts*, 2nd edn, (Blackwell Science, Oxford, 1997), at p. 202.
52 (1988) 39 BLR 106.

conditions are then encountered and the contractor is delayed by 4 weeks as a result. If this cause of delay is a ground for extension of time, it may be argued that as the conditions would not have been encountered if the contractor had finished on time, he should not be entitled to an extension. Now suppose that the cause of delay is not weather conditions but a variation issued during the period of culpable delay. Can the Architect extend time? If he has no jurisdiction then, on principles already discussed, time for completion *may* become at large. The word 'may' is emphasized because it is arguable whether completion time can become at large when it has already expired. Besides, the case law from which the principle of time being at large has been developed concerned delays occurring prior to the contractual completion date.

There is the further question whether, assuming time can be extended, the extension is to be 'net' or 'gross'. In the example a net extension would result in the new completion date being the end of the 104th week whilst the 124th week would be the case with a gross extension. Thus, a gross extension has the effect of exculpating the contractor for his own earlier delays. *Dicta* in *Amalgamated Building Contractors Company Ltd v. Waltham Holy Cross UDC*[53] suggests that there is jurisdiction to extend and that net extension would be the more appropriate method. In that litigation Denning LJ (as he then was) said *obiter*:

> Take a simple case where the contractors, near the end of the work, have overrun the contract time by six months without legitimate excuse. They cannot get an extension of time for that period. Now suppose that the works are still uncompleted and a strike occurs and lasts a month. The Contractors can get an extension of time for that month. The architect can clearly issue a certificate which will operate retrospectively. He extends the time by one month from the original completion date, and the extended time will obviously be a date which is past.

On the question of the Architect's jurisdiction to extend time, the Guidance Notes to Amendment 4 of the JCT 80 suggest that there is such a right. However, Murdoch[54] argues that the wording of Clause 25.3.3 of that Contract may not bear such an interpretation. The Commercial Court considered these chestnuts in *McAlpine Humberoak v. McDermott International Inc. (No. 1)*[55] Responding to the argument that variations during culpable delay had the effect of rendering completion time at large, Lloyd LJ said:

> If a contractor is already a year late through his culpable fault, it would be absurd that the employer should lose his claim for unliquidated damages just because, at the last moment, he orders an extra coat of paint.

The implication of that judgment for the JCT 98 is that the Architect would be authorized to extend time on a 'net' basis if the Contractor is delayed during a period of his culpable delay by a event for which the Employer is responsible. These questions were again put to the courts in *Balfour Beatty Building Ltd v. Chestermount Properties Ltd.*[56], a case which arose from a contract substantially in the JCT 80 form of contract with approximate quantities. When the Contractor failed to complete the works by a new completion date fixed by the Architect (9 May 1989), he issued a Certificate of Non-

53 [1952] 2 All ER 452.
54 Murdoch, J., 'Contractual Overruns and Extensions of Time' (1992) 8 Const. LJ 32.
55 (1992) 58 BLR 1.
56 (1993) 62 BLR 1.

Completion under Clause 24.1. By agreement between the Contractor and /
the Architect then issued variation orders between 12 February and 12 Jul'
these variations into account, the Architect finally fixed the complet'
November 1989 (note that this date is before the issue of the var
completion was achieved on 25 February 1991. Upon the issue of a Ct.
Completion in respect of the new completion date, the Employer sought to ιι.
liquidated damages.

In the litigation that ensued, two questions were put before the court. Did the issue of a
variation order within the Contractor's culpable delay have the effect of rendering
completion time at large? On this question, the Contractor argued that on the proper
interpretation of Clause 25 a variation was a Relevant Event only if it was issued before
the contractual completion date. It was further claimed that, as the variations were issued
after the contractual completion date, time for completion was at large with the result that
the Employer was not entitled to liquidated damages. The arbitrator's rejection of the
argument was upheld by Mr Justice Colman. The judge explained that from the general
scheme of Clause 25, in the absence of clear words to the contrary, it has to be inferred
that a variation after the contractual completion date is a Relevant Event.

12

Claims for damages and contractual claims: an overview

(See also: Chapter 13: Contractual claims under JCT 98.)

12.1 Introduction

The term 'claim' has no precise meaning and conjures up a variety of emotions in employers, their professional teams, and contractors. In some sectors of the construction industry 'claims' are an unwelcome concept. They are often associated with aggressive predatory commercial practice by contractors. Experience of claims has even led some building clients to remove contractors from tendering lists solely on the grounds that they had received claims from those contractors on earlier projects. Even contractors sometimes distance themselves from claims, and market themselves as being not 'claims conscious'. Some view claims as no more than a request for a contractual right, including payment for a variation;[1] for others a claim is a last resort, often borne out of frustration at not being able to recover what they think is due to them – a last stage before formal proceedings. For the purpose of this book a 'claim' is any request by the Contractor for payment, other than an application under Clause 30. Such claims are often categorized as:

1. loss and/or expense arising from matters provided for expressly in the contract, (principally Clause 26 where JCT 98 applies),
2. damages arising out of a breach of contract, or a breach of duty in tort (e.g. negligence),
3. claims in Restitution,[2]
4. other matters.

Category One is commonly called a 'contractual claim' or 'loss and expense claim', whereas Category Two is often described as a 'common law claim' or 'damages claim'. These two categories are dealt with in the following sections.

---

1 See General Conditions of Contract for Water Industry Plant Contracts Form G/90.
2 See Chapter 1: Section 1.8.2, sub-section: 'Quasi-contract or restitution?'

Category Three is dealt with briefly in Chapter 1,[3] and is outside the normal authority of the Architect.

Category Four comprises mainly 'moral' or 'sympathy' claims, in which the Contractor feels hard done by, but for which there is no breach or other actionable wrongdoing by the Employer. Thus there is no entitlement to reimbursement either under the contract or as damages. An example might be a claim for extra costs incurred by the Contractor entirely at his own risk and choice, but which in hindsight benefits the Employer. In such cases the Employer, out of gratitude, might contribute towards the Contractor's costs as a goodwill gesture. The Architect has no authority to deal with these matters unless expressly empowered by the Employer. Such payments are often described as *ex gratia* payments.[4]

12.2 'Contractual' and 'common law' claims: similarities

A claim for loss or expense made under Clause 26 of the Contract, is broadly the contractual equivalent of damages for breach of contract. In *Wraight Ltd v. P. H. & T. Holdings*[5] it was said:

> there are no grounds for giving to the words 'direct loss and/or damage caused to the Contractor' any other meaning than that which they have, for example, in a case of breach of contract

Many of the 'matters' described in Clause 26, giving rise to an entitlement under the contract, are matters which would also describe breaches by the Employer. For example, Clause 26.2.1 provides for the possibility of recovery by the Contractor in the event that the Architect issues information late; in the absence of such provision, late supply of information would be a breach of the contract, giving rise to damages at common law. Similarly under Clause 26.2.6, failure by the Employer to allow access would also be a breach of contract. However, some matters which create entitlement under Clause 26 would not be breaches of contract. The most common example is probably the effect of Variations. Whilst a late Variation might be hindrance by the Employer (i.e. a breach of contract), the issue of other Variations is not in itself a breach, because the Contract allows it; but the effect constitutes grounds for extra payment under Clause 26.2.7.

12.3 'Contractual' and 'common law' claims: differences

12.3.1 Architect to ascertain

Whilst both common law claims and contractual claims are made against the Employer, the procedure for dealing with common law claims is not dealt with in the Contract. One of the principal characteristics of contractual claims is the ability of the Architect to ascertain the amount due. His authority and duty is set out in the Contract (see Chapter 13: Section 13.2.2); under JCT 98 it does not extend to dealing with claims from the

---

3 Ibid.
4 *Ex gratia*: as a favour: *Osborn's Concise Law Dictionary* 7th edn (Sweet & Maxwell, 1983).
5 (1963) 13 BLR 26.

Contractor for the Employer's breach of that contract, unless the Employer expressly gives him power as an agent. This can create confusion when a Contractor submits to the Architect a claim which describes itself as both a request for loss or expense and/or a claim for damages. In those circumstances the Architect should deal with the request to the extent that the claim is a *bona fide* request under Clause 26, and he should ignore the common law element other than advising his Client of the contents. On the other hand, the Contractor cannot expect the Architect to deal with such matters, and should submit his common law claim direct to the Employer.

12.3.2 Interim payment

A great advantage to the Contractor of framing his claim as a loss or expense claim under the Contract, as opposed to making a common law claim, is the entitlement to payment of loss and/or expense in interim certificates.

12.3.3 Notices

In return for the entitlement to recover the equivalent of damages as a right under the Contract, the Contractor is obliged to follow an administrative procedure in the form of notices to the Architect (see Chapter 13). The procedures apply only to contractual claims; they do not apply to common law claims. Indeed it is this point which normally prompts the Contractor to submit his claim in the alternative (i.e. as a contractual claim and as a common law claim); this is particularly likely when the Contractor realizes he has failed to comply properly with the requirement for notices of his loss or expense claim and anticipates rejection by the Architect.

12.4 'Contractual' and 'common law' claims: as alternatives

Some contracts do not have the equivalent of the loss or expense provisions in JCT 98. Under those contracts, the Contractor would need to pursue a common law claim if the Employer were to commit a breach of the contract. Most construction industry standard contracts, however, contain some form of provision to compensate for the Employer's breaches. Under those contracts the obvious and intended remedy is the use of the compensation provisions in the contract; indeed that might be the only remedy where the contract limits the rights of the parties to the contractual remedies.[6] That is not the case with JCT 98 which, at Clause 26.6, states that the provisions of Clause 26 are without prejudice to any other rights and remedies the Contractor might possess. That means the Contractor is not obliged to use Clause 26, so it does not necessarily prevent a common law claim instead of the contractual claim; nor does it preclude a common law claim as an alternative, or even addition to, the contractual claim. The position was explained in *London Borough of Merton v. Stanley Hugh Leach Ltd*[7] in relation to Clause 24 of JCT 63 (the forerunner of the modern Clause 26):

---

6   See Clause 44.4 of MF/1 (Rev 3) *Conditions of Contract*, published by the Institution of Electrical Engineers.
7   (1985) 32 BLR 51.

But the Contractor is not bound to make an application under Clause 24(1). He may prefer to wait until completion of the work and join a claim for damages for breach of obligation ... under the contract. Alternatively, he can make a claim under Clause 24(1) in order to obtain prompt reimbursement and later claim damages for breach of contract, bringing the amount under Clause 24(1) into account.

It is open to Contractor to choose which route he wishes to pursue,[8] or he may pursue both, but clearly he will not be entitled to recover the same damages twice.

12.5 Common principles

12.5.1 Cause and effect and global claims

It is tempting for a Contractor, when preparing a claim, to give notice that he is incurring both loss and expense, then simply to send a lengthy monetary calculation. The apparent basis is that he has incurred lots of additional costs (or damages), so it follows that the Employer must be liable for them. The effort is put into calculating compensation, rather than establishing entitlement.

A common flaw in claims is the absence of a causal link between the matters giving rise to the claim and the compensation claimed. Similarly there is often no link identified between the matters described and a breach of contract or a matter described in the contract as giving rise to entitlement. The principle which such claims ignores is the maxim 'He who alleges must prove'. The paucity of causal links is understandable. Proving that a particular cost results solely from one particular cause can be difficult. It can be particularly difficult if causes are closely related, such as the progressive late issue of a number of drawings. Each drawing may be a cause in its own right, and the task of showing which portion of each head of claim results from each delayed drawing is daunting to say the least. It may not even be possible; but the Contractor is required to try. Indeed the main task of the Contractor in a contractual claim is to provide sufficient information to enable the Architect to ascertain the amount due.

Likewise the Contractor may have difficulty when two or more causes of loss occur concurrently. The problem is often associated with delay, particularly when the contract expressly provides for reimbursement in respect of one concurrent cause but not in respect of another. One example would be delay caused by exceptionally adverse weather running concurrently with delay caused by variations. It may in those circumstances be that the start of one cause preceded the other, so if a variation pushed work back into bad weather loss may be recoverable under principles used in 'winter working' claims.[9] There are a number of alternative ways in which such delays can overlap, and which are dealt with in several specialist texts,[10] but the overriding principle is that damages are losses which would not have been incurred in any event.

---

8 *Merton v. Leach* on this point.
9 In *Ellis-Don Ltd v. The Parking Authority of Toronto* (1985) 28 BLR 98, an employer's delay preventing summer work being carried until the winter entitled the contractor to be paid extra for the winter working.
10 See Dunn, S., *The Law of Damages* (Net Law Books, 1999), Chapter 2 Causation, for concise analysis of alternatives; see also Egglestone, B., *Liquidated damages and Extension of Time in Construction Contracts* 2nd edn, paras 14.2, 14,5; *Keating*, 6th edn. pp. 193–96.

Whilst Contractors are often criticized for failing to demonstrate causation, Architects are often equally criticized for showing dogged resistance to any claim, and for demanding more and more information in the hope that it will not be forthcoming. This may seem a cynical view, and obviously it does not apply to all contractors and architects; but it is a common aspect of dealings on the battleground known as the 'global claim'.

Global claims gained notoriety amongst architects, and popularity amongst contractors, following the reporting of the decision in *Crosby v. Portland UDC*,[11] in which a 'rolled up claim' was said to be acceptable in some limited circumstances. Contractors seemed to take that to be a general rejection of the requirement to establish cause. After a number of cases looking into the extent of the need to identify each causal link, the position was neatly summed up by the court in the case of *Mid Glamorgan County Council v. J. Devonald Williams and Ptnrs*:[12]

1. A proper cause of action has to be pleaded.

2. Where specific events are relied upon as giving rise to a claim for moneys under the contract then any preconditions which are made applicable to such claims by the terms of the relevant contract will have to be satisfied, and satisfied in respect of each of the causative events relied upon.

3. When it comes to quantum, whether time based or not, and whether claimed under the contract or by way of damages, then a proper nexus should be pleaded which relates each event relied upon to the money claimed.

4. Where, however, a claim is made for extra costs incurred through delay as a result of various events whose consequences have a complex interaction that renders specific relation between event and time/money consequence impossible or impracticable, it is permissible to maintain a composite claim.

These guidelines were made in the context of litigation, and apply in principle to any claim made by the Contractor, whether as a contractual claim or as a common law claim and would certainly apply to a claim in arbitration. However, it must be remembered that contractual claims are intended to be paid in interim certificates. The Architect cannot expect to receive on a month-to-month basis, the same amount of information, records, and analysis as would be presented for formal proceedings. In short, the Architect is required to do what he can with the information available at the time, and the Contractor is required to provide only sufficient to enable the Architect to form an opinion. The practicality of the situation during the course of the project was recognized by the court in *London Borough of Merton v. Stanley Hugh Leach Ltd*:[13]

If application is made for reimbursement of direct loss and/or expense attributable to more than one head of claim and at the time when the loss or expense comes to be ascertained it is impracticable to disentangle ... the part directly attributable to each head of claim, then ... the architect must ascertain the global loss directly attributable to the two causes. ... To this extent the law supplements the contractual machinery which

---

11 (1967) 5 BLR 121.
12 [1992] 29 Con. LR 129; (1992) CILL 722; 17 September 1991.
13 (1985) 32 BLR 51.

no longer works in the way it was intended to work, so as to ensure that the contractor is not unfairly deprived of the benefit which the parties clearly intend he should have.

If the information provided by the Contractor is insufficient to form any opinion, the Architect may request more, stating what he needs. This extra information could be given orally and might need to be only brief, since the Architect 'is not a stranger to the work'.[14] Indeed, the judge continued, in some cases, 'the briefest and most uninformative notification of a claim would suffice ... for instance where the architect was well aware of the contractor's plans'. It seems the Architect cannot just sit back and wait for the Contractor to meet the highest standards of notification and analysis; he must act pragmatically. Nevertheless, the prudent Contractor will still keep proper and adequate records, and still give as much information as he can in the time available, to meet the monthly timetable of interim certificates.

12.5.2 Measure of recovery

The basic object of damages, whether recovered either as loss or expense under the Contract, or as damages for breach of the Contract, is to put the claimant back 'so far as money can do it ... in the same situation ... as if the contract had been performed'.[15] This is subject to any limitation agreed between the parties, such as express agreement that liability may be 'capped', or limited to 'costs', thus excluding 'losses'.[16]

Neither loss or expense, nor common law damages, are intended to enhance profits; nor are they to be treated like added turnover, unless the contract treats them as such. Some engineering contracts provide for profit to be added to costs in the manner of Variations,[17] but JCT 98 does not.

However, not all losses or costs are recoverable; the law allows only those amounts that are not too remote or extravagant, and that do not unnecessarily improve the claimant's position.

Remoteness: The rules are stated in *Hadley v. Baxendale*:[18]

the damages which the other party should receive in respect of such breach should be such as may fairly and reasonably be considered as either arising naturally, i.e. according to the usual course of things, from such breach of contract itself, or such as may reasonably be supposed to have been in the contemplation of both parties at the time they made the contract as the probable result of the breach. ... If special circumstances ... were communicated ... to the defendants ... the damages ... would be the amount ... which would ordinarily follow from a breach of contract under the special circumstances so known and communicated.

The two concepts emerging from this statement are known as 'general damages' (the first rule), and 'special damages' (the second rule). An example of the first rule in the context

---

14 *Per* Mr Justice Vinelott in *Merton v. Leach* (1985) 32 BLR 51.
15 *Robinson v. Harman* (1848) 1 Ex. 850, 855 (1991).
16 E.g. ICE *Conditions*, 6th edn (1991), which limits liability to 'costs' defined in Clause 1(5).
17 See *Conditions of Contract* MF/1 (rev 3) published by the Institution of Electrical Engineers, Clause 41.2.
18 *Per* Baron Alderson (1854) 9 Ex 341.

of a construction project would be site establishment costs, which everyone would reasonably expect to occur naturally as a result of site delay. An example of the second rule would be a claim for loss of profit suffered as a result of losing a particularly lucrative contract elsewhere.[19] A common example of putting special circumstances into contemplation at the time the contract is formed can be seen in the practice of advising a sub-contractor of the liquidated damages applying on the main contract. The forewarning enables a Contractor to claim from a sub-contractor (in principle), the damages paid to the Employer; it prevents the sub-contractor from alleging that the damages are 'special' and not in contemplation at the time of entering the sub-contract.

The rules in *Hadley v. Baxendale* have been refined on a number of occasions to deal with difficult cases, where the contemplation of likelihood of the resulting damages is questioned.[20] Each instance of claim has to be considered on its own merits, but the general principles of recovery have been summarized as:[21]

- One can only recover that part of the loss actually resulting as may be reasonably considered as arising according to the usual course of things from the breach.
- The matter has to be considered as at the time of making the contract.
- The question as to whether the loss was liable to occur does not have to be actually asked; it is sufficient if the party breaking the contract would as a reasonable person have concluded the loss to be liable to result, if he had considered the question.
- 'Liable to result' should be read in the sense conveyed by the expressions 'serious possibility' and 'real danger'.

Extravagance (the 'duty' to mitigate): Strictly there is no duty to mitigate losses, unless the contract expressly requires for a contractual claim. The injured party can incur whatever he wishes; the so called 'duty' simply means the wrongdoer cannot be expected to compensate the injured party for his extravagance. In mitigating his loss the claimant must take reasonable steps to minimize his loss, and not take unreasonable steps that increase it. However, it is permissible to spend money in order to save money, and claim the expenditure as mitigation. This can be so, even if the resulting losses actually turn out to be greater than the losses that the expenditure was intended to reduce, provided the decision was a reasonable decision to make at the time it was made.[22] In building terms, this approach can sometimes be seen in the use of acceleration measures to reduce time overrun and resultant prolongation costs.

Betterment: The general purpose of damages for a breach, is to put the injured party back in the position he would have been in, but he should not be enriched by an award of damages (see above). There are times when the correction of a breach either requires something different from the original intention, or provides an opportunity to change or improve the claimant's position.

An example of something different being required would include replacement of a

---

19 See *Victoria Laundry (Windsor) Ltd v. Newman Industries Ltd* [1949] 2 KB 528.
20 See *Victoria Laundry* case at n. 19; and *Czarnikow v. Koufos* [1969] 1 AC 350.
21 See *Bevan Investments v. Blackhall & Struthers*, Court of Appeal of New Zealand, 29 July 1977; (1979) 11 BLR 78.
22 See *Melachrino v. Nicholl & Knight* [1920] 1 KB 693

defective boiler with a new model when the original was no longer available; [23] The change is necessary, and there is no reasonable alternative but to install the new model. The general principle that the claimant should not be enriched may not apply where the 'enrichment' cannot reasonably be avoided. In *Harbutt's 'Plasticine' Ltd v. Wayne Tank & Pump Co Ltd*,[24] the full cost of rebuilding a factory was awarded without any allowance for the factory being new and more valuable; the factory owner had no alternative but to rebuild. However, in extreme cases betterment may be considered; in *Bacon v. Cooper (Metals) Ltd*[25] it was said that if absurdity resulted from ignoring betterment, then justice may require it to be taken into account.

An example of opportunity to improve would be replacement of a defective gas fired boiler with solar panels. Here the claimant is taking advantage of necessary work to improve his overall position, by changing the original specification to solar heating. He is not replacing new for old. This, arguably, is better categorized as a failure to mitigate.[26]

Whether costs actually incurred: It is not necessary for an injured party to have actually incurred cost to be able to maintain a claim. In *Forsyth v. Ruxley Electronics & Construction Ltd*,[27] the House of Lords considered the damages due to the building owner when a swimming pool was built nine inches shallower than the contract specification. It was held that in principle it was irrelevant what the plaintiff intended to do with his damages. However in that case, it was also held that where reinstatement of defective work is necessary, such reinstatement must be a reasonable thing to do; in short it would be wrong to spend large sums of money to obtain a new pool which gave no additional benefit over that which is replaced, simply to achieve the strict specification (it appears that if the pool had been required for diving competitions, it may then have been reasonable to reinstate). The situation in the *Ruxley* case should not be confused with the position when there is an existing liability to pay a third party which has not yet been discharged; the late payment of a cost will not remove entitlement to claim that cost as damages.

12.6 Employer's claims

Claims made by the Employer against the Contractor are governed by the same principles as those applying to Contractor's claims, i.e. under the rules of the contract where a claim is made under the contract, and under general damages principles where the claim is made in respect of a breach.

There are a several situations in which the Employer has entitlement to claim against the Contractor under the contract; but each may also give rise to alternative or additional rights in common law. The most common are:

---

23 *McGregor on Damages*, ed. Harvey McGregor, 15th edn (Sweet & Maxwell, 1988) at para. 17, discusses 'betterment' in terms of new for old, and only arising when there is an increase in value.
24 [1970] 1 QB 447.
25 [1982] 1 All ER 397.
26 This point was considered in *Skandia Property (UK) Ltd and Vala Properties B.V. v. Thames Water Utilities*, ORB; 30 July 1997, in which the obligation to give credit under 'betterment' principles was differentiated from the principles in 'mitigation' to act reasonably.
27 (1995) 73 BLR 1, HL.

- delay in completion of the Works
- failure to complete the Works
- failure to comply with an Architect's instruction
- the correction of defects.

12.6.1 Delay in completion of the Works

The Employer's remedy under the Contract lies in the deduction of liquidated and ascertained damages (see Chapter 11). The remedy is exhaustive.[28] When entering his liquidated damages sum in the appendix, the Employer is expected to have included all the costs and losses he might expect to suffer as a result of delay, so he cannot then deduct liquidated damages and add a claim for other delay related matters. However, he can still maintain a claim for damages for breach if the liquidated damages clause fails.[29] This can occur when the clause is challenged due to a legal defect in construction[30] or if the liquidated damages amount is successfully challenged as being a penalty (see Chapter 11 for principles of liquidated damages). Whether the Employer is entitled to recover more in general damages than he would have received under the failed liquidated damages clause is uncertain. It has been held that the figure mentioned as liquidated damages does not always bind a party from recovering a larger sum;[31] but any such claim would be subject to the rules on remoteness, so the level of liquidated damages could be taken into account when considering contemplation under the second rule in *Hadley v. Baxendale*.[32]

12.6.2 Failure to complete the Works

The Employer's contractual remedy in the event of failure to complete is the right to determination under Clause 27, to have the work completed by others, and to adjust the contract sum accordingly (see Chapter 16). Nevertheless, Clause 27.8 provides that such rights are without prejudice to any other rights or remedies that the Employer may possess. Such rights would encompass similar rights in common law and entitlement to damages, including the additional cost of completing the work plus any other damage suffered. In some circumstances the proper measure of damages might be depreciation of property value, or an amount for loss of amenity.[33] Such damages would be subject to the general rules on remoteness, mitigation, and betterment.

---

28 See *Temloc Ltd v. Errill Properties Ltd* (1987) 39 BLR 30, 12 Con. LR 109.
29 *Per* Lord Justice Croom-Johnson in *Temloc v. Errill*.
30 E.g. *Peak Construction Ltd v. McKinney Foundations Ltd* (1970) 1 BLR 111, in which it was held the Employer lost his rights to liquidated damages when an Employer's breach put time at large.
31 Per Scrutton LJ in *Cellulose Acetate Silk Co. Ltd v. Widnes Foundry (1925) Ltd* [1933] AC 20.
32 See Section 12.5.2. See also *Temloc v. Errill* in which liquidated damages were inserted as 'NIL' and the Employer was unable to recover any general damages.
33 In *Forsyth v. Ruxley*, see n. 28, it was held there was no diminution in value of the property caused by a swimming pool being nine inches shallower than the contract requirement, but it was an option for consideration. £2500 was awarded for loss of pleasurable amenity by the trial judge which was criticized by Lord Mustill in the House of Lords as being a large amount, although the quantum had not been challenged.

12.6.3 Failure to comply with Architect's Instructions

Under Clause 4.1.2 the Employer may employ others to carry out an Architect's instruction which the Contractor has refused to do after receiving a notice to comply. All costs incurred in such employment may be deducted by the Employer from sums due to the contractor, and are treated as a debt. In the absence of such a clause the Employer would have a similar right in common law.

12.6.4 Correction of defects: limitations and Employer's additional losses

The Contract, at Clause 17.2, provides that any defects resulting from work or materials not in accordance with the contract (i.e. a breach) appearing during the Defects Liability Period shall be corrected by the Contractor at no cost to the Employer. Many contractors look upon this arrangement as no more than an obligation, but it entitles the Contractor to put right his breach at his own cost, instead of the cost which he would have to bear if the Employer engaged others to do the work.

Limitations: If the Employer fails to allow the Contractor to correct his work, and instead has the work done by others, the Employer will be in breach; he cannot then recover all his losses. In *Pearce & High Ltd v. John P. Baxter and Mrs A. Baxter*[34] the Court of Appeal held that a notice required by a contract[35] to be given within a stipulated time, advising the Contractor of the need to correct defects, was a condition precedent to the Contractor's obligation. However, it was also stated[36] that the Employer does not lose his right to recover entirely. This confirmed the court's decision in *William Tompkinson & Sons Ltd v. The Parochial Church Council of St Michael-in-the-Hamlet*,[37] that if the sum claimed from the Contractor is greater than the amount which the Contractor would have incurred, had he been allowed to correct his own work, the claim will be limited to that lesser amount.

In some circumstances the Employer may claim diminution of property value, although he may also receive no more than nominal damages for loss of amenity. In *Earl Freeman v. Mohammed Niroomand*,[38] the Court of Appeal upheld the decision of the trial judge who had said:

> In some circumstances it might be the diminution in value of property. In other circumstances it might be the cost of remedial work. In this case neither is ... [applicable] ... since there is no evidence of any diminution ... and ... [no evidence of intention to correct the work]. Therefore the only remaining touchstone is the evidence of [the builder] that it would have cost him an additional £130, if at the time of the original construction, he had complied with the drawing.

34 (Unreported), 15 February 1999, 1999 CILL 1488; referring to JCT Agreement for Minor Building Works but equally applicable to JCT 98.
35 JCT Minor Works form.
36 *Per* Evans LJ (*obiter*), at 104.
37 (1990) Const. LJ 319.
38 (Unreported): CA: 8 May 1996: CCRTF 95/0660/C: BLISS 4 November 1996 p. IB 43/2.

Whether diminution or remedial costs apply will depend on individual circumstances, although the trend seems to be towards remedial costs, when remedial work is a reasonable course. If property is intended for sale or leasing, then any diminution value may be relevant. If remedial work would be unreasonable and provide no benefit, and if diminution is not present, then damages may be limited to a nominal amount for loss of enjoyment or amenity.

Employer's additional losses: The Contractor's obligation to correct defects is not an exhaustive remedy. It may be that, in allowing the Contractor necessary access to do the work, the Employer suffers other losses. For example, correcting a leaking pipe at high level over a production area, may cause disruption to the Employer's production schedule; major work in offices, or inhabited premises, may even require temporary accommodation. Addition costs such as these may be recovered by the Employer from the Contractor, in a damages claim. In *H. W. Nevill (Sunblest) Ltd v. William Press & Son Ltd* [39] the employer claimed both the cost of delays caused to a second contractor on site, when the first contractor returned to repair defective drains, and also the cost of late opening. It was argued by the contractor that the obligation to correct his own work was his only obligation. The court held that the remedies under the contract were not exclusive, that the defective work was a breach of contract, and that the Employer was entitled to recover his losses from the Contractor in addition to having the work corrected. The Employer's right to recover such addition losses will be a claim for damages and subject to the rules on remoteness, mitigation, and betterment.

12.7 Excessive claims: criminal liability

Whilst it is the duty of the Architect to ascertain loss and expense under the Contract, Contractors often produce applications in formal claim documents. Such claims are usually presented after the Contractor has become dissatisfied with the Architect's ascertainment. Consequently claims may be viewed by Contractors both as a last chance to 'get everything in' and also to create a negotiating position. Occasionally the latter can lead to the claim being inflated to a level higher than the Contractor really thinks it is worth. It was said by one distinguished judge in 1704:[40] 'Shall we indict one man for making a fool of another?' He had just decided it was not stealing when the defendant had pretended to be authorized to collect money on behalf of someone else. But contractors should be very wary; there are many Employers who do not take an eighteenth-century view of things, and who are not aware of what is seen by some as the unspoken ground rules adopted by building trades. To them, inflating the claim could be a criminal act, and the twentieth-century courts agree. Under the Theft Acts 1968 and 1978 it is an offence to obtain money or avoid payment by deception or false accounting. If a grossly excessive price is quoted and charged, where there is a relationship of trust between customer and

---

39 (1981) 20 BLR 78.
40 Holt CJ in *Jones* (1704) cited in Smith, J.C. and Hogan, B., *Criminal Law*, 5th edn (Butterworth, 1983) at p.507.

tradesman, an offence may be committed under the Theft Act 1968 s.15(1).[41] Likewise, it is an offence to fabricate or change records relied on to support a claim; this would extend to daywork sheets. Employers should also avoid overstated counterclaims made with the sole intention of cancelling a Contractor's claim.[42]

41 E.g. *R v. Silverman*; 31 March 1987, Times Law Reports 3 April 1987, 33; (CA).
42 See *Building*, 2 February 1997 p.33; Tony Bingham discusses a case in which the director of a building company received a prison sentence after his company avoided paying a sub-contractor's account by fabricating an excessive counterclaim for alleged defects.

13

Contractual money claims under JCT 98

(See also: Chapter 12: Claims for damages and contractual claims: an overview.)

13.1 Introduction

Loss or expense incurred by the Contractor as a result of deferred possession of the site, or disruption to the progress of the Works caused by specified matters, may be recovered by the Contractor from the Employer under Clause 26. It is for the Architect to ascertain the amount due, and to certify payment to the Contractor in interim certificates. The successful operation of the Clause 26 provisions relies on a procedure triggered by the Contractor in which he first makes application, together with details if he wishes, to the Architect, followed by information which the Architect may request. Recovery under JCT 98 is described as 'direct loss and/or expense',[1] that has been held to be the equivalent of damages for breach of contract,[2] i.e. to put the injured party back (so far as money can do it) into the position he would have been in had the contract not been broken. The general principles governing the level of damages recoverable apply equally to 'loss and/or expense' (see Section 13.2.2 (3) below).

13.2 Procedures

13.2.1 Application by Contractor (notices)

Clause 26.1 states that if the Contractor makes an application to the Architect that he has incurred, or is likely to incur, direct loss and/or expense for which he would not be reimbursed under other provisions, either as a result of deferred possession or as a result of specified matters, the Architect must then take action (see Section 13.2.2). The Contract does not require the Contractor to make application; it simply states the effect if he does.

---

1 Some construction contracts limit recovery, e.g. ICE *Conditions 6th Edition* which refers to defined 'costs'.
2 See Chapter 12 for comparison with claims for damages, and general principles applying to both loss or expense, and damages.

An application by the Contractor (often referred to as a 'notice') must conform to several conditions:

1. The application must be in writing and made to the Architect. An oral application would not be sufficient, even if made at a site progress meeting and recorded in the minutes.[3] Neither would a Contractor's valuation application for an interim certificate under Clause 30.1.2.2 constitute good notice, even if it identified the relevant matters, since such a valuation is submitted to the Quantity Surveyor.

2. The Contractor 'may give' his quantification of loss and expense if he wishes, but there is no obligation since it is for the Architect to ascertain the amount due.

3. Under Clause 26.1.1, the application must be made as soon as it becomes (or should reasonably become) apparent to him that regular progress of the work has been or is likely to be affected. It should be noted here that it is the recognition of disrupted progress that triggers an application, not the recognition of the monetary effect.

4. The Contractor is not obliged to provide information with his application, but must do so on the request of the Architect (Clause 26.1.2 and 26.1.3). The level of information is that which would reasonably enable the Architect (or in the case of loss or expense, the Architect or Quantity Surveyor) to form an opinion. This seems to require objective consideration by the Contractor, and dispute can arise over what is reasonable. It is not clear from the words whether it is the duty of the Architect or the Quantity Surveyor to make the request for details of loss or expense; it is simply the obligation of the Contractor to provide the information 'on request'. The prudent Contractor will ask the Architect in his application what information is required. Unfortunately, it is common for Architects to forget that the Contractor provides information on request. The presentation of claim documents often results from stagnation of a Contractor's application, with the Contractor and the Architect (or the Quantity Surveyor) each awaiting action from the other.

5. The Contractor's application under Clause 26.1 must be in respect of 'material' effect on progress. This would exclude disruption of a trivial nature. What is trivial, and whether a series of individual disruptions each trivial in nature but amounting to a significant disruption can be termed 'material', is unclear. However, Contractors are only likely to make application when a disruption is great enough to come to attention.

6. The effect on progress identified in the application must be caused either by deferred possession or by one or more of the matters listed in Clause 26.2. The words 'any one or more of the matters referred to' seems to allow the Contractor to roll up causative matters, at least in his application, in conflict with the general rules on causation (see Chapter 12, Section 12.5.1). Any matters outside those parameters are not things which the Architect is obliged to consider under this clause.

13.2.2 Architect to ascertain

The role of the Architect is set out in Clause 26.1: 'If the Contractor makes written application ... if, and as soon as, the Architect is of the opinion that the direct loss and/or expense ... (is incurred) ... due to ... (the matters set out in the application) ... then the

---

3 In the Scottish case of *John Haley Ltd v. Dumfries and Galloway Regional Council* (1988) 39 GWD 1599, the Court of Session held that minutes were not good notice of delay.

Architect from time to time thereafter shall ascertain, or shall instruct the Quantity Surveyor to ascertain, the amount of such loss'.

1. 'If the Contractor makes application': There is no common law obligation on the Architect to consider loss or expense in the absence of an application from the Contractor. Clause 26.1 states the procedure to be followed if the Contractor starts the process by making an application. It has been held that the Architect would be negligent if he certified loss and expense in the absence of a written application.[4]

2. 'If, and as soon as the Architect is of the opinion': There is no absolute obligation on the Architect to ascertain the amount due; the obligation is to consider the matter fairly and to form an opinion, if the Contractor starts the process by making an application. This issue was considered in a series of engineering cases[5] concerning the duty of an Engineer to certify interim amounts. In *Kingston-upon-Thames v. Amec Civil Engineering Ltd*[6] it was said:

> The use of the word 'opinion' ... implies that there may be a degree of latitude. ... It may not be practicable to produce an exact valuation in 28 days; and the ... valuation has to be made on the basis of the contractor's statement. ... On matters of opinion ... there may well be room for difference of opinion. ... it is manifestly implicit that in arriving at his opinion he must correctly apply the provisions of the contract.

The Architect may actually form an opinion that nothing is due; if he does then he may have discharged his duty under Clause 26, provided he has correctly applied the provisions of the Contract. However, if he fails to apply his mind properly or at all, he will put the Employer in breach and the Contractor will be entitled to damages.[7] Such a claim would be in common law and not within the Architect's normal authority.

3. 'direct loss and/or expense': Clause 26.1 removes any doubt about the nature of the compensation which may be claimed. The use of the word 'direct' prevents reimbursement of consequential damages.[8] Some construction contracts provide only for reimbursement of 'costs' which may be defined in the contract;[9] JCT 98 allows for both costs and losses, and to avoid pedantic argument, expressly allows either or both to be ascertained. Neither of the terms 'loss' or 'expense' are defined in the contract, but the Court of Appeal in *F. G. Minter v. W.H.T.S.O.*[10] held that 'direct loss and/or expense' is the same as damages

---

4 *Turner Page Music Ltd v. Torres Design Associates Ltd*, Judgment 12 March 1997, ORB 0237; 1997 CILL 1263.
5 Relating to ICE *Conditions of Contract 5th Edition*.
6 (1993) 35 Con.LR 39.
7 See *Croudace Ltd v. London Borough of Lambeth* (1985) 6 Con. LR 70, in which it was said 'it necessarily follows that Croudace must have suffered some damage as a result of there being no one to ascertain the amount of their claim'. In *British Sugar plc v. NEI Power Projects Ltd* (1998) 87 BLR 42, the Court of Appeal decided 'consequential loss' was loss coming under the second rule in *Hadley v. Baxendale*, i.e. special damage (see Section 12.5.2 dealing with remoteness).
8 'Direct damage is that which flows directly from the breach without intervening cause and independently of special circumstances, while indirect damage does not so flow': *Saintline v. Richardson Westgarth & Co.* [1940] 2 KB 99.
9 See ICE *Conditions 6th Edition*, Clause 14(8) dealing with delay caused by late information; see also definition of 'cost' in Clause 1(5).
10 (1980) 13 BLR 1.

arising naturally in the ordinary course of things as described in the first rule in *Hadley v. Baxendale*[11] (see also Chapter 12, Sections 12.2 and 12.5.2). The task for the Architect is to satisfy himself that the heads of claim under consideration are caused by the matters alleged, i.e. that they arise naturally without intervening cause.

4. 'loss and/or expense is (due to the matters set out in the Contractor's application)': The Architect does not have any discretion to consider matters which are not referred to in the application. Indeed he would probably be in breach of his contract with his client if he were to do so, unless the client had given express permission.

When forming an opinion the Architect is required to establish that the loss or expense is due to the matters set out in the application. In order to do this he needs to apply the normal rules of causation (see Chapter 12, Section 12.5.1).[12]

5. 'for which (the Contractor) would not be reimbursed by a payment under any other provision': The Contractor can only apply for additional amounts. This provision makes Clause 26 a final mopping-up clause. The main effect should be to remind the Architect that the Contractor may be receiving payment towards his site establishment costs and overheads through Variations (whether they be based on rates in the Contract Bills or dayworks), insurance claims, or as amounts in lieu of loss or expense in an accepted Clause 13A Quotation.

6. "the Architect from time to time ... shall ascertain, or shall instruct the Quantity Surveyor to ascertain, the amount of such loss': It is for the Architect to calculate the amount due; he is required to ascertain, to find out for certain.[13] Alternatively the Architect is empowered to instruct the Quantity Surveyor to perform the duty.

13.2.3 Relevance of extensions of time

The Contractor's entitlement to recover his loss or expense under Clause 26 is not directly linked to extensions of time under Clause 25. Both clauses independently set out events and matters giving rise to entitlement. An extension of time does not automatically bring with it the right to extra payment. The fundamental difference is that extensions are related to a delayed completion date, whereas loss or expense is related to disrupted, or prolonged, progress. In addition some grounds for extensions of time (e.g. 'neutral events' such as exceptionally adverse weather) do not appear as relevant matters for the purpose of Clause 26.

Nevertheless, under Clause 26.3 the Architect is obliged to notify the Contractor of any extensions of time granted in respect of specified events, i.e. adjustment of discrepancies (2.3), variations and provisional sums (13.2 and 13.3), postponement (23.2), opening up work not found to be defective (25.4.5.2), failure to meet dates on an Information Release Schedule (25.4.6), Employer's own work or supply (25.4.8), and failure to give ingress or egress (25.4.12). This information may assist the Contractor to identify the actual costs or

11 (1854) 9 Ex 341.
12 See summary in *Mid Glamorgan County Council v. J Devonald Williams & Ptnrs* (1992) CILL 722, cited in Chapter 12, Section 12.5.1.
13 *Alfred McAlpine Homes Northern Ltd v Property and Land Contractors* (1996) 76 BLR 59.

losses for which the Employer may be liable, but liability will still depend on a causal link between the matters cited in the application and the money damage (if any) suffered by the Contractor.

13.2.4 Nominated Sub-contractors

If a Nominated Sub-Contractor makes a proper application for loss or expense under its sub-contract, the Contractor must pass on the application to the Architect for his consideration. The Architect is required to deal with the substantive content of the application in a similar manner as he does the Contractor's application, including instructing the Quantity Surveyor if he wishes. As with Contractor's own loss or expense, the Architect must notify the Contractor of extensions of time given to the Nominated Sub-Contractor that are relevant, with a copy to the sub-contractor. The extension of time events relevant to the sub-contractor's loss or expense are the same in principle as those applying to the Contractor (see Section 13.2.3).

13.3 Matters giving rise to entitlement

13.3.1 Deferment of giving possession of the site

Clause 26.1 provides that the Contractor may apply to the Architect if possession of the site is deferred under Clause 23.1.2. That clause allows the Employer to delay the start of work for up to 6 weeks or such lesser period as is entered in the Appendix. However, it is important for the Architect to remember that Clause 23.1.2 is optional, triggered by deleting the relevant option in the Appendix. If the deferred possession option is not operative, or if possession is deferred longer than the operative period (i.e. the deferment is not within the parameters of Clause 23.1.2), then the Contractor is not entitled to loss or expense under Clause 26. That is not to say he has no entitlement at all, but his claim will be for common law damages arising out of the Employer's breach of contract. In those circumstances the Architect has no authority to deal with the matter unless expressly authorized by the Employer.[14]

13.3.2 List of matters affecting regular progress of the Works

Clause 26.2 identifies the matters which may affect the regular progress of the Works. The following is a brief summary with notes where necessary:

Clause 26.2.1: Failure by the Architect to comply with an Information Release Schedule, or failure to comply with Clause 5.4.2, i.e. provide information at a time necessary to enable the Contractor to complete by the Completion Date, having regard to progress if the Works are behind programme.

Unlike previous editions of this form, JCT 98 no longer makes requests in writing for information a condition of entitlement to loss or expense. However, a prudent Contractor

---

14 See Chapter 12, Section 12.3.

will probably still provide a list of information required, so that he can at a later date establish the effect on his progress in support of his Clause 26.1 application.

Clause 26.2.2: Opening up work which is found to be in accordance with the Contract.

Clause 26.2.3: Discrepancies in or divergence between documents.

Clause 26.2.4: Employer's own work or materials supply.

Clause 26.2.5: Instructions regarding postponement under Clause 23.2.

Clause 26.2.6: Failure to give ingress or egress.

This provision deals with situations where the site is both accessed over property in the Employer's possession and control, and such access is dealt with in the Contract Documents. These circumstances commonly arise where the work is an extension or refurbishment to an existing facility, on a site where the Employer's operations are continuing. If the Contract Documents provide for the Contractor to give notice requiring access, then he must give such notice to the Architect to maintain his entitlement. Failure by the Employer to comply with any agreements between the Architect and the Contractor regarding ingress or egress also gives rise expressly to entitlement. If the Contract Documents are silent as to access over the Employer's property, and the Architect makes no agreement, then the Contractor's claim will be in common law, rather than a claim under Clause 26.

Clause 26.2.7: Architect's instructions in respect of Variations (other than an accepted 13A Quotation), or undefined provisional sums (other than Contractor's work and sums for Performance Specified Work).

The Architect must constantly keep in mind that loss or expense may be reimbursed elsewhere, and in particular through the Contractor's Price Statement (under Clause 13.4.1.2.A7.1.1)[15] and Clause 13A procedures. In many cases where a 13A Quotation is accepted it will be difficult for the Architect later to separate the individual heads of claim relating to individual causes; in those circumstances he may need to resort to a simple abatement of Clause 13A against Clause 26 entitlement. Unfortunately, the complexity of this exercise may encourage the Architect to refrain from agreeing amounts in lieu of loss and/or expense in 13A Quotations.

Clause 26.2.8: Approximate Quantities in the Bills which are not a reasonably accurate forecast of the work required.

The nature of Approximate Quantities is that the work cannot be accurately determined, so the Contractor might expect some adjustment; but that is not to say it should be at his risk. However, there is no guidance as to what reasonable accuracy might mean, and it will be for the Contractor to show how the change in quantity affected his work.

Clause 26.2.9: Matters arising out of the Employer's obligation to ensure that duty-holders appointed by him (other than the Contractor) under the CDM Regulations perform their duties.

---

15 See Chapter 6, Section 6.11.6.

Clause 26.2.10: Suspension by the Contractor resulting from the Employer's failure to pay the full amount due by the Final Date for Payment, provided the suspension was not frivolous or vexatious.

The right to suspend was introduced into JCT contracts to comply with the Construction Act.[16] The entitlement to loss and/or expense as a result of such suspension is a matter also incorporated by the JCT but is not a requirement under the Act. Suspension may be valid under the Act even though the sums involved may be minor, but in order to maintain entitlement to compensation the Contractor must ensure his suspension is a deserving response to the payment default. What may be considered frivolous will depend on the circumstances. Whilst a Contractor would be entitled under the Act (and therefore under the Contract), to suspend for underpayment of ten pounds, it would be considered frivolous to incur thousands of pounds in extra costs in suspending a multimillion pound project for such an amount. In those circumstances he would be entitled to an extension of time, but not to compensation. The Contractor's claim would then be in common law, although he might have difficulty showing the expenditure was reasonable.[17] What might constitute vexatious suspension is unclear, particularly if the sum outstanding is considerable, and the Architect must not lose sight of the fact that the Contractor has a statutory right to suspend if money is not paid when it should be.

13.4 Heads of claim

13.4.1 Introduction

Whilst it is for the Architect to ascertain loss and/or expense, the Contractor will often feel moved to prepare a calculation to guide him. Thus the notion of the Architect (or the Quantity Surveyor) ascertaining the amount due tends to be a fiction, and it is far more likely in practice that the Architect will check the Contractor's figures.

Claims from Contractors are often split into delay (or prolongation) heads, and disruption (or uneconomical working) heads. Typical heads include:

- Site establishment costs
- Head office overheads
- Visiting head office staff
- Uneconomical working
- Uneconomical procurement
- Loss of profit
- Acceleration
- Third party settlements
- Inflation
- Financing other heads of claim
- Financing retentions
- Interest
- Cost of producing claim

16 Housing Grants Construction and Regeneration Act 1996, Part II, s.112.
17 See Chapter 12, Section 12.5.2 dealing with mitigation.

13.4.2 Site establishment costs

Sometimes misleadingly called preliminaries (from association with site establishment allowances in the Contract Bills), this head includes such items as hutting, electricity, standing plant, small tools, site supervision, and non-productive labour including cleaning operatives and those in attendance on domestic sub-contractors.

Hutting and other plant: Claims will sometimes be made (and paid) on the basis of Bill allowances, but that is not what the Contractor is entitled to recover. For a damages claim (or loss and/or expense) the entitlement is the extra cost incurred. This will normally be based on presentation of hire invoices, including charges from a sister company, when the item claimed is not owned by the Contractor. A difficulty arises when the Contractor owns his own huts and plant. It has been the habit of many Architects and Quantity surveyors, when faced with a claim for contractor-owned items, to pay a reasonable hire rate, i.e. the commercial rate which the Contractor would have paid if he had hired the equipment. This course is practical, particularly for inclusion in interim certificates, but it is not what the Contractor is entitled to receive and it is not what the Architect is authorized to certify. The position was clarified in *Alfred McAlpine Homes North Ltd v. Property and Land Contractors Ltd*[18] in which the Court considered an arbitrator's award based on reasonable hire charges:

> Ascertainment on the basis of hire charges might not have been questioned if there had been a finding that (the contractor) would have hired out this plant but there is no such finding. ... Only if there had been such a finding could the ... award have been justified as representing ... the valuation of lost opportunity. ... The question of law implicit in this part of the appeal [is] ... that in ascertaining direct loss or expense under Clause 26 of the JCT conditions in respect of plant owned by the contractor the actual loss or expense incurred by the contractor must be ascertained and not any hypothetical loss or expense that might have been incurred whether by way of assumed or typical hire charges or otherwise.

The costs which the Contractor may claim are depreciation, maintenance, and addition fuel (if any). Different companies write off plant in different ways, but there seems no reason to calculate the depreciation in any way other than that normally used by the Contractor.[19]

Supervisory staff and non-productive labour: The principles are the same as for plant. Hired staff could be claimed against invoices, but employed staff must be claimed on the basis of cost. Cost in this instance, taken from the Contractor's accounts and wages records, will include wages, benefits such as health insurance, and any statutory payments by the company in respect of employment.

13.4.3 Head office overheads

Head office overheads relates either to the cost associated with running the Contractor's business, or the contribution required by the Contractor from each of its contracts towards

---

18 (1996) 76 BLR 59.
19 In *McAlpine v. Property and Land* (see n. 13) the Court allowed the Contractor to rely on its normal manner of trading with only one company.

such cost. The former should be considered as expense, and the latter as loss. On JCT contracts the difference is immaterial,[20] since the entitlement is to loss and/or expense. Items falling within this head include offices, support overheads such as buying and accounts departments, rent, rates, heating and telephone bills, indeed anything going towards the cost of maintaining the business operation as a whole, as opposed to an individual project.

Claims for such costs are notoriously difficult to establish, the main problem for the Contractor being to show entitlement in principle. The basis of claim is that because of delay on the project, the company workforce was deprived of the chance of earning contribution or recovering its overhead costs from elsewhere. Small companies with limited staff resources can sometimes demonstrate by reference to correspondence that they have had to turn away new work; but it is difficult for a large national company to show that a delay on a single project had a significant effect on the whole company's ability to accept new work.[21]

Once a claim is established in principle, a contractor is then obliged to provide information to enable the loss or expense to be calculated. In practice the calculation is normally prepared by the Contractor for checking by the Architect. A popular means of calculation among contractors is the use of a formula. Formulae in common use are the 'Hudson formula', the 'Emden formula', and the 'Eichleay formula'.

The 'Hudson formula'[22] calculates loss as an average overhead and profit allowance per week, based on the contract period and percentage mark-up included in the Bills, then multiplied by the length of delay. The formula can be criticised on the grounds that it relates to tender allowances (i.e. value), rather than actual costs.

The 'Emden formula'[23] differs from the Hudson formula only in that the overheads are taken as an average percentage from the Contractor's accounts. As a means of calculation once entitlement in principle is established, the Emden formula has received some apparent approval in the courts.[24]

The 'Eichleay formula'[25] approaches the problem from a different direction, arguably based on a shortfall in contribution.[26] The average weekly contribution necessary to run the company is calculated from the company accounts, and is multiplied by the period of delay. The product, which represents the total contribution required from all income in order to run the business during the delay period, is reduced pro rata to reflect the share of the total contribution required from the project in delay to pay its way. The pro rata is made by comparing the turnover of the project in delay during the period with total turnover (value) on all projects during that period. This formula was used in *Alfred*

20 Under contracts such as those published by the ICE which refer only to 'cost', the difference may be significant.
21 In *Whittal v. Chester-le-Street D C* (unreported) 3 July 1984, it was found as a fact that work was available which could not be taken.
22 Set out in *Hudson's*, 11th edn, at para. 8.182.
23 Set out in *Emden's Construction Law*, 8th edn, rev. A. J. Anderson, S. Bickford-Smith, N. E. Palmer and R. Redmond-Cooper (Butterworth, 1990) 57, Aug. 1999, vol. 2, at p. N/46.
24 In *J. F. Finnegan Ltd v. Sheffield City Council* (1988) 43 BLR 124, the judge preferred the 'Hudson formula' to one of the Contractor's own making, although he then went on to describe a form of the 'Emden formula'.
25 Based on the first case in the USA to use the formula; see Duncan Wallace, I. N., *Construction Contracts: Principles and Policies in Tort And Contract*, (Sweet & Maxwell, 1986), paras. 8-30 to 8-33 for detailed comment.
26 This point was argued in *McAlpine v. Property and Land*, see n. 13.

McAlpine Homes North Ltd v. Property and Land Contractors Ltd,[27] but again the Court emphasized the need to establish entitlement before any form of calculation should be employed:

> there may be some loss as a result of the event complained of, so that in the case of delay to ... completion ... there will be some 'under recovery' towards the cost of fixed overheads as a result of the reduced volume of work ... but this state of affairs must of course be established as a matter of fact. If the contractor's overall business is not diminished during the period of delay ... (due to increased contribution from variations etc.), or if as a result of other work, there is no reduction in the overall turnover so that the cost of fixed overheads continues to be met from other sources, there will be no loss attributable to the delay.

Similarly, in *Amec Building Ltd v. Cadmus Investment Co. Ltd.*[28] is was said:

> it is for Amec to demonstrate, in respect of the individuals whose time is claimed that they spent extra time allocated to the particular contract. This proof must include the keeping of some form of record that the time was excessive and their attention was diverted in such a way that the loss was incurred. It is important ... that the plaintiff places some evidence before the court that there was other work available which but for the delay he would have secured ... thus he is able to demonstrate that he would have recouped his overheads.

If a formula claim for overheads is allowed, care must still be taken to remove any duplication with other heads, such as overlap with site supervision costs which may be included as overheads in the company accounts.

One alternative to calculation by formula is by identifying, where possible, the cost of managerial time spent in problem solving. An example can be seen in *Tate & Lyle Food and Distribution Ltd v. GLC*[29] in which a claim for damages, albeit in tort and not contract, was enhanced by the addition of 2.5 per cent to cover managerial and supervisory resources. It was held that the time spent was a proper head of damage, but in this case nothing was proved. It was not sufficient to add a percentage; in an organization such as Tate & Lyle there should be records available to demonstrate the loss:

> While I am satisfied that this head of damage can properly be claimed, I am not prepared to advance into an area of pure speculation when it comes to quantum.

> I have no doubt that the expenditure of managerial time in remedying an actionable wrong done to a trading concern can properly form a subject matter of a head of special damage. ... I would ... accept that it must be extremely difficult to quantify. But modern office arrangements provide for the recording of the time spent by managerial staff on particular projects.

It is clear that the 'Tate & Lyle method' is likely to produce a more accurate result than any of the formula methods, but it relies on accurate record-keeping. Contractors should remember that by giving a loss and expense notice to the Architect, they also put themselves on notice to keep whatever records are required to prove their case.

---

27 See n. 13.
28 (1997) 51 Con.LR 105; (1997) 13 Const.LJ No. 1 p.50.
29 [1982] 1 WLR 149.

13.4.4 Visiting head office staff

One head of claim that straddles site project costs and head office costs is the additional time spent by head office personnel visiting site, either in problem solving or in prolonged involvement. A typical example would be the Contractor's surveyor who may be required to visit the site regularly to deal with domestic sub-contractors and internal cost accounts. A disrupted site may result in greater involvement each week, whereas a delayed site may prolong the visits over a longer period. Claims of this nature fall squarely into the type considered in the *Tate & Lyle* case.[30] The difficulty for the Contractor is in establishing what resources would have been spent in any event, which may require evidence of the normal level of resource required for that type of project. Actual costs should be comparatively easy to identify, and include salaries, travelling expenses, car hire or depreciation, fuel, accommodation etc., under the same principles as those referred to in Section 13.4.2, Site Establishment.

13.4.5 Uneconomical working

Uneconomical working claims are often made under the heading of 'disruption'. Delay or disruption on site may manifest itself as loss of motivation leading to loss of productivity. The difficulty with disruption is that everyone can see the effect of its presence on site, but the type of information required to demonstrate quantification and causation are rarely available. However, if the Contractor is able to show that events caused uneconomical working and wasted costs, the amount of such costs may be calculated by comparison of work output during disrupted and undisrupted periods.

In *Whittal Builders Co. Ltd v. Chester-le-Street D. C.*[31] the value of work output per man week while the Employer's breaches continued was compared with the value of work output after the breaches had ceased. An indication of willingness to find an answer can be seen in the words of the judgment:

> It seemed to me that the most practical way of estimating the loss of productivity, and the one most in accordance with common sense and having the best chance of producing a real answer was to take the total cost of labour and reduce it in the proportion which those actual production figures bear to one another – i.e. by taking one third of the total as the value lost by the contractor.

The calculation of one third wasted costs in this case came from comparing disrupted value of £108 per man week with undisrupted costs of £161 per man week.

Contractors should have no difficulty in obtaining figures; costs or man weeks may be taken from accounts records, and value may be taken from interim valuation calculations. The pitfall comes in finding two periods in which the work is sufficiently similar to be compared like with like. It is tempting for Architects to resist ascertainment on the grounds that evaluation cannot be made scientifically and in minute detail; but it is worth

---

30 See Section 13.4.3.
31 See n. 21.

remembering the pragmatic approach in the *Whittal* case[32] and the judgment in *Chaplin v. Hicks*:[33]

> Where it is clear that there has been actual loss resulting from the breach of contract, which it is difficult to estimate in money, it is for the jury to do their best to estimate: it is not necessary that there should be an absolute measure of damages in each case.

13.4.6 Uneconomical procurement

If a project is delayed or disrupted, it is possible that procurement of materials may be affected. Such claims are rare but may be sustained if the purchase of materials in large quantities is prevented, or if materials on long-term delivery have to be accepted out of sequence. A claim could include the loss of special bulk discounts, storage costs, and funding capital expenditure until the materials are included in a certificate.

13.4.7 Loss of profit

Loss of profit claims fall under the same general principles as head office overheads. As with overheads it is for the Contractor to establish that he was deprived of the opportunity to earn profit elsewhere. It is not sufficient to add a percentage to the net value of other heads of claim.[34] If the Architect omits work from the contract in order to give it to someone else, the Employer will be in breach of contract. The damages will be the loss of profit which the Contractor included in his tender.[35]

13.4.8 Acceleration

Employers will sometimes require their finished project by the original Completion Date, or by some other fixed date, irrespective of the Contractor's entitlement to extensions of time. Contractors may then propose accelerating their work if they are reimbursed their additional costs or sometimes an Employer will offer inducement in a bonus; sometimes the Employer will simply influence the Architect (or the Architect will act on his own) to deny extensions of time in order to coerce the Contractor into finishing earlier than his entitlement; or the Contractor may simply decide to accelerate of his own accord, in both his own and the Employer's interests.

Only the last of these three categories fall within the authority of the Architect without additional express powers being given by the Employer.

***Agreement to accelerate/bonus*:** If an agreement is made between the Contractor and the Employer (or with the Architect if the Employer gives him the authority), whether

---

32 See also *Penvidic Contracting Co. Ltd v. International Nickel Co. of Canada Ltd* (1975) 53 DLR (3d) 748, in which the contractor was awarded the difference between what he tendered and what he would have tendered if he had foreseen the conditions caused by the Employer's breach.
33 [1911] 2 KB 786.
34 See Section 13.4.3, dealing with the *Tate & Lyle* case.
35 See Chapter 6, Section 6.3 referring to *Carr v. J. A. Berriman Pty Ltd* (1953) 879 Con. LR327; *Amec v. Cadmus*, see n. 28.

payment is in the form of acceleration costs, or bonus, the agreement needs to be considered carefully. The Employer is at risk if the Architect issues further Variation instructions under the Contract. If the Contractor fails to meet the deadline because of such Variations, or if the Architect in any other way prevents the Contractor from meeting his target, the Employer will be in breach of an implied 'non-hindrance' term. In *John Barker Construction Ltd v. London Portman Hotels Ltd.*[36] the damages awarded were one half of the promised bonus.[37] Whilst there was no certainty that the Contractor in that case would have met his target date, it was held he was deprived of the chance by being hindered; the hindrance was the issue of *bona fide* Variations given under the Contract. A claim by the Contractor under this head is not a claim under the Contract, but is a common law claim. If the Employer wishes such matters to be dealt with by the Architect, the acceleration/bonus agreement needs to clearly set out the administrative arrangements, together with the effect of further Variations.

Constructive acceleration: The Contractor is faced with a dilemma if he is entitled to an extension of time, but he is deprived of his entitlement either by the Employer's breach of contract in deliberately preventing the award of an extension, or by the Architect's failure to form an opinion. The Contractor has to decide whether to hope his entitlement will eventually be recognized (and to continue at a natural pace), or whether to avoid exposing himself to liability for liquidated damages (and accelerate to meet the current contractual date). Acceleration in these circumstances is said to be 'constructive'.[38] The Contractor's claim is for common law damages arising from an alleged breach, and is valid in principle. However, there are two major obstacles for the Contractor to negotiate. The first is in establishing a breach, and the second is satisfying the causation and remoteness requirements of the general principles applying to damages.[39]

Contractors tend to view any failure of the Architect to award extensions as a breach of duty, but it has been held that the Architect is under no duty to provide a completion date at which the Contractor can aim.[40] Clearly, a breach can be established if the Contractor can demonstrate that the Employer has interfered by instructing the Architect. However, intervention by the Employer is probably rare. It is more likely that the Architect has either failed to address his mind to the matter, or he has acted after forming an opinion, albeit that opinion might later be changed, or found by an arbitrator to be wrong. Unfortunately for Contractors, simply forming a wrong opinion is not in itself a breach.[41] The essential act is in forming an opinion as required by the Contract (see Section 13.2.2 (2)). Thus failure to consider the matter at all, provided the Contractor has fulfilled his obligations, could be a breach. In *Perini Corporation v. Commonwealth of Australia*[42] it was held that the certifier had to give his decision in a reasonable time in the absence of express periods. On this premise the Architect may be in breach for not forming his opinion at the proper

---

36 [1996] 50 Con. LR 43; (1996) 12 Const. LJ 277.

37 In *Bournemouth & Boscombe Athletic FC v. Manchester Utd FC*, The Times 22 May 1980, the whole bonus was awarded as damages when a footballer was prevented from scoring his target.

38 Where the law implies a right without reference to the intention of the parties. See *Osborn's Concise Law Dictionary*.

39 See Chapter 12: Section 12.5.1, Cause and Effect, and Section 12.5.2, Measure of Recovery.

40 See *Amalgamated Building Contractors v. Waltham Holy Cross UDC* [1952] 2 All ER 452.

41 See *S. Pembrokeshire D C v. Lubenham Fidelities and Wigley Fox Partnership* (1986) 33 BLR 39.

42 (1969) 12 BLR 82.

time; but it will still be for the Contractor to make application, and provide information as requested.

Acceleration in mitigation: Contractors entitled to an extension of time for reasons which also give entitlement to loss or expense, are required under general principles to mitigate their costs or losses if they intend to seek reimbursement (see Chapter 12, Section 12.5.2 dealing with extravagance and 'duty' to mitigate). Under those same principles the Contractor is entitled to spend money in order to save money. If the Contractor's estimation of the costs involved in a prolonged contract period could be reduced by working overtime or by introducing extra labour to complete earlier, then he is entitled to take such action. Provided the decision to 'accelerate' was a reasonable decision at the time it was made, the Contractor would be entitled to recover his acceleration costs; this would be so even if the eventual cost exceeded the estimated cost it was intended to reduce. There is no strict obligation on the Contractor to seek approval for such a decision, but the prudent Contractor would notify the Architect for evidential reasons, simply to show at a later date that the action came from a positive decision. Since acceleration in this form is claimed as mitigation of loss and/or expense, it is a matter which the Architect can (and must) consider.

The only real risk for the Contractor in this course seems to be the possibility of a poor decision to accelerate when he is later found to be not entitled to an extension; clearly the decision is still in his own interest if it is a good decision, whether or not he later receives an extension, since he will have minimized his own costs.

13.4.9 Third party settlements

Whenever a project is delayed or disrupted it is likely that the Contractor's sub-contractors will be affected. It is rare for Contractors to carry out all the work themselves, indeed it more likely that the majority of the work is carried out by sub-contractors. It follows that many of the costs or liabilities borne by the Contractor, in reality, are liabilities to sub-contractors. A difficulty arises for Contractors when a sub-contractor presses its claim long before the Contractor's entitlement is finally ascertained. Rather than face long legal battles the Contractor may reach a settlement with its sub-contractor, then turn to the Architect for reimbursement under Clause 26.

The Contractor will be probably be faced with argument that the settlement is irrelevant to the Contractor's entitlement, or may even be put to the task of proving the sub-contractor's claim. In this latter task he is likely to fail since he will not have the necessary records. However, the courts have considered the problem on a number of occasions,[43] and in *Oxford University Press v. John Stedman Design Group*[44] the following principles were identified (paraphrased):

- The law encourages reasonable settlement.

---

43 See *Biggin & Co. Ltd v. Permanite Ltd* [1951] 1 KB 422; [1950] 2 All ER 859; *Fletcher & Stewart Ltd. v. Peter Jay & Ptnrs* (1976) 17 BLR 42 (CA); In *The Royal Brompton Hospital H T v. Frederick Alexander Hammon*, 8 January 1999, 1999 CILL 1464, it was held that the principles in the *Biggin* case applied equally to cases where there were several defendants from whom contributions are sought.
44 (1993) 34 Con. LR 1; (1990) CILL 590.

- The cost of pursuing litigation is relevant in deciding whether a settlement was reasonable.
- A party who relies on the settlement as the basis for compensation must prove that it is reasonable; but he does not have to prove strictly the claim against him in all its particulars.
- In establishing the reasonableness of a settlement, it is relevant that it was made under legal advice, although the advice itself may not be relevant or admissible.

It is still for the claiming party to establish that the third party settlement was reasonable[45] (that does not mean simply showing it was reasonable to settle), but the Architect cannot avoid ascertaining the head of claim purely on the grounds of a demand for the Contractor to prove the case.

13.4.10 Inflation

If a non-fluctuating price project is delayed or disrupted due to an Employer's delay, the Contractor is entitled, in principle, to claim the costs resulting from work being carried out at a later time. The calculation should not be based on the Completion Date, but on the comparable dates of the various activities. Thus, the extra cost of carrying out brickwork in (say) week 50, instead of (say) week 30 is the cost claimed. The evaluation may be based on invoiced costs compared with anticipated costs from price lists, and a similar calculation for labour cost. Alternatively, a notional calculation may be made by calculating the costs which would have been incurred over the original contract period by using published indices, and comparing it with the extra costs incurred over the actual period, again using indices. In order to carry out this type of calculation it is first necessary to establish the anticipated 'cashflow' of the original project for application to the indices. The 'actual' indices are applied to the Quantity Surveyor's interim valuations.

13.4.11 Financing other heads of claim

In times of high borrowing rates, financing costs can be a significant part of the Contractor's total loss and expense claim. Financing is normally looked upon as an 'expense', incurred by way of interest charges paid by the Contractor on an increased overdraft, although it may also be a 'loss' equivalent to lost income on reduced capital invested. When a Contractor incurs additional expense for which he is entitled to reimbursement under Clause 26, he has to fund that expense until he is paid under an Architect's certificate. It is part of that extra funding which is claimed as financing cost, and it should not be confused with interest charged to the Employer for late payment of a debt (see Section 3.4.13).

The principles governing financing to be reimbursed as loss or expense are to be found in two cases. The first, *F. G. Minter Ltd v. WHTSO*,[46] established that financing was an entitlement as loss or expense, and the Architect was obliged to consider it if proper notice had been given. The second case is *Rees & Kirby Ltd v. Swansea City Council*,[47] which dealt with the method of calculation and timing. It was held that:

---

45 See *P&O Developments Ltd v. The Guy's and St Thomas' NHS Trust & others* [1999] BLR 3.
46 (1980) 13 BLR 1.
47 (1985) 5 Con. LR 34.

- The Contractor's notice of the primary loss (the head of claim being funded) must be given within a reasonable time of the loss being incurred.
- Financing costs cannot be considered parasitic to the primary loss, and therefore some mention of the further expenditure in financing must be made in the notice.
- Financing is due from the date of loss and expense being incurred.
- Financing is due up to the date of the last application made before the issue of the certificate in respect of the primary loss (i.e. the payment of the head of claim on which financing is incurred).
- Financing charges incurred during a period when an independent cause operated should not be recoverable (this would include delay in ascertainment caused by the Contractor's failure to make application or failure to provide information after being requested).[48]
- The date of practical completion is irrelevant to the calculation.
- The rate of interest to be applied is the actual borrowing rate paid by the Contractor, provided it was reasonable.
- The calculation should be made on the same basis as that used by the banks, i.e. simple interest compounded at quarterly intervals.
- Where the Contractor is 'cash rich', the rate is that earned by the Contractor on money placed on deposit.
- Account should be taken of payments made progressively.

The calculation of financing is not difficult once the relevant periods have been established. The most elusive of these periods are the dates of incurring the primary losses, and the dates on which application ought reasonably to be made. Since the losses are incurred progressively an interest calculation becomes a rolling calculation, and the cost of preparing the calculation can easily exceed the amount of the claim. A practical solution is to calculate financing separately on each head of primary loss, starting at the date on which the 'centre of gravity' of the primary loss falls; the 'centre of gravity' is found by adding together all the costs under the head of claim, and identifying the date on which the mid point of cost occurs. As to identifying the reasonable date of notice, the Architect will need to take into account the requirements of the Contract, i.e. 'Clause 26.1.1: the Contractor's application shall be made as soon as it has become, or should reasonably have become, apparent to him that the regular progress of the Works ... has been or is likely to be affected ... (to cause loss or expense)'. The Contractor may not be aware of disruption or possible loss or expense immediately, so notice given responsibly may be a little time after the first signs. Insistence by the Architect on notice on the off-chance that disruption or additional costs may occur, is likely to produce a standard pro forma response to all directions and instructions; such an approach is not to be recommended since it clouds the important issues, and prevents the Architect taking preventive action to minimize the Employer's risk.

13.4.12 Financing retentions

When the Contractor tenders he is deemed to have included for his risks under the Contract. He knows that he will be paid at certain times and he can make due allowance

---

48 In the *Rees & Kirby* case the contractor's delayed notice was not held to prevent recovery, since the contractor had relied on ongoing negotiations in not submitting notices in strict compliance with the contract.

for the cost of funding the project. One factor to take into account is that he will be obliged to finance the retention monies until practical completion, when he will receive an injection of funds as monies are released. The Contractor may accordingly claim the financing costs of later retention release, delayed to the extent that the Employer is culpable under Clause 26. The calculation follows the same principles as those described in Section 13.4.11 above.

13.4.13 Interest for non-payment

Where money is not paid on time the Contractor may claim interest. Prior to JCT 80 Amendment 18, and subsequently JCT 98, there was no entitlement to interest under JCT contracts. Then the Contractor's only recourse was to rely on an Arbitrator's discretion in allowing interest on an award; but to trigger that possibility the Contractor was obliged to commence proceedings. Times change, and from November 1998 contractors (or at least some contractors) generally have one of two new entitlements, i.e. either an entitlement to interest under the Contract, or a statutory right to claim interest on a debt.

Contractual right to interest: Under Clause 30.1.1 the Contractor is entitled to simple interest on late or non-payment of an Architect's certificate. He is not entitled to interest under that clause for late issue of a certificate. Interest under Clause 30.1.1 is not a matter for the Architect (see Chapter 15, Payment).

Statutory right to interest: Under the Late Payment of Commercial Debts (Interest) Act 1998,[49] and in the absence of a substantial contractual right to interest, from 1 November 1998, a 'small'[50] company supplier of goods and/or services has a right to claim interest on a debt from a 'large' company. The Act, expressly, does not apply where the contract contains provision for interest on a debt (see also Chapter 15, Payment).

Interest as damages: The contractor may claim part of his additional loss or expense as a common law claim for damages. He will normally replicate all or part of his contractual claim for loss and/or expense. In such a claim his contractual head of 'financing' may appear as interest claimed as damages. In *Wadsworth v. Lydall*,[51] the Court of Appeal held that if it can be proved that special damage[52] is incurred by way of interest payments on an overdraft, which in turn is caused by late payment of a debt, then it may be recovered. However, the principle does not extend to interest as a primary loss prior to the Employer's breach of contract.

13.4.14 Cost of producing claim

One head which invariably appears in a Contractor's claim is the cost of preparing the claim. The cause normally given is that the Contractor was put to the trouble and expense

---

49 After 31 October 2000 small companies can claim interest from other small companies, and after 31 October 2002, any company can claim interest from any other company regardless of size.
50 Under the Act a 'small' company is defined as one having 50 or less full-time employees; a part-time employee working half a week is counted as one half of an employee. 'Large' companies are those having more than 50 full-time employees.
51 [1981] 2 All ER 401.
52 See Chapter 12, Section 12.5.2, dealing with the rules in *Hadley v. Baxendale*.

of preparing a claim by the failure of the Architect to ascertain the loss and expense properly. In many cases the claim is no more than the Contractor is obliged to provide as information given under the procedures in the Contract. One case often mistakenly cited is *James Longley & Co. v. S W Regional Health Authority*,[53] in which a claims consultant's fees were reimbursed in part. However, the portion paid related only to the work done in preparing schedules to be annexed to the Points of Claim in arbitration. Contractors sometimes annotate their claim with a note to the effect that it has been prepared in contemplation of arbitration. This is an attempt to bring the costs of preparing the claim into the ambit of the *James Longley* case, but it is not a matter within the Architect's power, and is not an entitlement. It is no more than evidence for consideration by an arbitrator at a later date, when he applies his discretion in awarding costs.

If the Contractor can show that the Architect has failed to carry out his duty to ascertain,[54] and as a result the Contractor was obliged to prepare a detailed and fully documented claim as a result of a breach, he may be able to recover.[55] In particular, a small Contractor company with limited staff expertise and without specialist knowledge might be able to demonstrate that he bought in management expertise, and prove his expenditure. The *Tate & Lyle* case[56] is sometimes cited as authority for entitlement to recovery. However, it is unlikely that the cost of preparing a claim would fall into the category of managerial time in problem solving, and the principle of keeping records is not unique to *Tate & Lyle*; it is an evidential principle of general application. In *Milburn Services Ltd v. United Trading Group (UK) Ltd,*[57] the judge said:

'I do not regard the Tate & Lyle case as enshrining any principle of law ... beyond the principle that the burden is on the plaintiffs to prove their case, but it is an interesting approach to a factual matter by an experienced and respected judge.

Such costs including programming or forensic accounting resources, may be recoverable as a result of the Architect's breach, in which case records would be necessary to ascertain the loss and expense properly.

Claims made under this head are usually for breach of contract, and are outside the normal authority of the Architect unless he is given powers by the Employer.

---

53 (1983) 25 BLR 56.
54 See Section 13.2.2 dealing with the Architect's duty.
55 See Powell-Smith, V., Architect Must Tot Up Contractor's Losses', *Contract Journal*, 30 July 1992, p.7.
56 See Section 13.4.3.
57 ORB; 9 November 1995; 1993 ORB 534; (1995) CILL 1109.

14

Fluctuations

In construction contracts with long contract periods, tenderers are not normally expected to make allowances for inflation in their tender prices. They are usually required to tender on only costs current at the time, with provisions being made for adjustment of the contract price for inflation during the performance of the contract. Such contracts are referred to as 'fluctuating price' contracts whilst the term 'fluctuations' refers to the amount of adjustment to the contract price. This chapter examines the JCT 98 provisions on fluctuations. Although the term covers both increases and reductions in payment caused by variation in costs, for simplicity and clarity, the discussion is mainly in the context of increases in costs.

There are three alternative clauses on fluctuations published as a separate booklet: Clauses 38, 39, and 40. Clause 37 provides that fluctuations under the Contract are to be dealt with in accordance with whichever one of these is indicated in the Appendix. Clause 38 applies if no choice is indicated. The general approach in Clauses 38 and 39 is determination of fluctuations as the net amount arising from changes in certain types of costs and is sometimes referred to as the 'traditional approach' whilst that in Clause 40 is the 'formula approach'. The formula approach attempts to determine what the Contract Sum would have been if the Contractor, assuming the same pricing level, had taken into account in his tender the inflation actually experienced.

14.1 Fluctuation under Clause 38

Generally recovery of fluctuations under Clause 38 are limited to changes in the Contractor's tax liability and other liability of a statutory nature incurred in respect of the Works. The Contractor has to bear any additional costs for which he is not statutorily liable. For these reasons, contracts incorporating this clause are sometimes referred to as 'firm' or 'fixed' price contracts. The specific recoverable costs are in respect of:

1. changes in rates and types of contributions, levies and taxes on the employment of labour (Clause 38.1);
2. changes in duties and taxes payable by the Contractor as a consequence of the procurement or use of the materials required for the carrying out of the Works (Clause 38.2.2);
3. changes in Landfill Tax (Clause 38.2.3);
4. an optional percentage addition to each of the above fluctuations (Clause 38.7).

14.1.1 Statutory contributions, levies and taxes in respect of employees

The Contractor is deemed to have allowed in the Contract Sum for any statutory contributions, levies and taxes which he, as an employer, was obliged at the Base Date to pay in respect: (i) his 'workpeople' engaged upon or in connection with the Works either on or adjacent to the site; (ii) his workpeople engaged upon the production of 'materials and goods' for the Works on or adjacent to the site. 'Workpeople' is defined under Clause 38.6.3 as 'persons whose rates of wages and other emoluments (including holiday credits) are governed by the rules or decisions or agreement of the National Joint Council for the Building Industry (NJCBI) or some other wage-fixing body for trades associated with the building industry'.[1] In *Murphy & Sons Ltd v. London Borough of Southwark*[2] the Court of Appeal held that labour-only sub-contractors were not included in same definition of 'workpeople' in the JCT 63. It is submitted that the decision is equally applicable to the JCT 98. 'Materials' and 'goods' are defined in Clause 38.6.2 to include timber used for formwork and electricity but excluding other consumable stores, plant and equipment. Fuels are included only if so stated in the Contract Bills. For simplicity, materials, goods, electricity and fuels are hereafter referred to collectively as just 'materials'.

Any changes in the rates or types of these statutory liabilities from the position at the Base Date are recoverable or allowable by the Contractor. Similar entitlements apply in respect of other employees engaged upon or in connection with the Works but who do not fall under the definition of 'workpeople', e.g. secretarial, administrative and managerial staff (Clause 38.1.3). Employees of this type are hereafter referred to as 'ancillary staff'. However, for the purpose of recovering the fluctuations, ancillary staff are to be treated as craftsmen with the following provisos in Clause 38.1.4: (i) the employee must have worked on or in connection with the Contract for at least 2 whole working days in the week to which the fluctuation relates; (ii) periods of less than a whole working day must not be aggregated into a whole working day; (iii) the highest rate for the Contractor's (or domestic sub-contractor's) craftsmen is to be used.

Under Clause 38.1.5 the Contractor is deemed to have included in the prices in the Contract Bills any refunds of statutory contributions, levies and taxes and other statutory payments in respect of the relevant workpeople to which the Contractor, as an employer, was entitled at the Base Date. If there are any changes in the type or rate of refund or payment after the Base Date, the net amount representing the changes is to be paid to or allowed by the Contractor (Clause 38.1.6). For the purpose of calculating fluctuations, employees who have contracted-out of the state pensions scheme are to be treated as if they have not done so (Clause 38.1.8).

14.1.2 Duties and taxes in respect of materials

The Contractor is deemed to have priced in the Contract Bills for any duties and taxes payable at the Base Date on account of the procurement or use of materials for the Works and which are on a list of materials attached to the Contract Bills. The Contract Sum is to

---

1 This definition is repeated in Clause 39.7.3.
2 (1983) 22 BLR 41.

be adjusted for any changes in the duties and taxes so payable by the Employer (Clause 38.2.2). VAT is not included because it is dealt with elsewhere in the Contract.

14.1.3 Landfill tax

Under the Finance Act 1996 landfill tax is payable by operators of licensed landfill sites. Contractors will have to pay this tax as part of charges by the operators. This tax liability is not incurred by the Contractor as a consequence of being and employer. Neither is it payable 'on the import, purchase, sale, appropriation, processing or use' of materials required for the carrying out of the works. Amendment 18 of the JCT 80 introduced what is now Clause 38.2.3 to allow price adjustment on account of changes in the amount of landfill tax. Under that clause, any change in the Contractor's liability for landfill tax from the position at the Base Date is recoverable by the Contractor or allowable to the Employer, as the case may be. However, no payment is due the Contractor if the circumstances are such that the Contractor would reasonably be expected to have used a waste disposal method different from the use of landfill sites. As the landfill tax will be only an element of the total cost charged to the Contractor by the operator of a landfill site, the Contractor needs to make arrangement with operators to obtain proper receipts covering the tax component of such charges.

14.1.4 Percentage for additional payment

The Contractor is allowed to indicate in his tender a percentage to be applied to the net amount payable under the heads already discussed to arrive at an additional payment (Clause 38.7). The applicable percentage is to be stated in the Appendix. The general intention behind Clause 38.7 is to compensate the Contractor against cost increases not expressly recognized as subject to fluctuation.

14.1.5 Notices, evidence and calculations

By Clause 38.4 the Contractor must, as a condition precedent to recovery of fluctuation entitlements, give written notice of any changes in respect of which adjustment for fluctuation is applicable. He is also required to supply the Architect, or Quantity Surveyor, with all evidence and calculations reasonably necessary to determine the amount of fluctuations. Types of evidence commonly required to be submitted include:

- invoices from suppliers;
- invoices/receipts from bodies to which contributions, levies and taxes are payable;
- take-off of materials and labour;
- miscellaneous supporting calculations.

Furthermore, he must, on a weekly basis, provide a certificate of the validity of any evidence and calculations supplied in respect of employees who do not fall under the definition of workpeople.[3] These requirements must also be complied with in respect of fluctuations on the work of domestic sub-contractors.

---

3 Intentionally issuing a false certificate could amount to theft; see Section 12.7.

Clause 38.4.3 provides that the Quantity Surveyor and the Contractor may agree the amount of adjustment to make in respect of any events giving rise to fluctuation entitlements. In *John Laing v. County and District*[4] it was held that this type of provision did not authorize the Quantity Surveyor to ignore the provisions of the Contract such as giving notices of the events as a condition precedent to recovery.

14.1.6 Exclusions from fluctuations under Clause 38

Changes in the costs of certain items are expressly excluded from fluctuations under Clause 38.5. These are changes in costs of:

- dayworks (they are paid for on the basis of current prices);
- work executed by Nominated Sub-contractors;[5]
- materials and goods supplied by Nominated Suppliers;
- work covered by a PC sum which the Contractor has successfully tendered for under Clause 35.2;[6]
- changes in VAT.

14.2 Fluctuations under Clause 39

Where Clause 39 applies, fluctuations are recoverable under the following heads:

1. labour costs (Clause 39.1.2);
2. statutory contributions, levies, and taxes relating to 'workpeople' (Clause 39.2);
3. costs of materials and associated duties and taxes (Clause 39.3);
4. the above costs in respect of sub-contract works (Clause 39.4);
5. an optional percentage addition to the above items of fluctuations (Clause 39.8).

14.2.1 Labour costs

The Contractor is deemed under Clause 39.1 to have included in the Contract Sum wages, other emoluments and other expenses payable to workpeople[7] and the cost of related employer's liability and third party insurance. This deeming provision expressly covers (i) workpeople employed upon or in connection with the Works either on or adjacent to the site; (ii) workpeople employed elsewhere by the Contractor in the production of materials for the works, e.g. employees in an off-site workshop. The wages, emoluments and other expenses referred to above are only those determined at rates and prices governed by the rules, decisions, bonus scheme, or other agreements of the NJCBI or other appropriate wage-fixing body promulgated[8] at the Base Date. In *Sindall (William) v. N. W. Thames*

4   (1982) 23 BLR 1.
5   These are recoverable under the Nominated Sub-contract and passed to the Employer. This provision avoids the double recovery that would otherwise occur.
6   If the Contractor wishes to recover fluctuations on such work, appropriate provisions should have been included in the separate agreement.
7   For discussion of the definition of 'workpeople' see Section 14.1.1.
8   The Contractor is deemed to have allowed for decisions and the like promulgated as at the Base Date even if they were then not yet in force.

Regional Health Authority[9] the House of Lords held that increases in costs attributable to a voluntary bonus scheme operated by the Contractor were not recoverable under the equivalent provisions of the JCT 63. It is submitted that recovery of fluctuations under the JCT 98 would be similarly restricted.

Fluctuations after the Base Date in wages, emoluments and other expenses of 'workpeople' arising from alterations in the rules, decisions, and agreements already referred to are recoverable or allowable by the Contractor. Similar fluctuations apply in respect of ancillary staff (Clause 39.1.3). However, for the purpose of recovering fluctuations, such employees are to be treated as craftsmen with the following provisos under Clause 39.1.4: (i) the employee must have worked on or in connection with the contract for at least 2 whole working days in the week to which the fluctuation relates; (ii) periods of less than a whole working day must not be added up into a whole working day; (iii) the highest rate for the Contractor's (or domestic sub-contractor's) craftsmen is to be used. Consequential changes in the cost of employer's liability and third party insurance in relation to such ancillary staff also qualify as relevant fluctuations.

The Contractor is also to recover or allow fluctuations in respect of costs of transporting workpeople for the purposes of carrying out the Works (Clauses 39.1.5 and 39.1.6).[10] The baseline from which changes apply is a list of basic transport charges attached to the Contract Bills or figures determined from the rules of a recognized wage-fixing body. Where there is a change in the transport charges, or the wage-fixing body fixes new rates of reimbursement of fares to workpeople, they are recoverable or allowable by the Contractor.

Good practice requires the Contractor and sub-contractors to draw up and maintain up-to-date schedules of relevant workpeople and ancillary staff, their wages, other emoluments and related expenses.

14.2.2 Statutory contributions, levies, or taxes

Clause 39.2 broadly mirrors Clause 38.1. The Contractor is deemed to have allowed in the Contract Sum for any statutory contributions, levies and taxes which he, as an employer, is obliged to pay in respect of the wages, emoluments and other specified expenses paid to, or incurred in respect of, his workpeople (Clause 39.2.1). Any changes in the rates of these statutory liabilities from their figures at the Base Date are recoverable or allowable by the Contractor (Clause 39.2.2). The same principle applies to ancillary staff (Clause 39.2.3). There are similar provisions in Clauses 39.2.4 and 39.2.5 regarding changes in tax refunds and other payment to which the Contractor may be entitled in respect of his employees engaged in connection with the Works.

For the purpose of calculating fluctuations, employees who have contracted-out of the state-pensions scheme are to be treated as if they have not done so. However, there is an exception where the private pension scheme is one established and operated by the NJCBI or other recognized wage-fixing body. In such a case, the Contractor's contributions are to be treated as an element of wages, emoluments and other expenses under Clause 39.1 and, as such, recoverable or allowable as fluctuations.

---

9 [1977] ICR 294.
10 There is no provision for recovery of fluctuations in respect of ancillary staff.

14.2.3 Materials

The Contractor's tender is deemed to have been prepared on the basis of the market prices at the Base Date of materials. The Contractor is expected to attach a list of these market prices to his tender. This list is normally referred to in industry as the 'List of Basic Prices'. If during construction the Contractor incurs additional costs or makes savings on these resources because of changes in their market prices, the net difference is recoverable or allowable by the Contractor.[11] To qualify as fluctuations, any change in costs must satisfy two conditions. First, it must be in respect of items on the List of Basic Prices. Second, the change must be due solely to changes in market forces, including the effects of changes of statutory taxes and duties on these items.

14.2.4 Sub-contractors and suppliers

The Contractor is to include in domestic sub-contracts provisions that mirror the main contract fluctuation provisions: Clause 39.4. Fluctuations under the sub-contract are reimbursable under the main contract. The Nominated Sub-contract conditions, Conditions NS/C, contain equivalent provisions to the JCT 98 fluctuation clauses. Whichever of these applies is either selected by the Architect before sub-contract tenders are sought or left to the tenderers to decide. Irrespective of the option selected, the amount of fluctuation determined under the Nominated Sub-contract becomes payable to or allowable by the Contractor under the JCT 98. The amount of fluctuations payable to or allowable by a Nominated Supplier is a matter of the supply contract between the sub-contractor and the supplier. However, any price reductions or increases are passed to the Employer under the JCT 98.

14.2.5 Percentage addition

Where a percentage is stated in the Appendix against Clause 38.7 or 39.8, the net amount of fluctuations determined under the heads discussed above is to be increased by that percentage. This is designed to allow additional payment to compensate the Contractor against cost increases not expressly recognized as subject to fluctuation: construction plant, consumable stores, head office overheads and profit.

14.2.6 Exclusions from fluctuations under Clause 39

Clause 39.6 is the same as Clause 38.5, which is explained in Section 14.1.6.

14.2.7 Notices, evidence and calculations

These matters are covered in Clause 39.5 and are broadly similar to Clause 38.4 already discussed in Section 14.1.5. To avoid repetition, the reader is therefore referred to that section.

---

11 Strictly speaking, the difference is to be determined by reference to prices payable when the Contractor bought the relevant materials or fuel.

14.3 Fluctuations under Clause 40

Clause 40 provides for the calculation of fluctuations by a formula requiring the use of indices[12] that reflect movements in national wages and prices. Neither the costs upon which the Contractor built his tender nor the actual costs incurred by the Contractor are used in the calculation. There is therefore no need to specify the resources subject to fluctuation or their basic prices. Unlike the traditional approach, the formula method allows recovery of additional overheads and profit.

The formula method is incorporated into the 'With Quantities' and 'With the Approximate Quantities' variants of the JCT 98, in both Local authorities and Private versions, by selecting clause 40 in the Appendix (i.e. by deleting clauses 38 and 39). The formula method is based on a classification of general building work into 49 Work Categories listed in Table 14.1. An index for each Work Category is calculated from indices tracking variations in costs of labour, plant and materials and weightings of the resources required to carry out the work in that Work Category. Clause 40 incorporates into the JCT 98 the *Formula Rules*,[13] which sets out the formulae and defines their use. The primary source of the indices is a monthly bulletin although they are usually subsequently carried in the construction press. Indices for any month are provisional when first published. This is because the collection of the information to compile the indices takes considerable time and effort if they are to be accurate. Usually firm indices are published when sufficient information is available. The firm indices are then substituted for the provisional figures in the valuation following their publication as firm.

Clause 40 requires a schedule of fluctuations, which is a list of the following items, with the value of work in each item indicated against it:

- each of the Work Categories;
- Contractor's specialist work;
- Balance of Adjustable Work;
- Fix-Only work;
- work covered by provisional sums subject to formula adjustment;
- work excluded from formula adjustment.

The schedule is normally completed by a quantity surveyor, who is often but does not have to be the Quantity Surveyor under the Contract, and sent out as part of the tender documents to each tenderer. Sometimes, rather than include a schedule in the tender documents, each item in the tender Bills of Quantities is annotated with whichever of the above classifications it belongs to and the schedule prepared from the annotation after contract formation. Once the contract is entered into, the schedule or the annotation is binding upon both parties.

Specialist work includes electrical installations, heating, ventilating and air-conditioning installations, lift installations, structural steelwork, and catering equipment installations. For these types of work, there are no Work Categories except structural steelwork which may be treated as part of Work Category 27 (steelwork). However, there are specialist engineering formulae for dealing with them. These formulae are not

---

12 These indices have traditionally been referred to as the 'NEDO Series 2 Indices'.
13 JCT, *Consolidated Main Contract Formula Rules*, RIBA Publications Ltd, London, 1987.

Table 14.1 Work Categories

Work Category No.	Work In It
1	Demolition and alteration
2	Site preparation, excavating and disposal
3	Hardcore and imported filling
4	General piling
5	Steel sheet piling
6	Concrete
7	Reinforcement
8	Structural precast and prestressed concrete units
9	Non-structural precast concrete components
10	Formwork
11	Brickwork and blockwork
12	Natural stone
13	Asphalt work
14	Slate and tile roofing
15	Asbestos cement sheet roofing and cladding
16	Plastic coated steel sheet roofing and cladding
17	Aluminium sheet roofing and cladding
18	Built-up felt roofing
19	Built-up felt roofing on metal decking
20	Carpentry, manufactured boards and softwood flooring
21	Hardwood flooring
22	Tile and sheet flooring (vinyl, thermoplastic, linoleum, and other synthetic materials)
23	Jointless flooring (epoxy resin type)
24	Softwood joinery
25	Hardwood joinery
26	Ironmongery
27	Steelwork
28	Steel windows and doors
29	Aluminium windows and doors
30	Miscellaneous metalwork
31	Cast iron pipes and fittings
32	Plastic pipes and fittings
33	Copper tubes, fittings and cylinders
34	Mild steel pipes, fittings and tanks
35	Boilers, pumps and radiators
36	Sanitary fittings
37	Insulation
38	Plastering (all types) to walls and ceilings
39	Beds and screed (all types) to floors, roofs and pavings
40	Dry partitions and linings
41	Tiling and terrazzo work
42	Suspended ceilings (dry construction)
43	Glass, mirrors and patent glazing
44	Decorations
45	Drainage pipework (other than cast iron)
46	Fencing, gates and screens
47	Bituminous surfacing to roads and paths
48	Soft landscaping
49	Lead work

described in this book. Readers are referred to the *Formula Rules* for their details. 'Balance of Adjustable Work' covers that part of the Contractor's work which ranks for adjustment but which is not allocated to Work Categories or specialist work. Examples include preliminaries, water for the works, insurance, and attendance on Nominated Subcontractors. Where there is a significant value of work in respect of items of 'Fix Only' nature, it may be appropriate to create a weighted index covering such work. The weighted index is called a 'Fix Only' index. 'Fix Only' work items may also be dealt with by allocating them to appropriate Work Categories, or including them in the 'Balance of Adjustable Work'. The *Formula Rules* require that where the method of dealing with 'Fix Only' items has not been specified in the Contract, the method to be adopted should be agreed with the Contractor.

14.3.1 Calculation of fluctuations

There are different formulae for each type in the schedule of fluctuations. The appropriate formula is then applied to each type as next described. The net total of the fluctuation for all elements constitutes the adjustment for that valuation period. The net total of the fluctuations for all valuation periods then forms a component of the 'total value of work properly executed' under Clause 30.2.1.1.

14.3.2 Work Categories/Work Groups

The formula for work allocated to a Work Category is stated in the *Formula Rules* as:

$$C = [V(Iv - Io)]/Io,$$

where,

C = the amount of fluctuation for the Work Category to be paid to, or allowed by the Contractor;

V = the value of work executed in the Work Category during the valuation period;

Iv = the index number for the Work Category for the month during which the mid-point of the valuation period occurred;

Io = the Work Category index number for the Base Month.

An alternative form of this formula may be expressed as follows:

$$C_g = [V * I_v]/I_o$$

Where C_g is the gross amount due on the valuation, i.e. the sum of the value of work in the Work Category executed in the valuation period and the amount of fluctuation in respect of that Work Category; the other symbols have the same meaning as before.

Work Categories may be combined into a Work Group for which a weighted index number can be calculated and used in the manner already described for a Work Category. For example, where a priced Activity Schedule was provided, index numbers for each activity on the Schedule may be calculated. In the same way, if trade bills are used, suitable trade groupings may be established. The *Formula Rules* require that where this procedure is to be adopted the following should be complied with:

- it should be stated in the Contract Bills that Part II of the *Formula Rules* shall govern the formula adjustment;
- the Contract should define what Work Categories are to be included in each Work Group;

- the Base Month should be stated in the Contract Bills;
- the schedule of fluctuations should show the total value of each Work Category within each Work Group.

For the Balance of the Adjustable Work the formula is:

$$C \quad = \quad [Vo * Cc]/ Vc$$

where,

C = the amount of fluctuation for the Balance of Adjustable Work to be paid to, or recovered from the Contractor;

Vo = the value of work in the Balance of Adjustable Work in the Contract Bills;

Cc = the total amount of fluctuation for all other Work Categories to be paid to, or allowed by the Contractor for the Valuation Period;

Vc = the total value of work in the other Work Categories for the Valuation Period.

For the purpose of calculating fluctuations, a Valuation Period is defined as commencing on the day after that on which the previous valuation was done and finishing on the date of the succeeding valuation. The index numbers for the Valuation Period are those for the month in which the mid-point of the Valuation Period occurs. If the Valuation Period has an odd number of days, then it will be the middle day of the Period. However, if it contains an even number, the mid-point will be the middle day of the period remaining after deducting the last day. It should be clear from the description of the way the formula method works that it cannot be used without valuation of work completed within each Valuation Period. Without such valuation, it would be impossible to determine the work in each Work Category completed within each Period. For this reason, Clause 40.2 makes valuation before each Interim Certificate mandatory.

14.3.3 Fluctuations after Practical Completion

The formula for adjusting the value of work which is included in Interim Certificates issued after the Practical Completion is given in the *Formula Rules* as:

$$C \quad = \quad [V * C_t]/V_t$$

where,

C = the amount of the fluctuations to paid to, or recovered from, the Contractor;

V = the value of work executed;

C_t = the net total of the formula adjustment in all previous certificates excluding fluctuations on specialist engineering work;

V_t = the total value of work but excluding the Contractor's specialist work included in previous Certificates.

This formula therefore has the effect of applying the average fluctuation rate over the contract period to the additional work involved.

14.3.4 Imported articles

Rule 4 of the *Formula Rules* allows the Contractor to attach to the Contract Bills a list of articles to be imported for the Works indicating the market prices of their delivery to the site at the Base Date. Under clause 40.3, any change in these prices at the time of actual delivery is recoverable or allowable by the Contractor.

14.3.5 Delay, suspension and cessation of publication of the indices

Clause 40.6 deals with the eventuality of delay in, suspension, or cessation of publication of the indices. The Contractor and the Employer are to continue with the normal procedures for determination of the fluctuations, e.g. interim valuations of the value of work in each Work Category for each valuation. Whenever calculation of the fluctuations is due but the indices are not available for whatever reason, the Contract Sum is to be adjusted on a reasonable basis, e.g. using the last published indices increased by the monthly rate of general inflation in the economy. Whenever indices are published after such reasonable adjustment but before the Final Certificate, the estimated adjustment must be replaced with an adjustment by the formula in the next payment certificate.

14.4 Errors, certification and retention

From the description of the processes and calculations involved in determining the amount to be recovered on account of fluctuations, the possibility of errors either in the allocation of work to the Schedule of Fluctuations or in the calculations is only too real. To enable the Quantity Surveyor to agree fluctuation accounts speedily, well organized Contractors have well designed pro-formas and computer systems for the calculation of fluctuations. In examining the Contractor's statements of fluctuations under Clauses 38 or 39 for inclusion in valuations, the Quantity Surveyor will normally check that:

- any cost item included is allowed under the appropriate clause;
- all conditions precedents have been complied with, e.g. notices and proper evidence from the Contractor,
- the calculations are arithmetically accurate;
- quantities of items in respect of which fluctuations are claimed are not in excess of the requirements of the Contract.

The *Formula Rules* allow the Quantity Surveyor to correct the following types of error: (i) arithmetical errors in the calculation of the adjustment; (ii) incorrect allocation of the value of Work Categories to Work Groups; (ii) incorrect allocation of work as Contractor's specialist work, (iv) use of incorrect index numbers.

Fluctuations recovered or allowed under Clauses 38 or 39 are not subject to retention (Clauses 30.2.2.4 and 30.2.3.1) whilst those under clause 40 are subject to retention (Clause 30.2.1.1).

14.5 Extension of time, delayed completion and fluctuation

By Clause 38.4.7, if the Date of Completion (as stated in the Appendix or extended by the Architect) is overrun, the amount of fluctuations recoverable are frozen at levels which were operative at the Date of Completion. However, under Clause 38.4.8, if Clause 25 is amended, or the Architect fails to grant properly extension of time to which the Contractor is entitled, Clause 38.4.7 ceases to have effect. Clauses 39.5.7 and 39.5.8 contain corresponding provisions to the same intended effect. Under Clause 40.7, the formula is applied to the value of work executed after the Completion Date (or extended date of completion) but using the indices applicable to the month in which the date falls. This has the same effect as if all the work outstanding after the Completion Date was carried out during the month in which the Completion Date occurred. Similarly, Clause 40.7 does not apply where Clause 25 is amended or the Architect fails to grant properly extension of time to which the Contractor is entitled.

Two main implications may be drawn from these provisions. First, Clause 25 must not be amended without corresponding deletion of Clauses 38.4.8 or 39.5.8 or 40.7.2, whichever is applicable.[14] Second, they are yet another illustration of the general principle explained in Section 11.4.2, that an Architect does the Employer no favours by refusing to grant extensions of time in accordance with the Contract.

---

14 See Sections 1.4.3, 1.5.5 and 6.14.2 for general comments on amending the Contract.

15

Payment

In Article 1 of the Articles of Agreement, the Contractor undertakes to carry out and complete the project for a named sum: the Contract Sum. This sum, adjustable in defined circumstances, is to be paid to the Contractor at times and in a manner specified in the Conditions. This chapter covers the following relevant issues most of which are covered in Clause 30:

- Interim Certificates
- responsibility for valuation and certification
- valuation to determine the amount due
- unfixed materials
- failure to pay on a certificate
- the treatment of retention funds
- final adjustment of the Contract Sum
- the Final Certificate
- the Architect's liability for negligent certification.

15.1 Interim Certificates

Cashflow is the very lifeblood of the construction industry; its stoppage on any project will often bring it to a complete standstill.[1] Most construction contracts recognize this fact by allowing for the making of payment on account to the contractor before the works are complete. Under s. 109(1) of the Housing Grants Construction and Regeneration Act 1996, referred to hereafter as the 'Construction Act', a qualifying construction contract must provide for periodic payment to the contractor. Where there is a failure to comply with this requirement, a Scheme for Construction Contracts applies. This Scheme provides for periodic payment. Other sections of the Act make more detailed provisions on the parties' rights and obligations in respect of periodic payment. The main impetus towards the production of the JCT 98 was a need to redraft the JCT 80 to comply with the relevant provisions of the Act.

---

1   For examples of judicial notice of this feature of the construction industry see *Dawnays Ltd v. F. G. Minter* [1971] 2 All ER 1389, as *per* Lord Denning at p. 1393; *Gilbert-Ash (Northern) Ltd v. Modern Engineering (Bristol) Ltd* [1973] 3 All ER 195 (hereafter *Gilbert-Ash*), *per* Lord Diplock at pp. 215–216.

By Clauses 30.1.1.1 and 30.1.3, from time to time, the Architect is to issue Interim Certificates stating the amounts to be paid to the Contractor. Generally, the amount in an Interim Certificate is an instalment of the Contract Sum reflecting the accomplishment of the Contractor's obligations since the previous Certificate. As with every Certificate of the Architect, Interim Certificates are to be issued to the Employer, with copies to the Contractor (Clause 5.8). The amount in an Interim Certificate is to be determined by applying valuation rules in Clause 30.2 up to and including a date not more than 7 days before the date of the Certificate. This means that the amount does not include monies earned under the Contract in the preceding 7 days. Each Interim Certificate is to be accompanied with directions to the Contractor on monies due to Nominated Sub-contractors included in it (Clause 35.13.1.1). Furthermore, each Nominated Sub-contractor is to be informed of his entitlement (Clause 35.13.1.2).[2]

15.1.1 Timing of issue of Certificates

From the Date of Possession up to and including the date of issue of the Certificate of Practical Completion, the Architect is to issue the Interim Certificates at intervals of constant duration (Clause 30.1.3). This duration, referred to as the 'Period of Interim Certificates', is to be specified in the Appendix against Clause 30.1.3. The Appendix states after this entry that that a default period of one calendar month applies if none is specified.

The date of issue of the first Interim Certificate therefore determines the dates of subsequent Certificates. However, the Contract is silent on this date. It is to be noted that in deciding this date, it is often desirable that account is taken of any organizational and financial constraints of the Employer regarding payment. For example, if the Employer's organization issues cheques only within certain periods of the month, then the date must be set in such a way that payment on Certificates is not held up by any such restrictions. Failure to consider such factors may lead to delayed payment and, therefore, breach of the Contract by the Employer.

The rules governing the timing of issue of Interim Certificates are as follows:

1. Interim Certificates must be issued in compliance with the Period of Interim Certificates. The last of these regular certificates must be issued at the end of the period in which the date of issue of the Certificate of Practical Completion falls (Clause 30.1.3). In calculating the amount payable under this last Certificate, the percentage applied as retention is to be one half of the Retention Percentage of the Contract (Clause 30.4.1.3).

2. After the issue of the Certificate of Practical Completion, Interim Certificates are to be issued as and when further amounts are ascertained as payable to the Contractor by the Employer. Examples of such amounts include: claims for loss and/or expense accepted by the Architect, an adjudicator or an arbitrator; additional fluctuations payable as a consequence of publication of firm NEDO indices that are different from their provisional figures applied in previous certificates; final releases of retention in

---

2   See Section 9.6 for details on payment to Nominated Sub-contractors.

respect of parts taken over before the issue of the Certificate of Practical Completion;[3] and final accounts of Nominated Sub-contractors (see item 3 below). The flexibility in the timing of the issue of Interim Certificates after practical completion is subject to the proviso that the Architect is not obliged to issue an Interim Certificate within one calendar month of having issued a previous Interim Certificate.

3. Under Clause 30.7, the Architect is obliged to issue an Interim Certificate incorporating final accounts of all Nominated Sub-contractors.[4] This Certificate is to be issued as soon as practical and, in any case, not later than 28 days before the date of issue of the Final Certificate. As an exception to the general rule on the minimum interval between Interim Certificates issued during the Defects Liability Period, an Interim Certificate containing the final accounts of all Nominated Sub-contractors may be issued within a month of a previous Interim Certificate.

15.1.2 Payment on Interim Certificates

S. 110 (1)(b) of the Construction Act states that 'every construction contract shall provide for a final date for payment in relation to any sum which becomes due'. It is important to note that the Act only requires the parties to agree a final date and that it does not specify when this should be. The parties are therefore at liberty to fix whatever date they wish to apply to their contract.[5] Clause 30.1.1.1 implements the requirement for a final date in relation to payment on an Interim Certificate by providing that the final date is 14 days from its issue.

The payment procedures under the Contract entail two types of notice referred to in this book as 'Notice of Payment' and 'Notice to Withhold'. Both notices are required under the Contract to be in writing. The timetable for the procedures are summarized in Fig. 15.1.

15.1.3 Notice of payment

S. 110(2) of the Construction Act states:

> 'Every construction contract shall provide for the giving of notice by a party not later than five days after the date on which payment becomes due to him under the contract, or would have become due if:
>
> (a)   the other party had carried out his obligations under the contract, and
> (b)   no set-off or abatement was permitted by reference to any claimed to be due under one or more other contracts,
>
> specifying the amount (if any) of the payment made or proposed to be made, and the basis on which that amount was calculated'.

---

3   Regarding release of retention, provisions equivalent to those governing the retention for the whole contract apply to retentions on relevant parts where sectional completion is provided for, i.e. release of one half of the retention after practical completion of the relevant part and the remainder after the expiry of its Defects Liability Period or the issue of its Certificate of Completion of Making Good Defects whichever is the later.
4   See Section 3.8 for explanation of the rationale underlying the requirement for this Certificate.
5   This freedom has been the subject of some criticism of the Construction Act: it is feared that parties in strong bargaining positions, e.g. employers and main contractors, may impose unreasonably distant final dates for payment just to comply with the letter of the Act but to defeat its spirit, which includes speedy payment for work done.

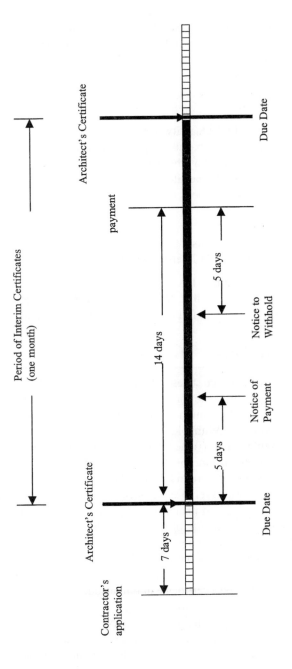

Fig. 15.1 Certification and Payment Timetable

This section is implemented in Clause 30.1.1.3, which requires the Employer to give this notice to the Contractor within 5 days after the issue of a payment certificate. This is the notice referred to in this book as a 'Notice of Payment'. As neither the Construction Act nor the JCT 98 prescribes the detailed contents of the notice, one to the effect that the payment is for work executed under the Contract and that it has been ascertained in accordance with Clause 30.2 for the relevant period would satisfy the requirement. However, it would be good practice to include the following:

- project title,
- project reference number,
- names of the parties,
- names of the Architect and the Quantity Surveyor,
- the total value of work properly executed by the Contractor since commencement,
- total loss and/or expense granted by the Architect,
- value of materials and goods on the site,
- value of relevant materials and goods off site,
- total adjustment for price fluctuation,
- total retention deducted,
- amount payable under the previous interim certificate,
- amount payable under the present Interim Certificate.

The Guidance Notes to Amendment 18, which introduced the requirement for Notices of Payment into the JCT 80, suggest that Notice of Payment is not necessary where the Employer intends to pay on time the full amount stated in a Certificate.

15.1.4 Notice to Withhold

S. 111(1) of the Construction Act states: 'A party to a construction contract may not withhold payment after the final date for payment of a sum under the contract unless he has given an effective notice of intention to withhold payment.' This is the notice referred to in this book as the 'Notice to Withhold'. S. 111(2) of the Act specifies the contents of an effective Notice to Withhold as: (i) the total amount to be withheld; (ii) the grounds for the withholding; and (iii) itemization of the total amount to be withheld by individual grounds if there are more than one ground. S. 111(1) also states that a Notice of Payment with the required content may also serve as a Notice Withhold. This means that if a Notice of Payment also contained the information required in a Notice to Withhold, the latter notice may be dispensed with. The use of the term 'withhold' in s. 111(1) is unfortunate because it has no defined legal meaning. To decipher the effect of the section, we must therefore apply its ordinary everyday meaning. The *Concise Oxford Dictionary* (7th edn) defines it as 'hold back'. The most common grounds upon which building owners and main contractors refuse to pay on a certificate in full include abatement and set-off[6] which, as a matter of strict law, are different from holding back. Technically therefore, a would-be payer may lawfully refuse to pay in full without serving prior notice. Acceptance of this argument will almost certainly defeat the purpose of the legislation. It is submitted therefore that the courts are likely to construe 'withhold' to include abatement and set-off.

---

6   See Section 15.1.6.

Under Clause 30.1.1.4, the Employer is to serve a Notice to Withhold not later than 5 days before the final date for the relevant payment. The use in the clause of the wording 'the Employer may give written notice to the Contractor which shall specify any amount proposed to be withheld and/or deducted' was probably intended to avoid this problem. However, it is doubtful whether such intention has been achieved completely. This is because the Employer is entitled to challenge that the amount stated is really 'the amount due' under the Contract and refer it to adjudication. Refusal to pay pending the outcome of the adjudication could not by any stretch of semantics be referred to as withholding or a deduction.

15.1.5 Contractual grounds to withhold

Many clauses, summarized in Table 15.1, expressly give the Employer the power to deduct certain sums from payment due or to become due to the Contractor. The Employer can therefore deduct such sums against payment required under a Certificate. It is to be noted that the Architect and the Quantity Surveyor are not to take account of these sums in valuations or Certificates. They are matters entirely for the Employer to deal with, although the Architect or the Quantity Surveyor may be under a contractual obligation to advise him accordingly.

Table 15.1   Express rights of set-off under the JCT 98

Clause	Set-off
Clause 4.1.2	the cost to the Employer of giving effect to Architect's Instructions with which the Contractor has failed to comply
Clause 21.1.3	the cost to the Employer of taking out insurance as a result of the Contractor's failure to take out or maintain insurances required under Clause 21.1.1.1
Clause 22A.2	the cost to the Employer of taking out insurance as a result of the Contractor's failure take out or maintain insurances required under Clause 22A.1
Clause 24.2.1	liquidated damages
Clause 27.4.4	the Employer's loss resulting from determination of the employment of the Contractor under Clause 27
Clause 31.6.1	the amount of any deduction made by the Employer under the Construction Industry Tax Deduction Scheme
Clause 35.13.5.3	the amount of direct payments made to Nominated Sub-contractors following the Contractor's failure to discharge their monies to them
Clause 35.18.1.1	the Employer's loss due to defects in the works of a Nominated Sub-contractor discovered after he has received final payment and which the Employer is unable to recover from the guilty Nominated Sub-contractor under the indemnity given under Agreement NSC/W
Clause 35.24.9	the Employer's loss arising from the need to appoint a replacement sub-contractor following the valid determination by a Nominated Sub-contractor of his own employment (the excess of the amount paid to the replacement sub-contractor over the price of the original nominated sub-contractor)

15.1.6 Other grounds to withhold

In *Gilbert-Ash*,[7] the House of Lords held that a payer of a certificate under a construction contract is entitled to raise any set-off available to him under the general law unless the

7   See, n. 1

contract itself expressly removes that right. The JCT 98 does not remove this right of the Employer in relation to his obligation to pay on Interim Certificates. It follows therefore that the Employer would be entitled to deduct from Certificates not only sums expressly authorized under the Contract but also set-offs available under the general law. An example of a right of set-off under the general law is set-off through the defence of abatement at common law.[8] This set-off involves the contention that the Contractor's performance is worth much less than he is contractually obliged to achieve, e.g. because of defects. This right is appropriate where the Employer wants to rely on the defects to make a deduction against the certified sum. Liquidated mutual debts may also be set against each other even if they arise from different contracts.[9] For example, at common law, the Employer would be entitled to deduct liquidated damages incurred on another contract. In equity, the Employer may raise a cross-claim (counterclaim) by way of set-off, e.g. damages for delay. Finally, under s. 323 of the Insolvency Act 1986, where the Contractor is insolvent, the Employer may be entitled to set-off the contingent liability of the Contractor as a result of the determination, e.g. estimated damages of completing the project with another contractor.[10]

The general right of set-off recognized in *Gilbert-Ash* was applied in *Acsim (Southern) Ltd v. Danish Contracting Ltd*,[11] which concerned withholding of payment by a main contractor under a sub-contract. The sub-contract expressly stated that the parties' rights to set-off 'are fully set out in these conditions and no other rights whatsoever shall be implied as terms of this sub-contract relating to set-off'. The Court of Appeal decided that the exclusion was not wide enough to exclude the defence of abatement at common law.

15.2 Responsibility for valuation and certification

The responsibility for issuing Interim Certificates is the Architect's but he may delegate the task of producing the valuation to the Quantity Surveyor whenever he considers it necessary (Clauses 30.1.1.1 and 30.1.2). Where fluctuations are to be recovered under Clause 40, valuations are mandatory (Clause 40.2).

Clause 30.1.2.2 allows the Contractor to make an application setting out the gross valuation according to Clause 30.2. The Contractor's application must include similar applications from Nominated Sub-contractors under the terms of Conditions NSC/C. If the Contractor makes such an application, the Quantity Surveyor must make an interim gross valuation. If the valuation of the Quantity Surveyor differs from the Contractor's, he must submit a copy to the Contractor. There is a further requirement that, in case of such difference, the valuation of the Quantity Surveyor must have the same level of detail as that of the Contractor and identify the areas of differences. This is really a formalization of the common practice whereby valuations are prepared jointly by the Quantity Surveyor and the Contractor's quantity surveyor.

---

8   *Mondel v. Steel* (1841) 8 M. & W. 858; (1976) 1 BLR 106; see also s. 53 of the Sale of Goods Act 1979; *Duquemin v Slater* (1993) 65 BLR 124; *Barrett Steel Buildings Ltd v. Amec Construction Ltd* (1997) 15 CLD-10-07; *Mellowes Archital Ltd v. Bell Projects Ltd* (1997) 87 BLR 26 (damages for delays cannot be the basis of abatement).
9   *Hargreaves (B) Ltd v. Action 2000 Ltd* (1992) 62 BLR 72.
10 *Willment Brothers Ltd v. North West Thames Regional Health Authority* (1984) 26 BLR 51.
11 (1989) 47 BLR 55; see also *GPT Realisations Ltd (in Administrative Receiver & in Liquidation) Ltd v. Panatown Ltd* (1992) 61 BLR 88.

The Architect is responsible for the correctness of the sum stated in Certificates although he is entitled to rely on the measurements and valuations of the Quantity Surveyor unless they are obviously wrong.[12] The Architect therefore has the power to adjust any valuation prepared by the Quantity Surveyor where he considers it necessary. As quantity surveyors are more expert in valuation, most Architects would be reluctant to embark upon any major adjustments of the valuations without good cause.

The Conditions do not expressly provide for the possibility of the Architect failing to issue an Interim Certificate when he should have done so. However, a number of cases involving other contracts which were also silent on the point have been before the courts.

Compania Panamena Europea Navigaion Ltd v. Frederick Leyland & Co. Ltd:[13] a contract for repair of a ship provided that the repairers were to be paid their expenditure certified by the owner's surveyor. The surveyor contended that his jurisdiction under the contract went beyond the quality of the work of repair to include questions of economy of the repairs. He then refused to issue a certificate unless the repairers produced full information on which he could consider the issue of economy. This stance was concurred to by the owners. The House of Lords held that the surveyor's interpretation of the contract was incorrect and that the repairers' action to recover the value of the repairs without a certificate should succeed. The basis for the decision was that it was an implied term of the repair contract that certificates would be issued in accordance with the contract. It was also suggested that where a certifier failed to perform his role, the employer has a positive duty to take action to ensure his due performance and even to dismiss and replace him if his defaults continue.

Croudace Ltd v. London Borough of Lambeth:[14] the Architect under a contract in the terms of the JCT 63 form was the Chief Architect of the Council. When he retired, the Council failed to appoint a replacement, with the consequence that the Contractor's claims for loss and/or expense could not be considered. The Court of Appeal held that the failure of the Council to appoint or to instruct anyone to assess the Contractor's claim was a breach of a implied duty to take reasonable steps to ensure that the Contractor's claim was ascertained.

It is submitted therefore that failure to issue an Interim Certificate under the JCT 98 puts the Employer in breach of contract with the Contractor. The Contractor's damages for the breach would include the amount that should have been payable if the Architect has certified. The Contractor can therefore enforce payment even in the absence of a Certificate from the Architect. In addition, depending upon the nature and extent of the failure, the Contractor may argue that this type of breach goes to the root of the contract and therefore that it entitles him to treat the contract as terminated and to sue for damages.

Failure to certify is different from wrong certification. In *Lubenham Fidelities and Investments Co. Ltd v. South Pembrokeshire District Council and Another*[15] the Court of

---

12 *R. B. Burden Ltd v. Swansea* [1957] 1 WLR 1167; the liability of the Architect in respect of certificates is discussed in Section 15.9.
13 [1947] AC 428 (hereafter *Panamena*); this was applied by the Supreme Court of New South Wales in *Perini Corporation v. Commonwealth of Australia* (1969) 12 BLR 82.
14 (1985) 6 Con. LR 70; (1986) 33 BLR 20 (hereafter *Croudace*).
15 (1986) 33 BLR 39 (hereafter *Lubenham Fidelities*).

Appeal decided that an Employer under the JCT 63 who became aware of shortcomings in the Architect's performance of his function as certifier was not under a duty to stop the shortcomings or tell the Architect how to do the job properly. In that case, in calculating the amount due under a certificate, the Architect made deductions not authorised under the Contract. It was also stated that, as applies to most standard construction contracts, an Employer under the JCT 63 does not warrant that the contract administrator will certify the correct amount due under the contract. Any errors are to be pointed out to the Architect for correction in the next certificate or resolved through arbitration. The Contractor's contention that he was entitled under the principle in *Panamena* to recover the amount that should have been certified was rejected. That case was distinguished on the grounds that, unlike the JCT 63, the contract in *Panamena* did not contain an arbitration clause. It is submitted that there is no material difference between the JCT 63 and the JCT 98 on the issue of the Employer's obligations in respect of the Architect's performance of his role as certifier, particularly where Clause 41B applies.

15.3 Valuation to determine the amount of a certificate

The amount stated as due under an Interim Certificate is to be quantified in accordance with Clause 30.2. Some of the calculations involved may be unnecessary where a schedule of stage payments is determined and incorporated into the Contract. In such a situation, Clause 30 would have to amended accordingly. In view of the difficulties involved in amending the JCT 98, due care has to be exercised.[16]

The amount to be certified is given by the expression:

$$C = G - R - A - P$$

where

C is the amount of the Certificate;

G is the gross valuation;

R is the Retention;

A is reimbursement due on any advanced payment;

P is the amount stated as due in the previous Interim Certificate.

15.3.1 Gross valuation

The gross valuation can be determined from the formula:

$$G = g_r + g_o - g_c$$

where

g_r = total of value of work subject to retention (see Section 15.3.2);

---

16 See Sections 1.4.3, 1.5.5 and 6.14.2 for discussion on the effect of Clause 2.2.1 in relation to amendments.

g_o = total value of work not subject to retention (see Section 15.3.3);

g_c = the total deduction in price to be allowed by the Contractor (see Section 15.3.4).

15.3.2 Sums subject to retention

By Clause 30.4.1.1, this percentage is to be 5 per cent unless a lower percentage is agreed by the Contractor and the Employer and inserted in the Appendix entry for 'Retention Percentage'. A footnote advises that if the estimated value of the contract is more than £500000, the percentage must not be more than 3 per cent. The Retention to be deducted is determined by applying the Retention Percentage to the total of the following:

1. the value of the Contractor's own work properly executed including variations (Clause 30.2. 1.1);[17]
2. fluctuations where the formula method is to apply (Clause 30.2.1.1)(details on this method of recovering fluctuations are in Chapter 14);
3. the total value of materials and goods on the site and intended for incorporation into the Works (Clause 30.2.1.2);
4. the total value of materials or goods off the site but intended for the Works (Clause 30.2.1.3);
5. the total amount payable to Nominated Sub-contractors which is subject to retention under clause 4.18 of Conditions NSC/C – the total of equivalents of items 1, 2 and 3 above (Clause 30.2.1.4);
6. the Contractor's profit on the total value of work properly executed by Nominated Sub-contractors less price reduction under the fluctuation clause (Clause 30.2.1.5).

At Practical Completion, the percentage to be applied is one half of the Retention Percentage (Clause 30.4.1.3). The result of the combined operation of Clauses 30.4.1.2 and 30.4.1.3 is that the remainder is to be released after expiry of the Defects Liability Period or the issue of the Certificate of Completion of Making Good Defects, whichever occurs later.

15.3.3 Sums not subject to retention

These are:

1. costs incurred by the Contractor in respect of:
    (i)   fees and charges payable to local authorities or statutory undertakers and any other statutory fees or charges subject to the limitations in Clause 6.2;
    (ii) compliance with an AI requiring the opening up of work for inspection or the carrying of any tests where the results of the inspection/tests showed that the work or materials were in accordance with the Contract (Clause 8.3);

---

17 This figure may be determined by: (i) physically measuring work items completed, multiplying the quantity of each item completed by its rate in the Contract Bills and summing up for all items; (ii) assessing the stage the Works have reached and taking the applicable payment from a stage payment schedule if it applies; (iii) where a priced Activity Schedule has been provided, assessing the percentages of the activities completed, taking the same percentage of the relevant price in the Schedule. The value of work in variations is determined in accordance with the relevant provisions in Clause 13 or 13A.

(iii) the Contractor's liability for infringement of royalties and patent rights (Clause 9.2);

(iv) the taking out and maintenance by the Contractor of insurance against damage to property from the risks defined under Clause 21.2.1 (Clause 21.2.3);

2. loss and/or expense under Clauses 26.1 or 34.3 (Clause 30.2.2.2) (this subject is discussed in detail in Chapter 13);

3. the value of work of restoration, replacement or repair of loss or damage to the Works from the insured risks under Clause 22B or 22C and disposal of debris (Clause 30.2.2.2);

4. final payment due to Nominated Sub-contractors (Clause 30.2.2.3);

5. increase in costs due to cost fluctuations recoverable under Clause 38 or 39 (Clause 30.2.2.4)(details of these methods of recovery fluctuations are in Chapter 14);

6. amounts payable to Nominated Sub-contractors, which are not subject to retention under Conditions NSC/C; these mirror the items above (Clause 30.2.2.5).

15.3.4 Deductions to be allowed by the Contractor

Clause 30.2.3 lists the following deductions to be made as part of the pre-certification valuation process:

1. deductions in respect of the Contractor's setting out errors which are not to be amended (Clause 7);

2. deductions in respect of work, materials, and goods not in accordance with the Contract but which are by agreement to remain (Clause 8.4.2);

3. deductions in respect of defects, shrinkages and other faults for which the Contractor is responsible but which, by agreement, are to remain after Practical Completion (Clause 17.2 and 17.3);

4. cost reduction due to fluctuations under Clause 38, or 39;

5. cost reductions due fluctuations under Clause 4A or 4B (equivalent to Clause 38 and 39 of the main contract, respectively) of Conditions NSC/C.

15.4 Unfixed materials and interim certificates

Materials intended for construction contract works but not yet incorporated into them fall into two categories. The first category covers those delivered to the site, and are usually referred to as 'materials on site'. The second category consists of all those materials intended for but not yet delivered to, the site, e.g. goods of mechanical and electrical installation and materials for temporary works stored in the contractor's workshops or on the premises of suppliers and sub-contractors. This second category is usually referred to as 'off-site materials'. The two categories are often referred to collectively as 'unfixed materials'.

Most construction contracts contain provisions allowing the contractor to be paid for unfixed materials in Interim Certificates. Such payment exposes the employer to the risk that the materials may never actually be used for the works. For example, the employer may be paying for materials ownership of which never did vest in the contractor in the first place. Holding companies of the contractor, suppliers and sub-contractors and their holding companies may still have lien in the materials after they have been paid for by an employer. In addition, the materials and goods may be stolen, lost, damaged or even diverted by the contractor to another project. The JCT 98 makes elaborate provisions designed to protect the Employer against these risks. They are contained mainly in Clauses 16, 19.4, 30.2, 30.3.

15.4.1 On-site materials and goods

In respect of materials delivered to the Site and intended for incorporation into the Works, the important provisions are as follows:

1. They are not to be removed from the Site for any purpose other than the carrying out of the Works without the written consent of the Architect. This consent must not be unreasonably withheld (Clause 16.1). A corresponding provision must be included in every domestic sub-contract (Clause 19.4.2.1).
2. Their total value must be included in interim certificates provided they are reasonably, properly and not prematurely delivered and are adequately protected against loss or damage (Clause 30.2.1.2).
3. After their inclusion in interim certificates and payment on the certificate by the Employer, the materials become the property of the Employer (Clause 16.1). A term is to be included in every domestic sub-contract to the effect that the sub-contractor concerned shall not deny the transfer of title to the Employer where the sub-contractor's materials are affected (Clause 19.4.2.2).
4. The Contractor is to include a term in every domestic sub-contract to the effect that where the Contractor pays for the materials before he is paid by the Employer ownership in the materials is to pass to the Contractor (Clause 19.4.2.3).
5. They must be insured against loss or damage from the risks covered by the Clause 22.2 definition of 'All Risk Insurance' (Clause 22A.1/22B.1/22C.2).
6. The Contractor is responsible for any loss or damage to the materials even after ownership in them has passed to the Employer (Clause 16.1).

It is to be noted that DOM/1 and Conditions NSC/C contain the terms which must be included in all domestic sub-contracts.

The primary intention behind the conditions that must be satisfied before unfixed materials are included in an Interim Certificate is to ensure that they do become the property of the Employer upon payment on the Certificate. Unfortunately, complete compliance with the conditions does not necessarily guarantee this end. This is because it is not uncommon for contracts for the supply of materials to contain stipulations to the effect that the materials remain the property of the supplier until paid for in full by the buyer or even until all debts owed by the buyer to the supplier are settled in full. Such clauses are referred to as 'Retention of Title' (ROT) clauses or *Romalpa* clauses after a famous case in which the court upheld the effectiveness of such clauses against purported transfer of title to third parties.[18] In any such situation, no matter the terms of JCT 98, the Contractor cannot effectively transfer title in the materials and goods to the Employer as required by some of its provisions. The Latin maxim *nemo dat quod non habet* best expresses this position of the law.

---

18 *Aluminium Industrie Vaasen BV v. Romalpa Aluminium* [1976] 2 All ER 552 (hereafter *Romalpa*).

This problem came to prominence in the construction industry through the litigation in *Dawber Williamson Roofing Ltd v. Humberside County Council*:[19]

> Main contractors on a JCT 63 contract for the construction of the Council's school sub-contracted the roofing to the plaintiff sub-contractors who then delivered 16 tons of roofing slates to the site. After receiving payment for the slates through a certificate issued under the main contract, but before paying the plaintiffs for them, the main contractor became insolvent. The sub-contract provided that materials brought onto the site by the sub-contractor remained the property of the sub-contractor until they were paid for by the Contractor. Relying upon a clause of the JCT 63 that unfixed materials and goods paid for under an Interim Certificate become the property of the Employer, the Council claimed ownership in the slates. It was held that ownership of the slates had not passed to the Contractor. The sub-contractors were therefore entitled to damages for wrongful detention and use of the materials by the Council.

It is important to note that *Dawber Williamson* applies to contests between an employer and the contractor's sub-contractors for ownership of unfixed materials and that such a contest between an employer and the contractor's supplier may well result in a different outcome. The difference arises from the fact that the Sale of Goods Act 1979 applies to supply of materials but not to sub-contracts, which are contracts for work and materials rather than sale of goods contracts. S. 25(1) of the Act states:

> Where a person having bought or agreed to buy goods obtained with the consent of the seller, possession of the goods or the documents of the title to the goods, the delivery or transfer by that person, or by a mercantile agent acting for him, of the goods or documents of title, under any sale, pledge or other disposition thereof, to any person receiving the same in good faith and without notice of any lien or other right of the original seller in respect of the goods, has the same effect as if the person making the delivery were a mercantile agent in possession of the goods or documents of title with the consent of the owner.

The effect of this section on the competing interests of the contractor's suppliers and the employer, where clauses in main contracts purported to transfer ownership of unfixed materials, was considered in *Archivent Sales & Development Ltd v. Strathclyde Regional Council*:[20]

> This case arose between an Employer under a JCT 63 contract and supplier of ventilators to the Contractor. The supply contract provided that 'until payment of the price in full is received by the company the property and the goods supplied shall not pass to the customer'. Clause 14 of the JCT 63, along the same lines as JCT 98 Clause 16, provided that unfixed materials must not be removed without the consent of the Architect and that ownership of materials passed to the Employer upon their inclusion in an Interim Certificate and payment on it by the Employer. After the Contractor was paid for the ventilators by the Employer but before the Contractor had in turn paid for them, the Contractor went into receivership. The sellers took proceedings against the Employer claiming ownership of the ventilators. The Scottish Court of Session (Outer

---

19 (1979) 14 BLR 70 (hereafter *Dawber Williamson*).
20 (1984) 27 BLR 98.

House) held that the effect of s. 25(1) of the Sale of Goods Act was to confer goods on the Employer.

An essential requirement for application of s. 25(1) is that the second buyer was not aware of any ROT clauses in the original supply contract. The principle in this case is therefore likely to apply to suppliers only where the Employer and the Architect are not aware of the terms of the relevant supply contracts. However, such ignorance is unlikely to be the case with Nominated Suppliers because, as explained in Chapter 10, the Architect must be involved where the supply contract contains a retention of title clause of the type in this case.

15.4.2 Off-site materials and goods

Clause 30.3 now distinguishes between uniquely identified items, e.g. a boiler from a specified supplier, and those not uniquely identified, e.g. bricks. Regardless of whether the item is uniquely identified or not, the Architect must include its value in Interim Certificates provided the following conditions are met:

1. The item must be included in the listed items annexed to the Contract Bills.
2. The item must be in accordance with the Contract.
3. Where materials belonging to that item are in the premises of their manufacture, assembly or storage, they must either be set apart or suitably marked to identify the Employer, the party who ordered them (either the Contractor or a sub-contractor) and that they are destined for delivery to the Works.
4. The Contractor has insured them for their full value against loss or damage from the Specified Perils under a policy which covers both the Employer and the Contractor and has provided the Architect proof of the cover. The period of cover is to be from when ownership in the materials passed to the Contractor until they are delivered to the Site. As explained in Section 7.3.4, upon delivery, they will be covered by the Contractor's All Risk Insurance of the Works and Site Materials.
5. The Contractor has provided the Architect with reasonable proof that ownership of the relevant materials is vested in the Contractor so that, as provided for in Clause 16.2, they become the property of the Employer after payment on the relevant Certificate. As a minimum, the Architect must check relevant supply contracts ensure that under their terms the Contractor has acquired ownership of the relevant materials.

In addition, in respect of items not uniquely identified, the Contractor must have provided the Architect with reasonable proof that he has procured a bond from a surety approved by the Employer. The terms of the bond are to be the form of bond in the Appendix unless, pursuant to the 7th Recital, the Employer required a different form of bond, in which case the Contractor must have procured a bond in the terms specified before contract execution. The form of bond annexed to the Appendix is of the on demand variety, i.e. provided the Employer makes a call on it strictly in accordance with its terms, the surety must meet the claim.[21]

---

21 See Issaka Ndekugri, 'Performance Bonds and Guarantees in Construction Contracts: A Review of Some Recurring Problems' (1999) ICLR 294, pp. 294-311.

The Employer should have specified two requirements of the bond either by completing the appropriate section of the form provided or stating them in the tender documentation, e.g. in the preambles section in the Bills of Quantities. These are the amount of bond and its longstop date. The amount of bond should be fixed by taking into account the likely maximum payment for off-site materials of this category. The longstop date should be fixed only after the most careful consideration because the bond expires after that date even if the Contractor failed to deliver the materials. The date of issue of the Certificate of Practical Completion would be prudent.

There are similar requirements for a bond in respect of materials that are uniquely identified listed materials but the obligation to procure it is conditional upon it being completed in the Appendix that such bond is required. The reason why the requirement for this type of bond is optional is probably that it is easier to check who the legal owners of uniquely identified items are, thus making the protection of a bond less essential.

15.5 Remedies for non-payment by the Employer

The remedies against failure of the Employer to meet his payment obligation include claiming interest on the amount not paid, suspension of performance, or determination. Some of these remedies are available concurrently. For example, the Contractor may suspend work and claim interest at the same time. Regardless of whether the Contractor has decided to suspend performance or determine his own employment, he would still be interested in ways of compelling the Employer to pay on the Certificate, including interest, if the Employer is solvent. This section therefore also considers the appropriateness of arbitration, litigation and adjudication. Whichever of these is most appropriate depends upon the dispute resolution provisions in the Contract and the reasons for the failure to pay.

15.5.1 Interest

If the Employer fails to pay by the final date without valid grounds for withholding payment of the whole or part of the certified amount, the Contractor is entitled to recover simple interest on the amount the Employer failed to pay for the period of non-payment beyond the final date. The applicable rate of interest is to be 5 per cent over the Base Rate of the Bank of England current when the payment became overdue.

If the Employer and the Contractor both entered into the Contract in the course of a business, the Late Payment of Commercial Debts (Interest) Act 1998 (LPCDIA) would apply.[22] This Act implies into any qualifying contract a term that any debt under it carries interest at a rate fixed by the Secretary of State. The current rate is 8 per cent above Base Rates. There may be a problem with the 5 per cent stipulated in the Contract because s. 8 (4) of LPCDIA provides that any contract terms are void to the extent that they purport to 'vary the right to statutory interest so as to provide for a right to statutory interest that is not a substantial remedy for the late payment'. S. 9(1) of LPCDIA provides that a remedy shall be regarded as a substantial remedy unless:

---

22 At present this can be enforced by only small businesses. The legislation is to cover all businesses from the year 2002.

1. the remedy is insufficient either for the purpose of compensating the supplier for the late payment; and
2. it would not be fair and reasonable to allow the remedy to be relied on to oust or vary the right to statutory interest that will otherwise apply in relation to the debt.

Under s. 9(3), the reasonableness test involves having regard to the benefits of commercial certainty, the relative bargaining positions of the parties, whether the term was imposed by one party to the detriment of the other, whether the supplier received an inducement to accept the term. It may be argued that only percentages over the current figure set by the Secretary of State would be considered substantial. However, it is submitted that, as the JCT 98 is an industry-wide standard form, the 5 per cent is likely to be considered a substantial remedy.

15.5.2 Suspension of performance

At common law, there is no implied right to suspend work on account of non-payment by an employer.[23] Ss. 112(1) to (3) of the Construction Act state:

> (1) where a sum due under a construction contract is not paid in full by the final date for payment and no effective notice to withhold payment has been given, the person to whom the sum is due has the right (without prejudice to any other right or remedy) to suspend performance of his obligations under the contract to the party by whom payment ought to have been made ('the party in default').
>
> (2) The right may not be exercised without first giving the party in default at least 7 days' notice of intention to suspend performance, stating the ground or grounds on which it is intended to suspend performance.
>
> (3) The right to suspend performance ceases when the party in default makes payment in full of the amount due.

Clause 30.1.4 implements the statutory right to suspend performance of contractual obligations. To avoid confusion, the clause provides expressly that any suspension for non-payment under the Contract is not to be treated as the specified defaults under Clause 27.2.1 (wholly or substantially suspending the carrying out of the Works without reasonable cause) or Clause 27.2.2 (failing to proceed regularly and diligently with the Works). The Employer is therefore not entitled to determine the Contractor's employment on grounds of suspension where the reason for it is the Employer's failure to pay on a Certificate by the final date for payment without a valid notice to withhold.

The Contractor's right to suspend performance ceases when the Employer pays in full. The Contractor must then, therefore, recommence the carrying out of the Works unless, in the meantime, his employment has been validly determined. Unfortunately, the Contract is silent on how soon after the payment in full the Contractor must recommence the performance of his contractual obligations. It is submitted that an obligation to recommence within a reasonable time of the payment would be implied. What is a reasonable time would depend on the surrounding circumstances of the relevant

---

23 *Lubenham Fidelities*, see n. 15; see also *Perini Corporation v. Commonwealth of Australia* (1969) 12 BLR 82 (a decision of the Supreme Court of New South Wales) (hereafter *Perini*); *Canterbury Pipelines v. Christchurch Drainage* (1979) 16 BLR 76 (a decision of the Court of Appeal of New Zealand).

obligation, e.g. the period of non-payment, whether the Contractor has removed his resources from the Site, and other work taken on as a result of the suspension. For example, where the Contractor stopped insuring the Works, it may be quicker to reinstate the insurance cover than to re-start physical work on site.

It is essential that the Contractor follows the specified procedures for suspending performance. Failure to do this will leave the Contractor open to charges of repudiation. The procedure is as follows. When the Contractor reaches a decision to suspend after the Employer has failed to pay in accordance with the Contract, he is to give the Employer written notice of his intention to do so for those reasons. A copy of the notice to suspend must be served on the Architect. The right to suspend crystallizes 7 days after the Contractor has given the notice. It is arguable that the Contractor has not 'given' notice until the communication has actually been received by the Employer.[24] In this respect, it is regrettable that the deeming provisions on receipt notices of the type in Clause 28.1 do not apply here. Prudence demands that the Contractor must not work on the minimum notice of 7 days unless he has established receipt of the notice with reasonable certainty.

15.5.3 Adjudication

Invoking adjudication would be the Contractor's best course of action where the Employer puts up any defence to his liability to pay, e.g. that there are defects in work included for payment, or that the Architect failed to apply the valuations rules under clause 30.2 correctly, or that he is entitled to set-off. The issue submitted to the adjudicator must be stated very carefully. If he is required to decide whether or not the amount is due and the adjudicator decides for the Contractor, the decision can be enforced against the Employer through the courts by a Summary Judgment application. This is because Article 7A provides that disputes about the enforcement of an adjudicator's decision are outside the JCT 98 arbitration agreement, thus giving rise to the effect that s. 9 of the Arbitration Act 1996 does not apply even if Clause 41B applies.

15.5.4 Litigation

The *Beaufort Developments (NI) Ltd v. Gilbert-Ash (NI) Ltd and Others*[25] decision reasserted the jurisdiction of the courts to open up and review and revise certificates and decisions of contract administrators. It has been suggested that litigation may, as a consequence of that decision, become more attractive to project participants. There are two powerful weapons available in litigation that the Contractor can apply for on commencement of proceedings: Summary Judgment and/or Interim Payment under the Rules of the Supreme Court.[26]

The essence of an application for Summary Judgment, also known as Order 14,[27] is that there is no defence to the claim for payment. It enables the applicant to obtain final judgment without a full trial, thereby saving time and costs. Usually both the evidence to

---

24 See *Holwell Securities Ltd v. Hughes* [1974] 1 WLR 155.
25 *Gilbert-Ash*, see n. 1.
26 These have been replaced by the Civil Procedure Rules.
27 This is now Rule 24 of the Civil Procedure Rules.

support the application and that required to rebut it are provided in the form of affidavits. These are sworn statements which may give rise to prosecution for the crime of contempt of court if they contain falsehoods.

Interim Payment, or Order 29,[28] is ordered where the court finds that a full trial is required but is nonetheless satisfied that, should the trial be held, the applicant would be awarded a substantial amount. In such circumstances, the court can order immediate payment of part of the claim clearly due pending the final trial. However, if the outcome of the trial is that the amount paid was not owed in part or at all, the court can order an appropriate refund.

The appropriateness of litigation depends upon whether or not Clause 41B, the JCT 98 arbitration agreement, applies. If it does, the Employer is likely to seek stay of any legal proceedings to arbitration. S. 9(4) of the Arbitration Act 1996 states that, on such an application, the court must stay the proceedings 'unless the arbitration agreement is null and void, inoperable or incapable of being performed'. This was construed by the Court of Appeal in *Halki Shipping Corporation v. Sopex Oils Ltd (The 'Halki')*[29] to imply that the court cannot exercise Order 14 jurisdiction unless the Employer admits that the sum is due. This means that if the Employer puts forward any argument against liability, no matter how flimsy, the court may have no choice but to stay the proceedings.

15.5.5 Arbitration

Arbitration is an option if the Appendix is completed to indicate that Clause 41B applies or the parties, after the dispute has arisen, agree in writing to resolve their dispute by arbitration. It is often argued that the current arbitration law gives arbitrators sufficient powers to provide summary remedies similar to Summary Judgment and Interim Payment. This would be helpful to the Contractor only if the appointed arbitrator is adequately skilled and is prepared to use his powers in that way.

15.6 Treatment of retention funds

Each Interim Certificate is to be accompanied by a statement specifying the retentions of the Contractor and of each Nominated Sub-contractor (Clause 30.5.2.1). This statement is to be prepared by the Architect but he may delegate this duty to the Quantity Surveyor. The statement is normally drawn up on a standard form of the RICS titled 'Statement of Retention and of Nominated Sub-contractor Values' and is to be issued to the Employer, the Contractor and each Nominated Sub-contractor whose work is covered by the relevant Interim Certificate. Apart from serving the purpose of satisfying the requirement of this clause, the Statement of Retention and of Nominated Sub-contractors' Values also serves as compliance with the Architect's duty under Clause 35.13.1 to specify the amounts in each Interim Certificate which must be discharged to Nominated Sub-contractors.[30]

A criticism of Clause 30.5.2.1 is that there is no reason for any Nominated Sub-contractor to be provided with information on the retentions of any other sub-contractor

28 This is now Rule 25 of the Civil Procedure Rules.
29 [1998] 1 WLR 726.
30 See Section 9.6 on details on payment to Nominated Sub-contractors.

and that the clause should only have required each Nominated Sub-contractor to be given only information on his own retentions. The sole counter-argument that these writers can identify is one of administrative convenience of the Architect and the Employer in sending the same document to all concerned. Some architects and employers argue that there is no particular reason for secrecy about the parties' payment entitlements.

The primary purpose of retention in construction contracts is to provide an Employer with security against latent defects and other failures of the Contractor to perform his obligations. However, many Employers use retention funds as working capital in the running of their businesses. Apart from the risk of misappropriation of the Contractor's retention by the Employer, there is also the risk that in the event of the Employer's insolvent liquidation or bankruptcy, the retention funds will have go into the pot for distribution among his creditors.[31] To be able to follow our examination of how these risks are addressed in the Contract, there is a need to understand the concept of a 'trust'.

15.6.1 Concept of a trust

The essence of a trust is that the legal owner (the trustee) of property is only holding it for the benefit of another (the beneficiary). There is therefore a separation between legal ownership and beneficial ownership. The trustee is said to stand in a fiduciary relationship, i.e. one of the utmost confidence, to the beneficiary. This means that the trustee must not derive any personal benefit from holding the property unless the terms of the trust provide for that benefit. In addition, in the trustee's performance of his responsibility, he must not put unacceptable risks on the beneficiary's interests in the property.

A trust can be created expressly by statute, the operation of the law or a declaration of the legal and beneficial owner (settlor) during his lifetime. We are concerned here with creation of trusts by declaration. Although no specific words are required, the words actually used must show that: (i) the settlor intended to create a trust (certainty of intention); (ii) the intended beneficiaries are identifiable (certainty of objects); (iii) the property subject to the trust is specific or ascertainable (certainty of subject-matter).[32]

15.6.2 Trusts of retention funds

Clause 30.5.1 states that 'The Employer's interest in the Retention is fiduciary as a trustee for the Contractor and for any Nominated Sub-contractor (but with no obligation to invest)'. This clause raises the question whether it creates a trust of the retention funds in favour of the Contractor and Nominated Sub-contractors. The effect of a similar clause in the JCT 63 was considered in *Rayack Construction Ltd. v. Lampeter Meat Co. Ltd*[33] Clause 30(4) of that form stated:

> The amounts retained by virtue of sub-clause (3) of this condition shall be subject to the following rules: (a) the Employer's interest in any amounts so retained shall be fiduciary as trustee for the Contractor (but with no obligation to invest).

31 See *Re Jartray Developments Ltd* (1983) 22 BLR 134.
32 *Knight v. Knight* (1840) 3 Beav. 148.
33 (1979) 12 BLR 30 (hereafter *Rayack v. Lampeter*); this was followed by the High Court of Hong Kong in *Concorde Construction Co. Ltd v. Colgan Co. Ltd* (1984) 29 BLR 125.

Mr Justice Vinelott decided that the effect of that provision was to impose on the Employer an implied obligation to set up the Retention as a separate trust fund in favour of the Contractor. Until the Employer does this by putting the retention into a separate bank account, there is no trust because the mixing of the Retention with other monies of the Employer which would otherwise occur contravenes the principle of certainty of subject matter.

Setting up a trust fund deprives the Employer of the opportunity of putting the Retention to other alternative uses more to his advantage. This means that, in the event of the Employer's insolvent liquidation, the Retention will be held by the liquidator upon trust for the benefit of the Contractor and Nominated Sub-contractors. The creditors cannot therefore touch it. It is important to note that for the Retention to be unavailable to the Employer's creditors in insolvency, the trust fund must have been effectively set up prior to the commencement of proceedings to put the Employer into liquidation or bankruptcy.

Imposition of an implied obligation to set up a trust means that, if the Employer fails to set it up, the Contractor can apply for a mandatory injunction compelling him to do so. In the *Rayack v. Lampeter* case, the Contractor successfully applied for such an injunction. For reasons relating to future business opportunities with the Employer, many contractors are reluctant to use this weapon unless there is clear evidence of imminent insolvency. However, it has to be cautioned that an application for a mandatory injunction after the commencement of liquidation will fail. In *Re Jartray Developments Ltd*,[34] which arose from a JCT 80 Contract, the application was made only after the appointment of a liquidator. The injunction was refused because to have held otherwise would have given unfair preference to the Contractor as against other unsecured creditors, a contravention of the *pari passu* (Lat. equal pace) principle of insolvency.

Case law suggests that the application may be rejected even if the Employer is only in informal insolvency, i.e. events other than formal winding up such as administration and administrative Receivership. In *Mac-Jordan Construction Ltd v. Brookmount Erostin Ltd*[35] the contract incorporated the JCT 81 form Clause 30.4.2.1 of which was equivalent to Clause 30.5.1 of the JCT 80. The Employer gave a floating charge over his assets including his interests in the building contract to a bank. After the bank had appointed administrative receivers of Employer's assets, the Contractor applied for an injunction to compel the Employer to put the retention into a separate account. This was dismissed on the grounds that, at the time of crystallization of the charge, no trust existed over the retention and that therefore the charge holders had already obtained good title to the retention by the time of the application.

In the Private version only of the JCT 98, Clause 30.5.3 gives the Contractor and Nominated Sub-contractors the power to direct the Employer to pay the Retentions into a separate bank account. If such a request is made, the Employer must comply at the date of payment under each Interim Certificate and certify such compliance to the Architect with a copy to the Contractor. The effect of the requirement in Clause 30.5.3 that the account should be 'so designated as to identify the amount as the Retention held by the Employer on trust as provided in Clause 30.5.1' is probably that the title of the account must make it clear that it is a trust account for the benefit of the Contractor and Nominated Sub-contractors. This

---

34 See n. 31; see also *GPT Realisations Ltd (in Administrative Receiver & in Liquidation) Ltd v. Panatown Ltd* (1992) 61 BLR 92 (hereafter *GPT Realisation v Panatown*).
35 (1992) 56 BLR 1.

should alert third parties such as receivers, liquidators and administrators that they have no access to those funds. If this is not done, there is always the risk that these third parties may, without any other notice of the trust position, lawfully make use of the funds. Although the Contract does not specify that the request to set up the trust account should be in writing, it would be good management practice to do so in writing. In *J. F. Finnegan Ltd v. Ford Sellar Morris Developments Ltd*,[36] which arose from the Clause 30.5.3 equivalent in the JCT 81 form, the court rejected the Employer's contention that the Contractor must make the request each time retention is deducted, e.g. after the issue of each Interim Certificate. It was stated that only one request is necessary and it may be made at any time.

As an injunction is a discretionary remedy, the court may refuse to grant it if a grant would have unfair consequences. For example, in *Henry Boot Building v. The Croydon Hotel*[37] the Contractor's application for a mandatory injunction to the Employer to set up retention monies as a trust fund failed because liquidated damages for non-completion were in excess of the retention monies. Also, the enforceability of Clause 30.5.1 against local authorities and other similar public bodies is doubtful. This is because, the risk of their insolvency being virtually non-existent, the court may take a view that it is not fair to put public bodies to such trouble. There is also the argument that the deliberate omission of Clause 30.5.3 in the Local Authorities version indicates an intention that the Employer should not be compelled to set up the trust. However, such an argument questions the point of leaving Clause 30.5.1 in the Local Authorities' version.

Some light was thrown on this argument in *Wates Construction v. Franthom Property*[38] which arose from a contract in the terms of the Private version of the JCT 80 but in which Clause 30.5.3 was deleted. The Court of Appeal rejected the Employer's contention that the deletion demonstrated a common intention that the Employer would not be under any obligation to open a separate trust account for the retention monies. The Contractor was therefore granted an injunction to enforce the setting up of the trust. An implication of this decision is that Clause 30.5.3 is probably redundant. The lesson is therefore that, where the intention is to avoid the obligation of the Employer to set up the trust, both clauses must be deleted.

15.6.3 Set-off against Retention

Clause 30.1.1.2 provides that the obligation of the Employer to hold retention in trust does not negate his right to exercise set-off allowed under the Contract against retention funds. Whenever the Employer makes deductions against retention funds, he is required to specify to the Contractor the total amount of the Contractor's retention to date and how much of it is set-off (Clause 30.5.4). There is also a requirement for the Contractor to be given similar details in respect of set-offs against the retentions of Nominated Sub-contractors. It is a requirement under Clause 35.13.5.2 that, where the money payable under an Interim Certificate is retention, in respect of set-offs on account of direct payments made to a Nominated Sub-contractor, the amount of set-off must not exceed the Contractor's own retention, i.e. direct payment to one Nominated Sub-contractor must not be at the expense of another Nominated Sub-contractor's retention.

---

36 (1991) 53 BLR 38 (hereafter *Finnegan v. Ford Sellars*).
37 (1985) 36 BLR 41.
38 (1991) 53 BLR 23.

15.7 Final adjustment of the Contract Sum

Although the JCT 98 is a lump sum contract, the Contract Sum is adjustable for a variety of events the effect, or even occurrence, of which cannot always be determined accurately at the pre-contract stage. Clause 30.6 describes in detail when and how the statement of the final adjustment of the Contract Sum is to be produced. This statement is referred to in the construction industry as the 'Final Account' and is normally drawn up as a formal document in a sectionalized format. The first section, usually on one side of A4, contains a summary of the account and states the Contract Sum and various additions and deductions. This section is followed by separate sections setting out the details of:

- variations
- loss and/or expense
- fluctuations
- provisional sums
- prime cost items
- miscellaneous items.

Valuations of variations agreed by the Contractor and the Employer under Clause 13.4.1.1 and amounts stated in accepted 13A Quotations may be additions to, or deductions from, the Contract Sum. However, adjustments for some matters are always additions to, whilst other are always deductions from the Contract Sum. Deductions are summarized in Table 15.2 whilst Table 15.3 covers additions. It is important to bear in

Table 15.2 Deduction from the Contractor

Clause		Deduction
Clause 30.6.2.1	(i)	all prime cost sums in the Contract Bills
	(ii)	amounts in respect of Nominated Sub-contractors (including the Contractor's profit) in the Contract Bills
	(iii)	value of defective work by a Nominated Sub-contractor whose employment has been determined and which has been paid or discharged by the Employer
	(iv)	the Contractor's profit on any of the above included in the Contract Bills
Clause 30.6.2.2	(v)	all provisional sums in the Contract Bills
	(vi)	the value in the Contract Bills of all work for which an Approximate Quantity is provided
Clause 30.6.2.3	(vii)	valuation of omissions under Clause 13.5.2
	(viii)	the amount included in the Contract Bills for work which has been so affected by variations that it has itself to be re-valued as a variation under Clause 13.5.5
Clause 30.6.2.4	(ix)	amounts in respect of the Contractor's setting out errors which the Architect has, pursuant to Clause 7, instructed to remain without amendment
	(x)	amounts in respect of materials or work not being in accordance with the Contract but which the Architect has, pursuant to Clause 8.4.2, instructed to remain without amendment
	(xi)	amounts in respect of defects, shrinkages, other faults after Practical Completion for which the Contractor is responsible but which the Architect has, pursuant to Clauses 17.2 or 17.3, instructed to remain without amendment
	(xii)	price decreases from the fluctuation provisions
Clause 30.6.2.5	(xiii)	any other amount which is required by the Conditions to be deducted from the Contract Sum

Table 15.3 Additions to the Contract Sum

Clause		Addition
Clause 30.6.2.6	(i)	the total of the finally adjusted Sub-contract Sums or Tender Sums of Nominated Sub-contractors
Clause 30.6.2.7	(ii)	the sum due to the Contractor in respect of work the Contractor successfully tendered for under Clause 35.2
Clause 30.6.2.8	(iii)	amount properly chargeable to the Employer in respect of Nominated Suppliers (discount of 5% is to be included but not VAT which qualifies as input tax)
Clause 30.6.2.9	(iv)	profit on items (i), (ii) and (iii) above
Clause 30.6.2.10	(v)	amounts payable to the Contractor by the Employer in respect of statutory fees and charges under Clause 6.2
	(vi)	amounts payable to the Contractor by the Employer in respect of testing and inspection under Clause 8.3
	(vii)	amounts payable to the Contractor by the Employer in respect of infringement of patents rights under Clause 9.2
	(vii)	premiums in respect of Clause 21.2.1 insurance of property
Clause 30.6.2.11	(ix)	the valuation of additional or substituted work under Clause 13.5
	(x)	the valuation of work which is not varied itself but has to be treated as a variation under Clause 13.5.5
Clause 30.6.2.12	(xi)	the amount of any disbursement by the Contractor in respect of an AI as to the expenditure of a provisional sum
	(xii)	amounts of the valuation of executed of work covered by an Approximate Quantity in the Contract Bills
Clause 30.6.2.13	(xiiii)	amounts of loss and/or expense under Clauses 26.1 and 34.3
Clause 30.6.2.14 (Private Version only)	(xiv)	insurance premiums paid or payable by the Contractor under Clause 22B/22C because the employer failed to maintain the relevant insurance cover
Clause 30.6.2.15	(xv)	price increases from the fluctuation provisions
	(xvi)	any other amount which is required to be added to the Contract Sum
Clause 30.6.2.16	(xvii)	additional amount determined by the Contractor under Clause 13.4.1.2 and accepted in lieu of ascertainment under Clause 26.1 (for disturbance to regular progress by a variation)

mind that, contrary to practice with the final accounts of civil engineering contracts, the work in a JCT 98 contract is not to be re-measured for final accounts, i.e. for each work item, the Contractor is to be paid only the amount for it in the Contract Bills unless it was affected by a variation, in which case, there could be additions or deductions as explained above.

15.7.1 Timetable for final accounts and the Final Certificate

The timetable governing the preparation of the final accounts and the issue of the Final Certificate are explained in Sections 3.7 to 3.9. As explained in Section 3.7, preparation of the final accounts is the responsibility of the Quantity Surveyor whilst the issue of the Final Certificate is the Architect's.

15.7.2 Effect of the Final Certificate

This subject is covered in Section 3.9.

15.9 Architect's liability for negligent certification

The law on the liability of contract administrators to employers for negligence in the performance of their administrative responsibility has been settled since *Sutcliffe v. Thackrah*[39] in which the House of Lords decided that an architect was liable for the employer's loss arising from negligent over-certification.

Whether the Contractor has a cause of action for negligent under-certification or other decision against the Architect is less clear. As there is usually no contract between the Contractor and the Architect, there is no contractual basis for such a claim. That leaves the Contractor with only a possible cause of action in tort. In a number of cases the courts answered in the affirmative. For example, in *Arenson v. Casson Beckman Rutley & Co.*[40] Lord Salmon said *obiter* that, considering the importance of cashflow to a builder, an architect who negligently certified less money than was payable could be successfully sued by the builder for the damage caused by being wrongly starved of money in that way. In *Lubenham Fidelities*[41] the Court of Appeal accepted the possibility of the special tort of procuring breach of a contract to which the defendant is a stranger arising from deliberate under-certification by the Architect. However, the court concluded that the proper remedy available to the Contractor is for him 'to request the architect to make the appropriate adjustment in another certificate or if he declines to do so, to take the dispute to arbitration'. However, in *Pacific Associates Inc. v. Baxter*[42] the Court of Appeal held that supervising engineers were not liable to contractors for under-certification. In that case the plaintiffs, contractors, were engaged by the Ruler of Dubai under a FIDIC (2nd edition) contract to carry out dredging work.[43] Condition 86 of their contract stated, *inter alia*, that the Engineer shall not be in any way liable to the Contractor for the performance of the Engineer's duties under the contract. There was also an arbitration clause typical of construction contracts. Claims for unforeseen conditions were rejected by Halcrow, the consulting engineers appointed to act as the Engineer. Part way through subsequent arbitration proceedings, the claim was settled at a fraction of its value by agreement between the plaintiffs and the Ruler. However, Pacific Associates sought to recover the balance of their claim from Halcrow in tort. The issue that the Court of Appeal was required to decide was whether Halcrow owed Pacific Associates a duty of care in tort to avoid causing that loss. The existence of the duty was argued on the voluntary assumption of responsibility principle and the proximity/foreseeability/policy test. It was held not to exist for two main reasons: (i) the effect of the condition exonerating the Engineer meant that there was no voluntary assumption of responsibility; (ii) because of Condition 86 and the arbitration clause, it was not fair, just or reasonable to impose a duty of care. According to Purchas LJ at p.53:

> where the parties have come together against a contractual structure which provides for compensation in the event of failure of one of the parties involved, the court will be

---

39 [1974] AC 727; see also: *Townsend v. Stone Toms & Partners* (1984) 27 BLR 26; *West Faulkner Associates v. London Borough of Newham* (1994) 71 BLR 1.
40 [1977] AC 405, at 437G.
41 See n. 23.
42 (1989) 44 BLR 33.
43 Fédération Internationale des Ingénieurs Conseils.

slow to superimpose an added duty of care beyond that which was in the contemplation of the parties at the time that they came together.

Pacific Associates was applied by the Appeal Court of British Columbia in *Edgeworth Construction v. Lea & Associates*[44] in deciding that design engineers were not liable in negligence to road contractors for their loss arising from design defects and errors in drawings. This was reversed by the Supreme Court of Canada.[45] It was suggested that had the engineers expressly excluded their liability to the contractors in the contract documents, the decision might have been different.

Does the Architect under the JCT 98 owe the Contractor a duty of care in his role as certifier? The Contract does not provide expressly that the Architect is not liable to the Contractor in respect of negligent performance of his duties under the Contract. It follows therefore that where Clause 41B does not apply, i.e. the Contract does not incorporate an arbitration agreement, the factors supporting the contract structure argument in *Pacific Associates* would not exist. *Pacific Associates* can therefore be distinguished. The issue would therefore have to be decided on the proximity/foreseeability/policy test. Since it is difficult to deny that the ingredients of proximity and foreseeability exist in this context, the issue would then be essentially one of policy. Unfortunately, opinions are divided here.[46] For example, while the Supreme Court of Canada in the *Edgeworth Construction* case stated that there is no policy reason for denying liability, the learned editor of *Hudson's* argues against putting contract administrators in a position where they will be 'shot at by both sides'.[47]

44 (1991) 54 BLR 11 (hereafter *Edgeworth Construction*).
45 (1993) 66 BLR 56.
46 For a review of case law from some common law jurisdiction see Duncan Miller, 'The Certifier's Duty of care to the Contractor – *Pacific Associates v. Baxter* Reconsidered' (1993) ICLR 173.
47 Para 1.302.

16

Determination

The ideal outcome of a building contract is for both parties to perform their obligations in accordance with their contract. Whilst this ideal is achieved in the majority of building contracts, a contract can sometimes be brought to a premature end by one of the parties. When this happens, the contract is said to be 'determined', i.e. terminated. A note of caution must be sounded here regarding terminology. The terms 'termination of a contract' and 'determination of a contract' are to be understood as shorthand for the ending of the primary obligations under the contract. These obligations consist of the contractor's obligation to carry out and complete the works and the employer's obligation to pay the contract price in accordance with the conditions of the contract. Strictly speaking, the contract itself does not come to an end because its secondary obligations, i.e. the contract breaker's liability for damages, remain unaffected.

A contract may be determined either under the common law or by exercising rights of determination expressly provided for in it. This chapter examines determination under the terms of the JCT 98. It will be seen that many of the rights to determine under the Contract are expressed to be without prejudice to any other rights or remedies that the determining party may possess. This means that the party concerned may choose to bring the contract to an end on common law grounds. Before the provisions in the Contract are examined, the general nature of this choice is therefore explained.

16.1 Determination at common law for repudiation

Repudiation, also sometimes referred to as a 'repudiatory breach', of a contract arises when an act or omission of a party to the contract is such a serious breach that the innocent party is entitled to treat it as evidence that the contract-breaker no longer intends to be bound by it. This situation can come about although the contract-breaker did his best to avoid the breach. When this happens, the innocent party has two choices. First, he can accept the repudiation and thereafter the party who repudiated can no longer, without the agreement of the innocent party, revert to the *status quo* before the repudiation. Not only is the innocent party discharged from further performance of his obligations under the contract but he can also sue for damages immediately.[1] Second, the innocent party may

---

1 *Heyman v. Darwins* [1942] AC 356; *Photo Production v. Securicor* [1980] AC 827.

choose to treat the contract as continuing despite the breach and claim damages instead. He is then said to 'affirm' the contract. Upon affirmation, the innocent party's right to accept that particular repudiatory breach is lost unless it is repeated or of a continuing nature. Affirmation is very readily assumed if there is delay in accepting the repudiation. As the court is unlikely to order a party to perform a contractual obligation,[2] affirmation works only where the party in breach is still willing to continue performance.

The most easily understood form of repudiation is renunciation, i.e. an express statement by a party to the effect that he no longer intends to perform any of his obligations under the contract. It can also arise from a breach of a condition of the contract as opposed to a breach of a warranty. In addition, persistent and nonchalant breaches of a warranty may also constitute repudiation. For example, in *Sutcliffe v. Chippendale & Edmondson*[3] it was stated that persistent poor quality work could be treated as repudiation.

16.1.1 Conditions and warranties

A term is a condition if it is so important that its breach by a party entitles the other to treat the contract as repudiated. For this reason, a breach of a condition is often referred to variously as a 'repudiatory breach', 'fundamental breach', or a 'breach that goes to the root of the contract'. For example, it would be a condition of any construction contract that the employer will be able to grant possession of site without undue delay[4] and, further, that he will not expel the contractor from the site of the works without reasonable cause. On the contractor's side, it would be a condition that he will not wholly abandon the works without lawful cause[5] or sub-let the entire works without the employer's consent. South African authority suggests that where a construction contract requires the contractor to procure a performance bond, failure to comply would constitute a breach for which the employer may terminate the contract at common law.[6]

A warranty is a term of less importance than a condition. Its breach does not entitle the innocent party to terminate the contract; there is only entitlement to damages. For example, an isolated delay by a few days by an employer to pay on a certificate would not entitle the contractor to treat the contract as terminated. Similarly, an isolated defect that can easily be put right by the contractor would not result in a right of the employer to treat the contract as terminated.

For a long time, the position of the law was that a term was either a condition or a warranty. However, in modern times, it has been realized that some terms cannot be categorized in this way because whilst one type of breach of such a term could have only very minor consequences another breach of the same term could be serious enough to deprive the innocent party of substantially the whole benefit of the contract. For such a term, the legal consequence of its breach depends on the nature of the events arising from it. Where the legal effect of the breach of term is best judged by examining the nature of the events arising from the breach, the term is referred to an 'innominate term' or 'intermediate term'.[7] It is submitted that failure to pay on a certificate falls into this

---

2 The court is more likely to award damages for the breach.
3 (1971) 18 BLR 149.
4 *Carr v. J. A. Berriman Property Ltd* (1953) 27 ALJR 273.
5 *Marshall v. Mackintosh* (1898) 78 LT 750.
6 *Swartz & Son (Pty) Ltd Wolmaransstad Town Council* (1960) 2 SARL 1.
7 *Hong Kong Fir Shipping Co. Ltd v. Kawasaki Kisen Kaisha Ltd* [1962] 1 All ER 474.

category. For whilst delay by a few days would be tolerable, delays for months without explanation would be serious enough to entitle the contractor to terminate the contract at common law.

16.1.4 Wrongful determination at common law

A court may decide that a term is a condition or a warranty because it has been categorized as such by statute or binding judicial precedent. In the absence of relevant statutory provision and precedent, the court must determine from the contract itself and the matrix of its surrounding circumstances which category was intended by the parties. Provided that the intention of the parties is clear enough that a term is to be a condition, the court will treat it as such even if the consequences of its breach are very minor.[8] As a corollary, a clear statement that breach of a very important term is not to give rise to a right to terminate would be accepted by the courts as evidence that the term is, by the common intention of the parties, a warranty. However, in view of the draconian consequences of a breach of a condition, the courts lean against deciding that a term is a condition unless an intention to that effect is very clearly stated. Thus, merely stating that a term is a condition may not be conclusive that the parties intended that the innocent party shall have a right to terminate the contract in the event of its breach.[9] This approach reflects recognition that the term 'condition' is often used to refer to terms in general without any intention that the terms shall be subject to the consequences of a condition in the legal sense. Indeed, many of the standard forms in the construction industry, the JCT 98 included, are labelled 'conditions' although only a small proportion of their terms are intended as conditions in the legal sense.

There is a need to exercise due caution when faced with what appears to be repudiation. This need arises from the fact that where party A determines for alleged repudiation by party B and the conduct relied upon is not accepted in law as repudiation, A would be guilty of wrongful determination. B may be entitled to treat the wrongful determination as repudiation by A which, if accepted by B, would result in an overturning of the tables whereby A is liable to B for damages. However, it was suggested in *Woodar Investment Development Ltd v. Wimpey Construction UK Ltd*[10] that a purported termination under the contract based on an honest but mistaken interpretation does not always amount to a repudiatory breach. Lord Wilberforce explained:

> So far from repudiating the contract, the appellants were relying on it and invoking one of its provisions, to which both parties had given consent. And unless the invocation of the provision was totally abusive, or lacking in good faith (neither of which is contended for), the fact that it has proved to be wrong in law cannot turn it into repudiation. . . . . Repudiation is a drastic conclusion which should be only held to arise in clear cases of refusal, in a matter going to the root of the contract, to perform.

---

8 *Lombard North Central plc v. Butterworth* [1987] QB 527.
9 *Schuler AG v. Wickham Machine Tool Sales Ltd* [1974] AC 235.
10 [1980] 1 All ER 571.

16.2 Common law versus the contract

It is a common law principle that contractual determination clauses will not preclude a party from determination at common law for repudiation by the other party unless the contract itself expressly or impliedly provides that it can only be determined by exercise of the contractual right.[11] Clauses 27.8 and 28.5 of the JCT 98 provide that the parties' rights to determine under the contract are without prejudice to any other rights and remedies that they may possess. The parties therefore have the choice to proceed under common law even where the contract may be determined by invoking its terms. Regarding this choice, a party contemplating determination will usually consider the following factors.

1. With determination at common law, assuming repudiation has really occurred, there are no special procedures to follow. A simple notice to the effect that the contract has been determined for stated reasons would be sufficient. By contrast, contracts often lay down elaborate procedures to be followed. For this reason, a party who has failed, or is unable, to comply with procedures laid down in the contract may elect to bring the operation of the contract to an end by exercising his common law rights.
2. After determination at common law, neither party has any obligations under the contract except the contract breaker's liability for damages and obligations under an arbitration agreement in the contract.[12] However, where the employment of the contractor is determined under the contract, both parties are still bound by the contract although most of its terms would not be applicable after the determination. A party entitled to determine may therefore opt for a route because it offers greater advantages regarding his post-determination rights.
3. Where the determination is at common law, the innocent party is entitled to damages assessed under common law principles, which are explained in Sections 12.2 to12.5. Most determination clauses state expressly what the innocent party may recover after the determination and how it is to be quantified, e.g. under Clauses 27.6, 28.4, and 28A.2. However, where the contractual remedy is not void, e.g. for being penal, the determining party is entitled to only that remedy.[13]
4. If the Contractor provided a performance bond, the Employer may make a claim on it upon determination at common law. With determination under the contract, as explained in Section 16.19, the Employer cannot call the bond until after completion of the outstanding work or, where the Employer decides to abandon the Works, after submission of the Employer's statement of accounts under Clause 27.7.1.

16.3 JCT 98 determination clauses: an overview

The JCT 98 provides for determination of the employment of the Contractor by the Employer or the Contractor himself in a number of defined situations. This is done in four clauses:[14]

---

11 *Lockland Builders Ltd v. John Kim Rickwood* (1995) 77 BLR 38.
12 For explanation of the doctrine of separability of an arbitration agreement see Section 17.4.1.
13 *Thomas Feather & Co. (Bradford) Ltd v. Keighley Corporation* (1953) 52 LGR 30.
14 A clause of this type and intended to be operated by an employer is commonly referred to as a 'forfeiture clause'.

- 22C.4.3: determination by either party as a result of loss or damage to the Works from certain insured risks;
- 27: determination by the Employer in defined situations; 5 specified defaults of the Contractor (Clause 27.2.1), the Contractor's insolvency (Clause 27.3), and corruption (Clause 27.4);
- 28: determination by the Contractor in defined situations; 4 specified defaults of the Employer (Clause 28.2.1), suspension beyond the period of suspension in the Appendix caused by the specified suspension events (Clause 28.2.2), and the Employer's insolvency (Clause 28.3.1);
- 28A: determination by either party on the occurrence of specified events causing suspension of the Works beyond the relevant period of suspension in the Appendix.

Determination is such a drastic step that, if it is contested, the courts tend to lean against the party seeking to determine. It is therefore absolutely essential that the procedure for determination spelt out in the Contract is followed to the letter. Any attempt to determine without compliance with the stipulated procedure may amount to termination at common law but only if the default relied upon is one for which there is entitlement to determine at common law.[15] Two requirements are of paramount importance. First, the determining party must comply meticulously with the timetable. Second, the notice must be clear as to what is being notified. Case law[16] suggests that a notice in general terms but which clearly directs attention to what is amiss is sufficient. However, it is recommended practice not only to state clearly the default in question but also to specify the applicable clauses of the Contract. Ideally, the notice should adopt the words used in the contract to describe the default.

Whether a defective notice of termination under the contract is capable of being rectified or replaced depends on the circumstances. Such remedial action would not be effective where the receiver is entitled to treat the notice as a repudiatory breach and he chooses to accept the repudiation before the remedial action is taken. Immediate remedial action must therefore be taken as soon as it is realized that a notice was defective. It follows from the above discussion that where the default relied upon in a notice also entitles the innocent party to determine at common law, it may be prudent practice to serve the notice in the alternative. This way, if subsequently it is found to be invalid under the contract, it can take effect at common law. For similar reasons, where the default complained of entitles the innocent party to determine under the contract but he prefers the determination to take effect at common law, service in the alternative may be advisable unless it is clearly a default for which determination at common law is available.

The determination procedures laid down in the Contract require giving of certain notices. It is stipulated in Clauses 27.1, 28.1, and 28A.1.1 that notices required under specified clauses should be in writing and served by actual delivery, registered post or recorded delivery. It is further provided in those clauses that if a notice is sent by registered post or recorded delivery, subject to proof to the contrary, it is deemed to have been received 48 hours after the date of posting (excluding Saturdays, Sundays and public holidays).

---

15 *Architectural Installation Services v. James Gibbons Windows* (1989) 46 BLR 91 (hereafter *Architectural Installation*).
16 *Hounslow London Borough v. Twickenham Garden Developments Ltd* [1970] 3 WLR 538 (hereafter *Hounslow v. Twickenham Garden*); *Supamarl Ltd v. Federated Homes Ltd* (1981) 9 Con. LR 25.

16.4 Determination for loss/damage to the Works

Clause 22C is intended for use where the Works entail alteration of, or extensions to, existing structures and their contents. By Clause 22C.2, the Employer is required to take out and maintain insurance against loss and/or damage to the Works from the risks covered by 'All Risks Insurance' as defined in Clause 22.2.[17]

Clause 22C.4.3.1 entitles either party to determine the employment of the Contractor in the event of the occurrence of loss or damage to the Works from the insured risks. However, there is an important proviso that it must be 'just and equitable' to determine. There is no indication anywhere in the Conditions as to what is 'just and equitable'. It is submitted that useful considerations on this point include:

- the extent to which the nature of the Contract has been changed;
- the extent to which the financial commitments of the Employer have increased;
- the extent of loss or damage to the existing structures;
- the extent to which the Contractor can be adequately remunerated for carrying out the changed Works by the variation provisions in the Conditions.

It is likely that the type of loss and damage contemplated would be such as to frustrate the contract at common law.

There are four key requirements to the procedure for determination under this clause.

1. Upon discovery of the loss/damage, the Contractor must forthwith give notice in writing to both the Employer and Architect.
2. Notice of the determination must be served on the other party within 28 days of the occurrence of the loss/damage.
3. The notice must be sent by registered post or recorded delivery. This requirement on method of delivery is only directory rather than mandatory, i.e. notice actually delivered by other methods within the time limit would be valid.[18] It is unfortunate that the 'deeming' provisions in Clauses 27.1 and 28.1 regarding the time of receipt of notices do not appear to apply to notices under this clause.
4. Where the determination is to be contested, there is a limit of 7 days from receipt of the notice within which either party may invoke the relevant procedures applicable to the resolution of disputes in order that it may be decided whether such determination is just and equitable.[19]

16.5 Determination by the Employer for specified defaults

Clause 27.2.1 identifies 5 grounds upon which the Employer may determine the employment of the Contractor. These are referred to as the Contractor's 'specified defaults'. They are if, before Practical Completion, the Contractor:

---

17 See Section 7.3.4
18 *Goodwin v. Fawcett* (1965) 175 EG 27; *J. M. Hill & Sons Ltd v. London Borough of Camden* (1980) 18 BLR 31.
19 For discussion on whether this limitation of the right to challenge determination contravenes the Construction Act see Section 17.2.

1. wholly or substantially suspends the carrying out of the Works without a reasonable cause (Clause 27.2.1.1);
2. fails to proceed regularly and diligently with the Works (Clause 27.2.1.2);
3. refuses or neglects to comply with an AI requiring him to remove work, materials or goods and by such refusal or neglect the Works are materially affected (Clause 27.2.1.3);
4. assigns rights under the Contract or sub-lets any part of the Works without the required consent (Clause 27.2.1.4);
5. fails to comply with his contractual obligations in respect of the CDM Regulations (Clause 27.2.1.5).

It should be noted that some of these defaults might not be serious enough to constitute repudiation at common law. It is also to be noted that although the Contract does not limit the time within which the Architect may give notice of default, it would be implied that any such notice must be given within a reasonable time of the default.[20]

16.5.1 Suspension of the Works by the Contractor

This default is stated in Clause 27.2.1.1 as 'without reasonable cause he wholly or substantially suspends the carrying out of the Works'. To qualify as a valid ground for determination, any suspension relied upon must therefore satisfy two conditions. First, the suspension should be a total or substantial cessation of work on the whole of the site. Second, there must be no reasonable cause for the suspension.[21] In practice, it is difficult to prove that these conditions apply in any situation unless the Contractor is clearly minded to abandon the works. As long as the Contractor does not move all his resources off the site, he cannot be said to have wholly suspended the carrying out of the Works. However, it may amount to substantial cessation of the work. Whether that is the case in any given situation is a matter of fact and degree.

16.5.2 Failure to proceed regularly and diligently

Under Clause 23.1.1 the Contractor undertakes to proceed regularly and diligently with the Works until completion. Failure to do so is therefore a breach. Under the general law, it is doubtful if this type of breach goes to the root of the contract. However, by Clause 27.1.2, the Employer is expressly entitled to determine the employment of the Contractor. In a number of cases, very scathing judicial comments were made on the vagueness of the term 'regularly and diligently'. For example, in *Hounslow v. Twickenham Garden*[22] Megarry J described the same phrase in the JCT 63 as 'elusive words on which the dictionaries help little'. This remark was supported wholeheartedly by O'Connor J in *Lintest Builders Ltd v. Robert*.[23] In *West Faulkner Associates v. London Borough of Newham*[24] Judge Newey QC described what amounts to proceeding regularly and diligently in the following terms:

---

20 *Architectural Installation*, see n. 15.
21 Under Clause 30.1.4 the Contractor is entitled to suspend performance of his obligations on account of the Employer's failure to pay on an Interim Certificate by its final date for payment. See Section 15.5.2 for discussion of the Contractor's right to suspend.
22 See n. 16.
23 *Lintest Builders Ltd v. Roberts* (1978) 10 BLR 120; affirmed by Court of Appeal: (1980) 13 BLR 38.
24 (1993) 9 Const. LJ 233, at 249; (1994) 71 BLR 1, at p. 13.

contractors must go about their work in such way as to achieve their contractual obligations. This requires them to plan their work, to lead and to manage their workforce, to provide sufficient and proper materials and employ competent tradesmen, so that the works are fully carried out to an acceptable standard and that all time, sequence and other provisions of the contract are fulfilled.

When the case got to the Court of Appeal,[25] although the general approach was supported, Simon Brown LJ said that Judge Newey's definition of proceeding regularly and diligently could not be accepted in its entirety. He added that attendance and effort were not enough unless there was some measure of accomplishment. Pointing out that it would be unhelpful to seek to define the words 'regularly' and 'diligently' separately, he then offered an alternative definition in these terms:

> Taken together the obligation upon the contractor is essentially to proceed continuously, industriously and efficiently with appropriate physical resources so as to progress the works steadily towards completion substantially in accordance with the contractual requirements as to time, sequence and quality of work. Beyond that I think it impossible to give useful guidance. These are after all plain English words and in reality the failure of which clause 25(1) (b)[26] speaks is, like the elephant, easier to recognise than describe.

He concluded that, whichever of the definitions was applied, the Architect was not only entitled to give the notice but could not reasonably do otherwise than give it. It would therefore appear that failure of the Contractor to comply with the master programme is some, although not conclusive, evidence of failure to proceed regularly and diligently unless compliance with the programme is a term of the Contract.

Taking the judicial comments as a whole, it is concluded that the question whether the Contractor is proceeding regularly and diligently is a matter of fact and degree to be decided taking into account the master programme, the adequacy of the resources deployed for the purpose of performing the Contract, actual progress, outstanding work and productivity trends.

16.5.3 Refusal to remove defective work and materials

Refusal or neglect by the Contractor to comply with an AI requiring the removal of work, materials and goods not in accordance with the Contract is a breach of the Contractor's duty to comply with all valid AIs imposed by Clause 4.1.1. To constitute a valid ground for determination under this Contract, the refusal or neglect must materially affect the Works.[27] Apart from the inherent vagueness of this qualification, it seems that determination on the ground that the Works are likely to be affected at a future date may not be valid.

It has to be pointed out that a more appropriate course of action might be for the

---

25 (1995) 11 Const. LJ 157; (1994) 71 BLR 1.
26 The equivalent of JCT 98 Clause 27.2.1.2.
27 The Architect must have specifically required removal of the work or materials; condemning them as non-complying is not enough: *Holland Hannen & Cubitts (Northern) Ltd v. Welsh Hospital Technical Services Organisation* (1981) 18 BLR 80.

Employer to employ third parties to carry out the removal and to recover the cost of that course of action from the Contractor under Clause 4.1.2.

16.5.4 Assignment and sub-letting

Under Clause 19 the Contractor undertakes not to assign the Contract without the written consent of the Employer (19.1.1)[28] or sub-let any portion of the Works without the written consent of the Architect (19.2.2). Breach of any of these terms is a ground for determination by the Employer.

16.5.5 CDM Regulations

Under Clause 6A.2, while the Contractor is also the Principal Contractor, he must properly discharge the duties of a Principal Contractor under the CDM Regulations. If a different Principal Contractor is appointed, the Contractor must comply with all reasonable requirements of the Principal Contractor to the extent that such requirements are necessary for compliance with the CDM Regulations. The Contractor is to supply information reasonably required by the Planning Supervisor in writing for the purpose of preparing the Health and Safety File and to ensure that all sub-contractors do the same. Failure to comply with any of these obligations is a specified default for which the Employer may determine. It is submitted that any attempt to determine for a default giving rise to only very minor consequences would be caught by the requirement that notice of determination is not to be given unreasonably and vexatiously.

16.6 Determination for the Contractor's insolvency

Clause 27.3.1 lists a number of insolvency procedures in relation to the Contractor that could occur. Prior to July 1992, the month of publication of Amendment 11, the JCT 80 provided that on the occurrence of any of the insolvency events 'the employment of the Contractor under this Contract shall forthwith be automatically determined'. This blanket automatic determination for insolvency failed to recognize a very material distinction between the position of ordinary receivers, administrative receivers, or administrators on the one hand and liquidators and trustees on the other. This distinction is that whilst the former are bound by the contracts of the companies to which they are appointed, liquidators and trustees are not so bound. It was finally realized that in some situations it may be more advantageous to the Employer to leave the contractual obligations in force whilst the Employer attempts to agree a way forward with the Insolvency Practitioner acting for the Contractor. Accordingly, some of the events now only result in an option to the Employer to determine the Contractor's employment.

The effects of the insolvency procedures under the JCT 98 are as illustrated in Table 16.1. For purposes of making the provisions on their effect easy to understand, the procedures are here classified into 'constructive procedures' and 'non-constructive procedures'. The constructive procedures are those designed to improve the solvency of

---

28 See Section 8.2.1 for discussion on the meaning of 'assign a contract'.

the Contractor whilst the non-constructive procedures are designed to put the Contractor into bankruptcy or liquidation in an orderly manner. It can be observed from Table 16.1 that the effect of the non-constructive procedures is automatic determination[29] of the employment of the Contractor whilst that of the constructive procedures is that the Employer acquires an option to determine by notice.

In most insolvency procedures, an administrative receiver, administrator or liquidator, hereafter referred to collectively as an 'Insolvency Practitioner', would be acting for the Contractor. Any agreement with the Contractor would therefore be with that person.

Table 16.1   Effect of the Contractor's insolvency

	Insolvency Event	What the Contractor must do	Effect
(i)	Bankruptcy	The Contract is silent on this	Automatic determination of the employment of the Contractor but it may be reinstated by agreement between the Employer and the Contractor
(ii)	The making of a composition or arrangement with his creditors	To notify the Employer in writing immediately	Optional determination
(iii)	The making of a proposal for a voluntary arrangement for a composition of debts or scheme of arrangement to be approved in accordance with the Companies Act 1985 or Insolvency Act 1986 or any amendment or re-enactment of those Acts	To notify the Employer in writing immediately	Optional determination
(iv)	The appointment of a provisional liquidator	The Contract is silent on this	Automatic determination of the employment of the Contractor but it may be reinstated by agreement between the Employer and the Contractor
(v)	The making of a winding-up order	The Contract is silent on this	Automatic determination of the employment of the Contractor but it may be reinstated by agreement between the Employer and the Contractor
(vi)	The passing of a resolution for voluntary winding-up (except for the purposes of amalgamation or reconstruction)	The Contract is silent on this	Automatic determination of the employment of the Contractor but it may be reinstated by agreement between the Employer and the Contractor
(vii)	The appointment under the Insolvency Act 1986 of an administrator or administrative receiver	The Contract is silent on this	Optional determination

---

29 The validity of automatic determination clauses has been questioned, e.g. Powell-Smith, V., and Sims, J., *Determination and Suspension of Construction Contracts* (Collins, 1985), p. 54; Duncan Wallace, I.N., QC, 'Strict Cannons in the Court of Appeal, Not "Business Commonsense" ' (1996) 12 Const. LJ 178, at p. 184.

To be become a qualified Insolvency Practitioner, one must be a member of a professional body recognized by the Secretary of State for the Department of Trade and Industry and licensed to act in that capacity. To acquire such recognition, a professional body must satisfy the Secretary of State that its rules and regulations ensure that anybody licensed by it to practise as an Insolvency Practitioner has appropriate qualifications and training. Most Insolvency Practitioners are partners in the large legal and accountancy professions.

16.6.1 Automatic determination

The Contractor's employment may be reinstated by agreement between the Employer and the Contractor. In practice, reinstatement is usually agreed only where either a novation can be arranged or it is clear that the Contractor, under the management of the Insolvency Practitioner, will be able to complete the project. If the project is near Practical Completion, the Insolvency Practitioner may well choose this course of action in order to recover retentions and avoid or minimize liquidated damages. The Employer may be attracted to such an arrangement to avoid the disruption of getting other contractors in.

16.6.2 Optional determination

From the date the Employer acquires the option to determine for insolvency, the Contractor's obligation to carry out and complete the Works and the Employer's obligation to make payment in accordance with the Contract are suspended until a 'Clause 27.5.2.1 agreement' is concluded or the Contractor's employment is determined. A 'Clause 27.5.2.1 agreement' is defined as an agreement on a continuation contract, novation, or a conditional novation.

If the Employer wishes to exercise his option to determine, he is required to do so by written notice. There is a proviso that optional determination is not to be exercised where the Employer has made an agreement on a novation, conditional novation or continuation contract with the Contractor: Clause 27.3.4. It is a criticism that the Conditions do not address the fact that, after the Contractor's creditors have properly approved voluntary arrangements and schemes, the Employer would be bound by them under the companies or insolvency legislation. The Employer may therefore lose his rights to determine.

Clause 27.5.2.1 agreements
A continuation contract entails the Employer entering into another contract whereby the Contractor, under the management of the Insolvency Practitioner, undertakes to complete the Works. Such a continuation contract is often in effect a variation to the original Contract, i.e. new terms are introduced. For example, the Employer may agree to advance payment or a new completion date. However, it is possible for the continuation contract to require the Contractor to continue and complete, as he was originally obliged to. The provision for a continuation contract contemplates a situation where the Employer, after examining the circumstances of the insolvency situation and alternative courses of action available to him, comes a conclusion that his interests are best served by such an arrangement. A major factor that the Employer would usually consider is that the Insolvency Practitioner is personally liable for any non-performance of the continuation contract. This offers considerable protection since Insolvency Practitioners are normally

from the major legal and accountancy firms and are therefore usually covered by appropriate professional indemnity insurance cover. On the part of the Insolvency Practitioner, he must also come to a view that such an arrangement is more advantageous to his appointor or the general body of creditors than determination.

In this context, an agreement on a novation is an agreement involving the Employer, the insolvent Contractor and a Substitute Contractor that the latter has replaced the insolvent Contractor as the Contractor under the Contract. Thereafter the Substitute Contractor is governed by the Contract as if that has been the case from the start whilst the insolvent Contractor ceases to have any further obligations or rights under the Contract.[30] A novation is feasible only where the contract is profitable and the Substitute Contractor is willing to share the surplus in the completed work with the Insolvency Practitioner. The surplus is the total value of the work executed less total payment made under the contract. A conditional novation is a novation in which the terms of the Contract are varied by the agreement. This is a course of action that the Employer may consider where the Contract is not attractive enough to other contractors to bring about a pure novation. Practice Note 24 gives, as examples, a new Completion Date, exclusion of the Substitute Contractor's liability for defects in the pre-novation work, and a new Contract Sum.

Before Clause 27.5.2.1 agreement or determination

Under Clause 27.5.4, subject to any relevant Clause 27.5.2.1 agreement, the Employer may, after he first acquired the acquired the right to serve notice of determination, take such measures as are reasonable to protect work executed and Site Materials. The cost of such measures are recoverable from the Contractor as set-off under the Contract or as debt. The Contractor and the Employer may also agree any interim arrangement for work to be carried out. The Employer must make payment due under such an arrangement. The only set-off allowed against such payment is set-off in respect of the cost of reasonable measures to protect work and Site Materials. However, where the Contractor is in administration, to the extent that the measures entail taking into the possession of the Employer goods in the possession of the Contractor, they would be a breach of s. 11 of the Insolvency Act 1986 and, therefore, a contempt of court unless leave of the administrator or the court is first obtained. Also, measures amounting to a right to use goods owned by the Contractor are generally invalid against insolvency law.

16.7 Determination for corruption

This ground is applicable 'if the Contractor shall have offered or given or agreed to give any person any gift or consideration of any kind': Clause 27.4. The acts likely to amount to corruption are very wide because of the use inclusively of the phrase 'consideration of any kind'. The only controlling factor is the requirement that it is done by a person representing the Contractor as an inducement or reward for improper conduct in relation to the execution or performance of contracts between the Employer and the Contractor. It is to be noted that acts of the Contractor's employees without his sanction or knowledge are also covered. Also of significance is the provision that the clause covers similar acts committed on other contracts between the Employer and the Contractor.

---

30 Practice Note 24 contains a Specimen Novation Agreement.

16.8 Procedure for determination under Clause 27

The procedure varies depending upon the grounds for the determination.

16.8.1 On account of the Contractor's specified defaults

The procedure to be followed for determination on account of the specified defaults is shown in Fig. 16.1. The Architect is to set the ball rolling by giving notice to the Contractor that the Contractor has committed a default for which the Employer is entitled to determine his employment. Determination of the Contractor's employment is then conditional upon the Employer serving notice of determination within the relevant period from receipt by the Contractor of the Architect's notice of default. If the Contractor ceases a specified default within the 14 days but repeats it thereafter, the Employer may determine the Contractor's employment within a reasonable time of the repetition (Clause 27.2.3). The use of the phrase 'within a reasonable time' to qualify when the Employer may determine for repetition of a previous specified default creates unnecessary uncertainty where certainty is very necessary. There is a similar right to determine for a repeated default after the Employer failed to determine within the 10 days. Whilst the intended effect of Clause 27.2.3 is probably as stated, there is some amount of uncertainty. The wording 'the Contractor repeats a specified default' instead of 'the Contractor repeats the specified default' suggests that the Employer may also be entitled to determine for repetition if the second specified default is different from the first.

The two-tier nature of the notices required is to be noted. The Employer must be particularly careful not to issue the second notice too early. He must also be certain that the Contractor has either continued the default for the relevant period or repeated it. It was explained in *Hill v. Camden*[31] that the Architect is better placed than the employer to issue the notice of default not only because of his independent status but also because of his greater expertise and knowledge in recognizing any occurrence of the defaults. However, the Architect would be liable for the Employer's loss if his right to determine is lost because of his failure to give the appropriate notice.

West Faulkner Associates v. London Borough of Newham:[32] the plaintiffs performed the role of Architect on a contract let on the JCT 63 form. Under the contract, the defendants, the employer under the contract, were entitled to determine the employment of the contractors if they failed to proceed regularly and diligently with the work. Before this right could be exercised, the Architect was required to issue to the contractors notice that they were failing to proceed regularly and diligently. During the course of work, the defendants, being concerned at the fact that the contractors were making very slow progress, asked the Architect whether they could issue the appropriate notice. On a part of the project programmed to be completed in 9 weeks, the contractors took 28 weeks. The Architect responded that, although progress was slow, it did not amount to failure to proceed regularly and diligently. The defendants eventually replaced the contractors and terminated the Architect's contract of

---

31 See n. 10.
32 (1992) 61 BLR 81.

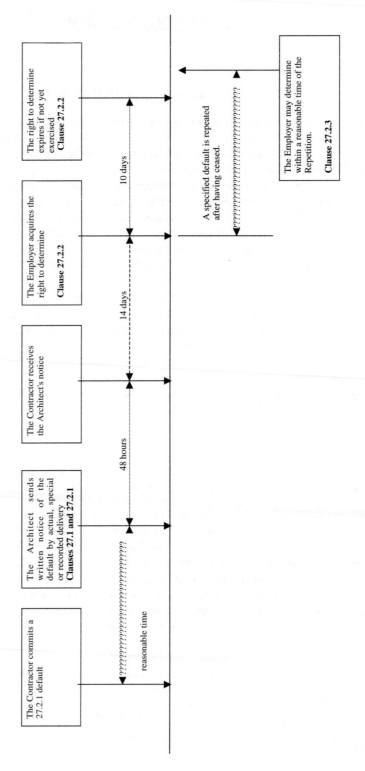

The Contractor commits a 27.2.1 default

The Architect sends written notice of the default by actual, special or recorded delivery
Clauses 27.1 and 27.2.1

The Contractor receives the Architect's notice

The Employer acquires the right to determine
Clause 27.2.2

The right to determine expires if not yet exercised
Clause 27.2.2

A specified default is repeated after having ceased.

The Employer may determine within a reasonable time of the Repetition.

Clause 27.2.3

??????????????????????????????
reasonable time

48 hours

14 days

10 days

???

Fig. 16.1 Procedure for Determining on account of the Contractor's Specified Defaults

engagement. When the Architect sued for their fees, the defendants counterclaimed damages for negligence in failing to issue the notice. Judge Newey held that there was no doubt that the contractors were failing to proceed regularly and diligently and that the Architect should have issued the notice. This decision was upheld by the Court of Appeal.[33]

16.8.2 On account of the Contractor's insolvency

There is no express requirement on the Employer to give any notice that the Contractor's employment has been automatically determined. However, it is not only good management practice but also prudent for the Employer to inform the Contractor or the Insolvency Practitioner because otherwise it could be claimed that silence amounts to reinstatement of the Contractor's employment. Optional determination requires only written notice to the Contractor, which takes effect upon its receipt (Clause 27.3.4). The deeming provisions in Clause 27.1 apply to this notice.

16.8.3 On account of corruption

The Contract is silent on the mode of notifying the Contractor that his employment has been determined for corruption. However, good practice demands that the Employer informs the Contractor in writing.

16.9 After determination under Clause 27

The effect of determination is that the rights and obligations are limited to those stipulated in Clauses 27.6 and 27.7.

16.9.1 Contractor's remaining obligations

1. He must remove from the Works any temporary buildings, plant, tools, etc., as and when required by the Architect in writing (Clause 27.6.3).
2. Where the reason for the determination is not insolvency,[34] the Employer, or the Architect on his behalf, may, within 14 days from the date of determination, request the Contractor to assign to the Employer, without any charge, the benefit of any agreement for the supply of materials or the execution of work entered into for the purposes of the Contract. The Contractor is obliged to comply with such a request only to the extent that the benefit of the agreement in question is assignable (Clause 27.6.2.1).

---

33 (1994) 71 BLR 1.
34 Where the Contractor is going through a formal insolvency procedure other than a company voluntary arrangement or scheme of arrangement, the *pari passu* principle of insolvency law prevents such assignment. This principle requires creditors in the same class to treated with an even hand, e.g. each is paid the same percentage of the insolvent company's indebtedness to him where the amount available for distribution is insufficient to cover all the debts.

16.9.2 Employer's rights regarding completion

There are a number of rights that the Employer may exercise after the determination. Although they are all designed to promote completion of the Works, whichever of these he chooses to exercise depends on the circumstances with which he is faced on the particular project.

1. The Employer may complete the Works by employing other persons. In doing this, he is entitled to use, or authorize the other persons he has employed to use, the Contractor's temporary buildings, materials and equipment. However, where these resources do not belong to the Contractor, the Employer must obtain the consent of their actual owners before such use (Clause 27.6.1). The enforceability of the right to use the Contractor's resources is doubtful for at least two reasons. First, if the Contractor were in administration, leave of the court or the administrator would be required. Second, if the Contractor is in liquidation or bankrupt, the liquidator or trustee would be entitled to take into his possession any asset owned by the Contractor for the benefit of the general creditors.

2. He may suspend payment to the Contractor under the Contract until the Works are complete. However, where the ground for determination is an insolvency event, certain payments which should already have been made by the Employer are payable to the Contractor, the receiver, liquidator, or trustee, as the case may be. These are payments which accrued 28 days or more before the date of determination (applies to automatic determination) or the date the Employer could first give notice to determine (applies to optional determination) (Clauses 27.5.1 and 27.6.4.1).

The Works are complete when Practical Completion is achieved. The Employer is not therefore entitled to refuse to discharge payment due under certificates after Practical Completion issued before determination.[35] The moratorium on payment by the Employer applies only if he intends to complete the project.[36]

However, case law suggests that the Employer may be able to resist successfully enforcement of accrued payment. In *Willment Brothers Ltd v. North West Thames Regional Health Authority*[37] the Employer issued a cheque to cover payment overdue on a certificate. After learning that a liquidator of the Contractor had been appointed, the Employer stopped the cheque. The Court of Appeal decided on what is now s. 323 of the Insolvency Act 1986 that the Employer was entitled to set-off against the certificate the contingent liability of the Contractor to the Employer as a result of the determination. This case appears to be giving a green light to Employers to delay paying on certificates on the slightest suspicion of insolvency and thereby to bring about the situation where otherwise it could have been avoided. Whether the Employer's right of set-off will defeat an adjudicator's decision requiring full payment on a certificate remains to be seen.[38]

---

35 *Emson Eastern v. E. M. E. Developments* (1991) 55 BLR 114.
36 *Tern Construction Group v. RBS Garages* (1992) 34 Con LR 137.
37 (1984) 26 BLR 51.
38 See Section 17.2.8

3. Except in insolvency events involving bankruptcy or winding-up,[39] the Contractor must comply with any request of the Employer, or of the Architect on his behalf, within 14 days of the determination date to assign to the Employer the Contractor's rights under supply contracts and sub-contracts relating to the carrying out of the Works. The Employer may also pay sub-contractors and suppliers directly for work executed or materials and goods supplied to the site for which the Contractor has not paid. Any such direct payment can be deducted from payment due, or to become due, to the Contractor. Where there is no payment due, or to become due to the Contractor, the Employer can recover it from him as debt (Clause 27.6.2). The right to assignment is certainly unenforceable against an administrator. It is also doubtful whether, in the light of the rule in the *British Eagle*[40] case, the right to make direct payment is enforceable where the Contractor is insolvent.

4. If the Contractor fails to remove his temporary plant, equipment of materials within a reasonable time of being requested by the Architect in writing to do so, the Employer can remove and sell any of these items. The proceeds of the sale are to be held to the credit of the Contractor after the Employer has deducted the expenses of the sale (Clause 27.6.3). Here too, the right to sell may not be enforceable against an administrator or liquidator. In *Re Cosslett (Contractors) Ltd*[41] the Court of Appeal held that a similar power to sell the Contractor's site plant and materials under the Institution of Civil Engineer's *Conditions of Contract*, 5th edn, was a floating charge which was ineffective against an administrator for lack of registration of the charge as required by s. 395 of the Companies Act 1985.

16.9.3 Where the Works are abandoned after determination

After determination, the Employer may abandon the carrying out of the Works (Clause 27.7.1). If he chooses to do so, he must so notify the Contractor in writing within 6 months of the determination. Thereafter, he must provide the Contractor with a statement of their final settlement within a reasonable time of the date of the notification. If the Employer abandons the Works but fails to notify the Contractor within the 6 months, the Contractor may request the Employer to confirm the position and require the Employer to prepare the statement of settlement if he has, in fact, abandoned the Works.

16.10 Financial settlement after determination under Clause 27

The financial settlement, which may be in the form of a statement by the Employer or a certificate issued by the Architect, must be prepared within a reasonable time of completion of the Works and the making good of defects. Clause 27.6.5 itemizes the

---

39 From the *British Eagle International Airlines v. Compagnie Nationale Air France* [1975] 1 WLR 758 decision it would be illegal for the Employer to pay creditors of the Contractor undergoing bankruptcy or winding-up. The Latham Report's recommendation that the principle in that case be abolished by legislation has not been acted upon.
40 Ibid.
41 (1997) 85 BLR 1.

settlement to include: (i) the amount of 'expenses properly incurred by the Employer' as a result of the determination; (ii) 'direct loss and/or damage caused to the Employer' as a result of the determination; (iii) the amount of any payment made to the Contractor; (iv) the total amount which would have been payable for the Works in accordance with this Contract (referred to hereafter as the 'Notional Final Account').

16.10.1 Expense, loss and/or damage

There is the question whether 'direct loss and/or damage', as used under Clause 27.6.5.1 is the same as 'direct loss and/or expense' under Clauses 26 and 34.3. It is suggested in *Keating*[42] in respect of the same problem in the JCT 80 that they describe the same entitlement. On the authority in *F.G. Minter v. Welsh Hospital Technical Services Organisation*,[43] this entitlement is damages under the first limb of *Hadley v. Baxendale*.[44] Allowable items include:

- amount payable to the completion contractor;
- additional professional fees payable on account of the determination;
- legal costs of the determination procedures;
- cost of managerial time expended in dealing with the determination;
- cost of work done to protect the uncompleted Works;
- cost of general site security;
- cost of disposing of the Contractor's plant, temporary buildings, etc. (with credit for proceeds);
- cost of insuring the works for the period before the start of the completion contract;
- additional amount payable to the completion contractor as a consequence of the determination;
- cost of additional finance;
- cost of inflation (double recovery through fluctuation payments to the completion must be avoided);
- damages for delayed completion.

16.10.2 Notional final account

As explained in Section 16.10, the term 'notional final account' refers to 'the total amount which would have been payable for the works in accordance with this Contract' stated in Clause 27.6.5.3. Many items would therefore have to be valued twice: first, in accordance with the completion contract and, second, in accordance with the original Contract. In the highly unlikely event of there being no variations and disruptions for which the Employer is responsible, this amount is very easy to determine. It is to be quantified in accordance with Clause 30 of this Contract and on the assumption that the original Contractor carried out and completed the Works. Unfortunately, the Contract fails to give any guidance on how the amount is to be determined in situations where there are variations and loss and/or expense items in favour of the completion contractor in circumstances where the same

---

42 At p. 663.
43 (1980) 13 BLR 1.
44 *Hadley v. Baxendale* (1854) 9 Ex. 341.

would have happened had the original Contractor completed the work. As an illustration of the uncertainty, consider a variation that would have been issued and for which the completion contractor furnished an accepted 13A Quotation. There are several ways of determining the amount that would have been payable to the original Contractor in respect of the variation. The obvious and easiest way is to accept the amount in the quotation. An alternative is to use the estimates in the quotation but apply the original Contractor's unit prices for resources required in the estimate. As the Architect is usually not privy to such prices, this refinement is hardly workable. A third possibility is to price the variation in accordance with Clause 13.5.1 and the original Contract Bills. There are similar questions surrounding what to do with loss and/or expense under Clause 26. Entitlement to recovery under that clause is expressed to be conditional upon notice of disruption. Perhaps, the fiction that the original Contractor carried out the remainder of the Works is to be applied to Clause 26.1 notices served by the completion contractor. Some practitioners favour a *pro rata* approach to variations, loss and/or expense and the like. This involves determining the two final accounts, ignoring variations and loss and/or expense items. The original contractor is then credited for these items with sums that bear the same proportion to the completion contractor's figures as the two final accounts bear to each other.

16.10.3 Calculation involved

The amount payable to, or payable by the Contractor, D, is to be calculated as follows:

$$D \quad = \quad a_0 + a_1 + a_2 - A \text{ (applicable if the Employer completes the work); or}$$

$$D \quad = \quad a_0 + a_2 + a_4 - a_3 \text{ (where the Works are abandoned);}$$

where

a_0 =   total amount paid or otherwise discharged to the Contractor before the determination;

a_1 =   the amount of expenses properly incurred by the Employer in completing the Works;

a_2 =   direct loss and/damage caused to the Employer by the determination;

a_3 =   the total value of work executed by the Contractor before determination;

a_4 =   the amount of expenses properly incurred by the Employer before the abandonment as a result of the determination;

A =   total amount that would have been payable to the Contractor if he had completed the Works in accordance with the Contract (Notional Final Account).

If D is positive the amount is to be paid by the Contractor. However, if it is negative it has to be paid to the Contractor.

16.10.4 Contingent liability claims

The final financial settlement upon completion may be years after the inception of the insolvency procedure. There is therefore the danger that the Insolvency Practitioner might have concluded his role and left. To counter this risk, a projected liability of the Contractor, referred to in industry as a 'Contingent Liability Claim', is usually prepared and submitted to the Insolvency Practitioner pending the final accounts of the completion contract. As explained in Section 16.9.2, the Employer is also entitled to set-off for a contingent liability claim against payment due.

16.9.4 Recoverability of liquidated damages

The general question whether a liquidated damages clause survives determination is difficult to answer from relevant case law. However, the correct starting point is that the answer depends on the terms of the particular contract. Where the contract provides expressly, clearly and unambiguously that the contractor is to be answerable for liquidated damages for the overall delay in completion regardless of who achieved the completion, it will be given effect.[45] Where the contract is silent on the issue, the general principle is that rights that accrued before determination are enforceable.[46] A contractor would therefore be liable for liquidated damages applicable to delay suffered before determination. On the issue of recoverability of liquidated damages for delay beyond the date of determination, the authorities conflict to some extent. On the one hand, it has been decided that liquidated damages for delay beyond the date of determination are recoverable only if there are express provisions in the contract to that effect.[47] However, as detailed in *Hudson's*,[48] different approaches have been followed in some common law jurisdictions.

The issue in relation to JCT 98 is clouded because it falls between the two extremes of clear express provisions and contractual silence. Two clauses may be relied upon in support of the proposition that the Employer is entitled to liquidated damages for the whole period of actual delay regardless of determination by the Employer under Clause 27. Clause 27.8 provides that all the other determination provisions under the clause are 'without prejudice to other rights and remedies which the Employer may possess'. Some commentators[49] have treated this as effectively maintaining the employer's entitlement to liquidated damages regardless of the time of determination. There is support for this view in *Re Yeadon Waterworks Co. and Wright*[50] in which the employer was given the right to determine the contractor's employment 'but without thereby affecting in any respects the liabilities of the said contractor'. That form of words was treated as retaining the employer's right to recover liquidated damages for the overall delay. In addition, Clauses 27.6.5 and 27.6.6 provide that in the settlement of accounts between the Employer and the Contractor, the Contractor is to be credited with 'the total amount which would have been payable for the Works in accordance with this Contract'. It is argued that the amount which would have been payable must reflect the Employer's right to set-off for liquidated damages for delay under Clause 24.2.1. It is submitted that both routes to recovery of liquidated damages are equally tenable. However, care must be exercised to avoid double

---

45 See the New Zealand case of *Bayliss v. Wellington City* (1886) 4 NZLR 84.
46 *Bank of Boston Connecticut v. European Grain and Shipping Ltd* [1989] AC 1056.
47 *Re Yeadon Waterworks Co. and Wright* (1895) 72 LT 538; *British Glanzstoff Manufacturing Co. Ltd v. General Accident Fire & Life Assurance Corporation Ltd* [1913] AC 143;
48 Paras 10.047 to 10.453.
49 Powell-Smith and Sims, *Determination and Suspension of Construction Contracts* at p. 62; see also n. 19 in para. 10.047 of *Hudson's*.
50 See n. 47.

recovery of liquidated damages: as 'direct loss and/or damage' under Clause 27.6.5.1 and in the preparation of the Notional Final Account.

16.11 Determination by the Contractor for specified defaults

The specified defaults of the Employer for which the Contractor may determine his own employment are listed under Clause 28.2.1 as if the Employer:

- fails to discharge to the Contractor any amount stated as payable under a certificate and/or VAT payable on the amount by the final date for payment (28.2.1.1);
- interferes with or obstructs the issue by the Architect of any certificate (28.2.1.2);
- assigns the Contract without the Contractor's consent (28.2.1.3);
- fails to discharge his contractual obligations in respect of the CDM Regulations (28.2.1.4).

16.11.1 Failure of the Employer to pay

By Clause 30.1.1.1 the Employer must pay to the Contractor the amount stated on any Interim Certificate within 14 days of its issue. This is subject to the Employer's right to make deductions.[51] Failure of the Employer to pay the amount properly due on a Certificate entitles the Contractor to determine his own employment. A less drastic course of action open to the Contractor is suspension under Clause 30.1.4.

This specified default refers to failure by the Employer to pay by the applicable final date 'the amount properly due to the Contractor in respect of any certificate'. It is argued in Section 15.1.6 that the Employer is entitled to make against the amount stated on the face of a certificate set-offs either under the Contract or under the general law. The Employer commits this default only where he makes unlawful set-offs or fails to follow the right procedure. It is suggested in *Keating's*[52] that the 'amount properly due' phraseology probably puts on the Employer an obligation, where there is a palpable error in the computation of the amount stated, to pay the amount that would have been stated had the error not been made. In view of the policy of the courts to discourage litigation on the quantum of interim certificates, except only in the most glaring and substantial cases, the application of this principle is likely to be rare if at all.

16.11.2 Interference/obstruction by the Employer

For this condition to apply, two facts must be established. First, the Employer, or his agents, must have had the intention either to prevent the Architect from performing his certification duties or to influence unduly the Architect's judgement in the performance of such duties. Second, there must have been actual interference or obstruction. In *R. B.*

---

51 See Sections 15.1.4 to 15.1.6 for discussion of the Employer's right to withhold and/or deduct from sums otherwise payable.
52 At p. 669.

Burden Ltd v. Swansea Corporation[53] it was said that inadvertent errors, negligence, or omissions of agents of the Employer who, at the request of the Architect, assisted in the certification process would not usually amount to interference or obstruction.

16.11.3 Assignment

Clause 19.1.1 prohibits any of the parties from assigning the contract without the consent of the other.[54]

16.11.4 CDM Regulations

Under Clause 6A.1, the Employer owes the Contractor a duty to ensure that the Planning Supervisor carries out the duties of a planning supervisor under the CDM Regulations. The same obligation applies in respect of the Principal Contractor but only where the Contractor is not also the Principal Contractor. The Employer commits a determination default if they fail to discharge their duties properly or he fails to appoint their replacements when they cease to act in those capacities. The point made in Section 16.5.5 about the effect of the Contractor's default in relation to his CDM obligations also applies here.

16.12 Determination for the specified suspension events

The Contractor is entitled to determine his employment if, before Practical Completion, the carrying out of the Works, or most of the Works, is suspended for a period exceeding the maximum period of suspension (Clause 28.2.2). Though the recommended maximum period of delay is one month, the Employer must, at the pre-contract stage, consider very carefully the possibility of determination on this ground in deciding the period most appropriate to the particular contract and complete the Appendix appropriately. For any suspension under this clause to apply, it must arise from one or more of the following:

1. an AI under Clauses 2.3 (discrepancies or divergences between documents), 13.2 (Variations and expenditure of provisional sums), or 23.2 (postponement of any part of the work) provided the AI was not necessitated by a default of the Contractor or his domestic sub-contractors;[55]
2. failure of the Architect to supply information in accordance with the Information Release Schedule or the requirements of Clause 5.4.2;[56]

---

53 [1957] 1 WLR 1167.
54 See Section 8.2.1 for discussion on 'assigning a contract'.
55 There is a right to determine if the AI is due to the default of a Nominated Sub-contractor, persons engaged by the Employer, statutory undertakers or local authorities carrying out their statutory obligations. In effect, therefore, the Contractor is entitled to determine if the Architect is compelled to issue a postponement instrument as the result of a default by a Nominated Sub-contractor. This position has been criticized in Duncan Wallace, I. N., QC in *Construction Contracts: Principles and Policies in Tort and Contract* (Sweet & Maxwell, London, 1986).
56 See Section 11.3.6 on this duty of the Architect.

3. delay by the Employer or his other contractors and licensees in the execution of work not forming part of the Contract (Clause 29 work) or failure to execute such work;

4. delay by the Employer in supplying materials and goods that he agreed to supply or failure to supply them;

5. failure to give ingress to or egress from the site.[57]

16.13 Determination for insolvency of the Employer

Clause 28.3 lists insolvency situations in which the Employer may find himself. These are exactly the same as those in Clause 27.3. The immediate effect of the occurrence of any of these situations is that the Contractor's obligation to carry and complete the works is suspended. This provision gives the Contractor the opportunity to stop and monitor the situation, without leaving himself open to the risk of determination by the Employer on grounds already explained, before deciding whether or not he wishes to continue with the Contract. If he decides to determine he must give notice to the Employer to that effect.

16.14 Procedure for determination under Clause 28

The procedure for determination on grounds other than for insolvency is as follows. The Contractor gives the Employer notice specifying the default or suspension. It is important that the Contractor states expressly that he is giving a preliminary notice of determination. If the Employer continues the default or the suspension for 14 days from the receipt of the notice then the Contractor may within a further 10 days give a final notice determining his own employment. The right to determine expires after the 10 days. However, if the Employer's default or the suspension is repeated, the right becomes available again but it can only be exercised within a reasonable time after the repetition. The procedure is shown in Fig. 16.2. With determination for insolvency only one notice is required. The determination takes effect upon its receipt.

16.15 Determination under Clause 28: consequences and financial settlement

These consequences are detailed under Clause 28.4. The Contractor is to remove his temporary buildings, plant, materials and the like from the site with reasonable despatch and to afford sub-contractors the facilities to do the same. The Contractor's liability in respect of personal injury, death and damage to property under Clause 20 applies until the removal is complete.

The Employer must refund to the Contractor within 28 days of determination all the retentions deducted before determination. However, the Employer is entitled to make against the retention any deduction allowed under the Contract. The Contractor's total payment entitlement consists of:

---

57 See Section 11.3.14 for a discussion of the Employer's responsibility for ingress and egress.

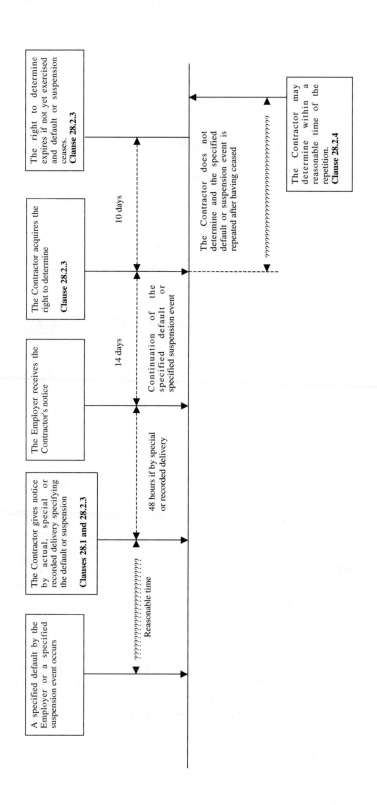

Fig. 16.2 Procedure for Determining for the Employer's Specified Default or Specified Suspension

- the total value of work complete at the date of the determination (the method of determining this amount is the same as that used for the valuation for interim payment under Clause 30.2);
- the value of work begun but not completed (the amount is determined in accordance with the rules for valuing Variations under Clause 13.5);
- the Contractor's loss and/or expense under Clauses 26 and 34.3;
- cost of materials and goods ordered for Works and for which the Contractor has incurred liability to pay (on payment they become the property of the Employer and the Contractor must not remove them);
- reasonable cost of removal temporary buildings, plant, etc.;
- any direct loss and/or damage caused to the Contractor or any Nominated Sub-contractor (e.g. profit on uncompleted work).

The Contractor is to prepare a statement showing these figures with reasonable despatch upon determination of his own employment. The difference between their total and total payment already made on account represents the final settlement. If it is positive, it is payable to the Contractor and vice versa.

16.16 Determination by either party under Clause 28A

Either party may determine the employment of the Contractor if, before Practical Completion, the carrying out of the Works, or substantially the whole of the Works, has been suspended for a continuous period exceeding the appropriate maximum stated in the Appendix and such suspension is caused by any of the following:

- *force majeure*[58]
- loss or damage to the Works from the Specified Perils[59]
- civil commotion
- breaches by statutory authorities
- hostilities (whether war declared or not)
- terrorist activities (Clause 28A.1).

This is subject to the proviso that where the suspension arises from any of the Specified Perils, the Contractor has no right to determine if the Specified Peril was caused by the negligence or default of the Contractor, his servants, agents, or anybody engaged by him upon the Works (Clause 28A.1.2). The Employer, his servants and agents, local authorities and statutory undertakers are expressly excluded from the proviso. The absence of the Nominated Sub-contractors in the list of exclusions must mean that the Contractor has no right to determine if a Nominated Sub-contractor caused the Peril.

The procedure for determination under this clause is as follows. The party wishing to determine gives notice to the other that if the suspension is not terminated within 7 days of receipt of the notice, the Contractor's employment is to be determined. If, after receipt of this notice, the suspension continues beyond that period, the determination takes effect after its expiry. Fig. 16.3 shows the procedure for determination under Clause 28A.

---

58 See Section 11.3.1 for discussion on the meaning of this term.
59 See Clause 1.3 for a definition of this term.

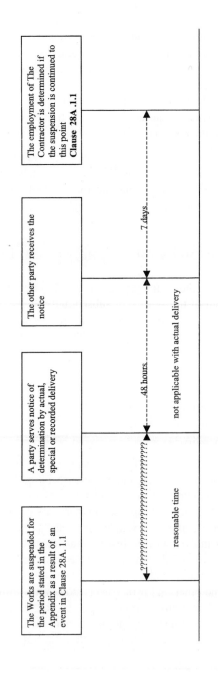

Fig. 16.3 Procedure for Determining for Suspension caused by the Clause 28A.1.1 Events

16.17 Financial settlement: Clauses 22C/28A determinations

Upon determination under Clause 28A or Clause 22C.4.3.1, the provisions under Clause 30 on payment and release of retention no longer apply. They are replaced by a new regime as follows:

1. One half of the Retention deducted by the Employer must be paid to the Contractor within 28 days of the determination: Clause 28A.4.
2. Not later than 2 months after the date of the determination, the Contractor must provide the Employer with all documents necessary for the preparation of the financial settlement between the parties. This settlement is referred to in the Contract as 'the account'. It must include: (i) the total value of work properly executed determined in accordance with the terms of the Contract; (ii) any direct loss and/or expense due under Clauses 26 and 34.3; (iii) the reasonable costs of removal of the Contractor's and Sub-contractor's resources from the site; (iv) costs of materials ordered towards carrying of the works that the Contractor is obliged to pay for. In addition, but only if the determination was for damage from the Specified Perils caused by the negligence of the Employer or of people for whom the Employer is responsible, it must include 'any direct loss and/or damage' caused to the Contractor by the determination. In *Wraight Ltd v. P. H. & T. (Holdings) Ltd*[60] it was accepted at first instance that the phrase 'direct loss and/or damage' allowed recovery of overheads and profit allowance in the Contractor's tender.
3. The Employer is to prepare the account with reasonable dispatch subject to receipt of the necessary documentation from the Contractor.

16.18 Meaning of 'unreasonably or vexatiously'

The consequences of many of the grounds for determination vary in gravity. For example, the effects of sub-letting without consent may vary from very trifling to disastrous, depending upon the nature of the work sub-let and the calibre of the sub-contractor. The policy in the Contract is to provide a filter in the form of Clauses 27.2.4 and 28.2.5 for minor defaults for which determination, although available technically, is not to take place. However, whilst the need for such filters is clear, the use of the phrase 'unreasonably and vexatiously' deprives the provision of the required certainty. In *Hill v. Camden*,[61] which concerned notice by a contractor of determination for failure of the Employer to pay on a certificate, the Court of Appeal considered the meaning of that phrase. On the subject, Ormrod LJ said:

> I imagine that it is meant to protect an employer who is a day out of time in payment, or whose cheque is in the post, or perhaps because the bank has closed or there has been a delay in clearing the cheque or something – something purely accidental or purely incidental so that the court could see that the contractor was taking advantage of the

60 (1968) 13 BLR 26.
61 See n. 18.

other side in circumstances in which, from a business point of view, it would be totally unfair and almost smacking of sharp practice.

In *J. Jarvis Ltd v. Rockdale Housing Association*[62] the Court of Appeal considered the same phrase used in Clause 28.1 of the JCT 80. Bingham LJ agreed with the views of Ormrod LJ as to the meaning of 'unreasonably'. He also suggested that it might be helpful to compare the benefit to the Contractor of determining against the burdens to the Employer of that action. The Contractor's exercise of his right to determine would not be unreasonable unless there is a gross disparity between the benefits and burden. The use of 'vexatiously' is equally troublesome. According to Bingham LJ in the *Jarvis v. Rockdale* case, it suggests 'an ulterior motive to oppress, harass or annoy'.

16.19 Performance bonds and guarantees

A performance bond, also sometimes referred to as a 'performance guarantee', is an agreement by deed between an employer and a third party (bondsman), usually a bank, insurance company, or a specialist bonding company, that if the contractor defaults in the performance of his obligations under the construction contract with the employer, the bondsman will pay the resultant loss of the employer up to a stated maximum sum. This is typically 10 per cent of the contract price. The contractor is usually required under the construction contract to obtain the bond for the employer. To procure the bond, the contractor pays a premium ranging typically from 1–3 per cent of the bond amount, depending upon the bondsman's assessment of the risk of default by the contractor. As a contractor would normally include this cost in his tender, the premium is ultimately paid for by the employer.

Although the JCT 98 does not require the Contractor to procure a performance bond in favour of the Employer, it is often amended to include such a requirement. Two particular problems with bonds have been highlighted in litigation arising from JCT 80 or similar contracts.[63] First, as the obligation of the bondsman to pay under the bond mirrors that of the Contractor under the construction contract, the bondsman does not have to meet any payment until the Contractor is obliged to do so. This means that in the event of determination, the bondsman will not have to pay until the work has been completed by alternative means, thus imposing upon the Employer the need to find additional funds to bring the project to completion.[64]

The second problem concerns a form of bond in which the Employer's right to make a claim on it is expressed to be conditional on the Employer's service to the bondsman of notice of the Contractor's default under the construction contract. In *Perar BV v. General*

---

62 (1986) 36 BLR 48 (hereafter *Jarvis v. Rockdale*).
63 For a full review of the problems see Issaka Ndekugri, 'Performance Bonds and Guarantees in Construction: A Review of Some Recurring Problems' (1999) ICLR 294.
64 *Trafalgar House Construction v. General Surety* (1995) 73 BLR 32; see also *Paddington Churches Housing Association v. Technical and General Guarantee Company Ltd* [1999] BLR 244 in which His Honour Judge Bowsher held that, following determination under the JCT Standard Form of Building Contract With Contractor's Design 1981 edition for the Contractor's insolvency, the surety's obligation to pay on a performance bond did not arise until the Employer had not only completed the Works but also prepared the statement of accounts required under the equivalent of Clause 27.6 of this Contract.

Surety and Guarantee Co. Ltd,[65] which concerned a bond of this type, the Employer gave the notice only after the Contractor's insolvency had resulted in automatic determination of his employment. The Court of Appeal held that the Employer could not claim at all on the bond because the effect of automatic determination was that there were no further obligations from which the Contractor's default could arise. This decision had the most unfortunate result that, on the occurrence of the very event against which the Employer had sought the protection of a bond, the bondsman was able to escape liability while keeping his premium. It is important to note that although this case arose from a JCT 81 contract, the principle applies to the JCT 98 where the insolvency event concerned is to result in automatic determination. It would also apply to those which do not result in automatic determination but only give the Employer an option to determine. This is because, under Clause 27.5.1, in those circumstances, the Contractor has no obligation to continue with the works.

The decisions suggest that, where a bond is required, care must be exercised with its drafting to ensure the following:

- that a claim can be made on the bond in the event of determination for the Contractor's insolvency;
- that it gives the Employer the right to immediate recovery of the estimated additional costs to complete but allowing final settlement of accounts between the Employer and the bondsman on actual completion.[66]

16.20 Proactive strategies

From the contents of this chapter, the importance of early detection of impending insolvency cannot be over-emphasized. Not only would the solvent party be able to determine on grounds other than insolvency, but also he could start formulating contingency plans towards minimization of the disruptive effects of the determination. In his planning, the solvent party must be mindful not to precipitate the insolvency by unnecessarily publicizing the impending problem. According to Powell-Smith and Sims,[67] the specified defaults under Clauses 27.2.1 and 28.2.1 are often the tell-tale signs of impending insolvency. They suggest the following additional warning signs:

- sudden disappearance of plant or materials from the site;
- high turnover in site management;
- excuses about late deliveries of materials;
- complaints by sub-contractors about non-payment;
- general lack of diligence in the carrying out of the works.

---

65 (1994) 66 BLR 72; see also *Oval (717) Ltd v. Aegon Insurance Co. (UK) Ltd* (1997) 85 BLR 97.
66 For a sample of this type of bond see *The Use of Performance Bonds in Government Contracts*, a report published in 1996 by the Construction Sponsorship Directorate of the Department of the Environment, Transport and the Regions.
67 Powell-Smith, and Sims, *Determination and Suspension in Construction Contracts*, pp. 54–55.

17

Dispute resolution

Certain aspects of the construction process make the performance of construction contracts intrinsically prone to disputes, e.g. new materials and construction techniques, multi-organizational participation, uncertainty in the physical and commercial environments within which they have to be performed, variability in human performance, and low margins. This reality of disputes is recognized in the JCT 98 by way of provisions on how they are to be resolved if they arise. The aim in this chapter is to explain these provisions.

The resolution techniques expressly recognized in the Contract are litigation, arbitration and adjudication. Generally, the court has the power to settle disputes between individuals. Furthermore, the parties often need the court to enforce settlements reached by the other techniques. Article 5 provides that any dispute arising under the Contract may be referred to adjudication in accordance with Clause 41A.[1] By Article 7A, failure to delete 'Clause 41B applies' in the Appendix has the effect that the parties have entered into an agreement in the terms of that Article to resolve their disputes by arbitration. The combined effect of Article 7B and Clause 41C is that if the phrase is deleted, disputes are to be resolved by litigation, i.e. there is no right or obligation to resolve any dispute by arbitration unless the parties enter a separate agreement to that effect. The parties may also decide to resolve their disputes by negotiation or other techniques referred to collectively as 'Alternative Dispute Resolution' although the Contract does not mention them.

17.1 Ambit of the dispute resolution clauses

As Articles 5, 7A and 7B stipulate that certain disputes or differences are to be resolved by the applicable resolution technique, a party may resist any of these procedures by denying that there is a 'dispute' or 'difference', thus raising the question of what these terms mean. In *Hayter v. Nelson*,[2] which is now considered the leading authority on the question, Saville J (as he then was) stated that they bore the same ordinary meaning. This is now supported by s. 82(1) of the Arbitration Act 1996, which states that the term

---

1 It is also to be noted that if the form is to be used on a project not subject to the adjudication legislation, appropriate amendments must be made unless the parties wish to be bound by the adjudication provisions.
2 *Hayter v. Nelson* [1990] 8 Lloyd's Rep. 265.

'dispute' includes 'difference'. According to the case law,[3] the ordinary meaning of 'dispute' includes any situation where a claim has been put forward by one party and is rejected by the other. Saville J emphasized that, except in the simplest and clearest cases, there can still be a dispute even if it can be easily determined without any shadow of doubt that one party is right and the other is wrong. It follows, therefore, that adjudication or arbitration cannot be commenced unless the other party has been given an opportunity to consider the claim and has rejected it.[4] This principle is illustrated by *Cruden Construction Ltd v. Commission For The New Towns*.[5]

> The plaintiff was a contractor on a JCT 63 contract dated 22 May 1980 for the construction of dwellings for Central Lancashire New Town Development Corporation, who contracted to sell some of the dwellings to a housing association. On discovery of structural defects, the housing association made a claim against the Corporation. The defendants, who were the Corporation's successors in title, wrote to the plaintiff on 7 October 1993 informing it of the claim. On 11 October 1993, Cruden replied requesting details of the claim. The same day the defendant gave notice of arbitration. The court granted the plaintiff's application for a declaration that there was no dispute capable of being referred to arbitration, explaining that, as a general principle, where a party is unaware of the basis of a claim or he seeks information about the claim, there can be no dispute.

It is therefore important, before embarking upon adjudication or arbitration, to ensure that a dispute has crystallized, i.e. there is evidence[6] that the other party has been given the chance to respond to the claim and has rejected it. There is case law also suggesting that where the other party receives the claim and ignores it[7] or prevaricates,[8] it may amount to a rejection.

Another source of contention concerns types of dispute covered by the relevant dispute resolution clause, as the other party can resist the proceedings by arguing that, although there is a dispute, it is outside the ambit of the appropriate clause. Regarding the ambit of the JCT 98 dispute resolution clauses, there is some guidance from decisions in cases involving dispute resolution agreements with some similarities. For example, where the clause includes disputes 'arising under the contract', disputes concerning whether the contract has become illegal to perform or is otherwise frustrated are covered.[9] However, that form of words would not cover disputes concerning negligent misstatement under the principle in *Hedley Byrne Co. Ltd v. Heller & Partners Ltd*[10] or innocent misrepresentation under the Misrepresentation Act 1967.[11] Before Amendment 18, the ambit of the JCT 80

---

3   See *Monmouth County Council v. Costelloe & Kemple Ltd* (1963) LGR 429; *M. J. Gleeson plc. V. Wyatt of Snetterton Ltd* (1994) 72 BLR 13 and *Halki Shipping Corporation v. Sopex Oils Ltd* [1998] 1 WLR 726 (hereafter *Halki*).
4   The court may be used for situations where there is no dispute as such, e.g. enforcement of payment admitted to be due.
5   (1994) 75 BLR 134.
6   For example, in correspondence or authenticated minutes of meetings.
7   *Tradax International SA v. Cerruhogullari T.A.S., The Eregli* [1981] 2 Lloyd's Rep. 169.
8   *Ellerine Bros (Pty) Ltd v. Klinger* [1982] 1 WLR 1375.
9   On illegality see *Mackender v. Feldia AG* [1967] 2 QB 590; on frustration see *Kruse v. Questier & Co. Ltd* [1953] 1 QB 669 and *Government of Gibraltar v. Kenney* [1956] 2 QB 410.
10 [1964] AC 465 (hereafter *Hedley Byrne*.
11 *Fillite (Runcorn) Ltd v. Aqua-lift* (1989) 45 BLR 27 (CA).

arbitration clause was stated as 'any dispute or difference as to the construction of this Contract or any matter or thing of whatsoever nature arising thereunder'. The Court of Appeal in *Ashville Investments v. Elmer Contractors*[12] considered the meaning of these words. It was stated that 'any matter . . . arising thereunder' included disputes about the interpretation of the terms of the contract, implying that the words 'the construction of this Contract' in that agreement were redundant. That probably explains the omission of those words from the JCT 98 arbitration agreement. It was also stated that 'any matter . . . in connection therewith' covered disputes concerning rectification of the contract, negligent misstatement and misrepresentation. It is doubtful whether the JCT 98 arbitration agreement covers other claims in tort.

Several clauses provide either that certain matters are not subject to review at all or that the relevant decisions are final. The provisions are as follows:

1. Under Clause 4.2, the Contractor may challenge the validity of an AI by requesting the Architect to name the clause from which his power to issue it arises. As explained in Section 5.3.1, if the Architect names a clause, the instruction is deemed to be authorized under the Contract after the Contractor complies with it before its validity is formally challenged by invoking the appropriate dispute resolution method. This means that the Arbitrator or court would not have jurisdiction to decide whether or not such an instruction was in accordance with the Contract.
2. Clause 22C.4.3.1 entitles any party to determine the employment of the Contractor in case of loss or damage to the Works from the insured risks provided it is 'just and equitable' to do so. The other party may contest any purported determination on the grounds that it is not just and equitable to do so. There is a limit of 7 days from receipt of the notice of the challenge within which either party may invoke an appropriate dispute resolution mechanism to determine the issue. The effect of this limitation clause is that, after the 7 days, the matter cannot be a dispute under the Contract.
3. Clause 30.9.1 provides that the Final Certificate is final and conclusive evidence of a number of issues in 'any proceedings under or arising out of or in connection with this Contract' unless the issue is properly challenged before expiry of 28 days after the issue of the Certificate. It is submitted that this provision clearly covers adjudication, arbitration or litigation proceedings.[13]
4. Clauses 38, 39 and 40 provide in detail how the amount of fluctuations under those clauses is to be determined. Clauses 38.4.3, 39.5.3 and 40.7 provide that, as an alternative to calculating the amount of adjustment strictly in accordance with those provisions, the Quantity Surveyor and the Contractor can agree the amount of adjustment. Provided the Quantity Surveyor acts within his authority, any such agreement reached with the Contractor is not subject to review.[14]

17.1.1 EDI Agreement and dispute resolution

As explained in Section 1.4.1, the parties must enter an EDI Agreement if it is completed in the Appendix that *The JCT Supplemental Provisions For EDI* apply. Clause 2 of these Provisions states:

12 (1989) 37 BLR 55
13 See Section 3.9 for detailed discussion on these issues.
14 See Sections 14.1.5 and 14.2.7 for further explanation.

The procedures applicable under this Contract to the resolution of disputes or differences shall apply to any dispute or difference concerning these provisions or the exchange of any Data under the EDI Agreement and any dispute resolution provisions in the EDI Agreement shall not apply to such disputes or differences.

The rationale behind Clause 2 is probably that standard forms of EDI Agreements are unlikely to have been drafted with the special needs of the JCT 98 in mind. The displacement of any dispute resolution provisions in the EDI Agreement in favour of the procedures incorporated by Articles 5, 7A and 7B (adjudication, arbitration or litigation) therefore avoids the risk of inconsistency between the two sets of procedures. The use of the wording 'any dispute or difference concerning these provisions or the exchange of any Data under the EDI Agreement' suggests that claims for negligent misstatement under the *Hedley Byrne* principle arising from electronic exchange of information within the EDI Agreement would be subject to the adjudication provisions. However, the same dispute arising from exchange of the same information in the traditional way is not a dispute 'under the contract' and is therefore outside the adjudication provisions.

17.2 Adjudication

Adjudication was first introduced into a UK standard form of construction contract in 1976 when it was incorporated into the then current DOM/1, the form of domestic sub-contract intended for use with the JCT 63. The ambit of adjudication under the sub-contract covered only disputes between the main contractor and a sub-contractor about set-offs against monies otherwise payable to the sub-contractor. The process involved submission of the dispute to an independent third party, the adjudicator, appointed either in the sub-contract or after the dispute had arisen. The adjudicator was required to make a very speedy decision which would be binding unless and until reviewed by a court or an arbitrator after practical completion. Similar clauses were thereafter incorporated into other sub-contracts and the works contracts of the JCT family of contracts. From the 1980s wider adjudication clauses were introduced into standard forms of main contract. For example, the JCT's Standard Form of Building Contract with Contractor's Design contained an optional adjudication clause in respect of disputes arising in connection with a wide spectrum of matters referred to as 'The Adjudication Matters'. The New Engineering Contract required any dispute to be referred to adjudication before practical completion.

Acting on the Latham Report's recommendation that dispute resolution by adjudication should be incorporated into all relevant standard forms in the construction industry, the Government passed the Housing Grants, Construction and Regeneration Act 1996 (hereafter the 'Construction Act'). Ss. 108(1) to (4) of the Act require such qualifying construction contracts to contain, as a minimum, the following provisions on adjudication:

1. A party may give notice of his intention to refer any dispute arising under the contract to adjudication at any time.
2. The timetable in the contract has the object of securing the appointment of the adjudicator and referral of the dispute to him within 7 days of the notice to refer.
3. The adjudicator must make his decision within 28 days of the referral or such longer period as agreed by the parties after the dispute has arisen.

4.   The adjudicator is allowed to extend the period for his decision by 14 days with the consent of the party who referred the dispute.
5.   The adjudicator is under a duty to act impartially.
6.   The adjudicator has authority to take initiatives to determine facts and law applicable to the issues in dispute.
7.   The decision of the adjudicator is binding until the dispute is finally determined by arbitration or litigation proceedings or by agreement of the parties.
8.   The adjudicator and his employees and agents are not liable to the Contracting Parties for actions and omissions in furtherance of the adjudication unless they were in bad faith.

Examination of Clause 41A of the JCT 98 indicates that these minimum provisions have been incorporated into the Contract. However, the qualification, in both the Construction Act and the JCT 98 itself, of disputes to be referred for adjudication by the phrase 'disputes under the contract' leaves open to debate the validity of complying with the letter of ss. 108(1)–(4) of the Act but avoiding adjudication on some issues by also stipulating in the Contract itself that those issues, which may qualify as disputes in the normal sense as explained in Section 17.1, are not disputes under the Contract or that the relevant decisions are final. Some matters that qualify as disputes in the ordinary sense of the term are either not subject to review at all or must be reviewed within specified time limits. Technically, therefore, there is compliance with the Act although it is arguable that the parties do not have the right *at any time* to refer *any dispute* for adjudication. The validity of clauses of this kind[15] is still moot. On the one hand, it is arguable that upholding such clauses is likely to drive the proverbial coach and horses through the spirit of the Construction Act. However, it is submitted that the courts are unlikely to impose a blanket ban on such clauses and that they are likely to decide each case as a question of fact and degree, regarding the extent to which the clause undermines the purpose of the legislation. On this basis, the JCT 98 is likely to be found compliant with the Act in both letter and spirit.

Leaving aside the question of compliance, there is the related question of the effect of non-compliance with the stipulated minimum contractual content. S. 108(5) provides: 'If the contract does not comply with the requirements of sub-sections (1) to (4), the adjudication provisions of the Scheme for Construction Contracts [SCC] apply'. The effect of this section has been a matter for some debate among construction lawyers. The debate has concerned whether the effect of failure to include the minimum contractual content is to incorporate the SCC in its entirety, thereby displacing other adjudication provisions in the Contract with which the Scheme is inconsistent. It is submitted that a literal construction of the section supports such effect. However, it can lead to very harsh consequences. To illustrate this problem, consider a contract in which it is provided that the parties agree that the adjudicator shall have jurisdiction to adjudicate at the same time on more than one dispute under the contract. Alleging defects, the employer refuses to pay in full on a certificate. Upon the contractor invoking adjudication to enforce full payment, the employer wishes the issue of defects to be considered by the same adjudicator. This is objected to by the contractor, who points out that their contract does not provide for immunity of the adjudicator and of his agents as required by s. 108(4) and that, therefore,

---

15 See also Clause 66 of ICE *Conditions*, 6th Edition, which provides that until the Engineer decides any difference between the Employer and the Contractor, it is a matter of dissatisfaction and not a dispute under the Contract.

the adjudication provisions of the Scheme apply in their entirety. The failure to provide for the immunity was inadvertent. On the literal construction of s. 108(5), the Scheme's requirement[16] for the consent of the contractor to concurrent adjudication would override the provision in the contract that the adjudicator shall have the power to do so. An alternative approach would be to imply the required immunity so as to uphold the parties' agreement. Only time will tell which approach prevails.

17.2.1 Appointment of the Adjudicator

As explained in Section 17.1, there can be no adjudication without a dispute under the contract. A party desiring adjudication should therefore ensure that the dispute has crystallized. The process is then commenced by the referring party serving notice upon the other party that their dispute, which must be identified in the notice, be referred to adjudication: Clause 41A.4.1. The parties may reach an agreement with the intention of appointing an adjudicator and referring the dispute to him within 7 days of the notice to refer: Clause 41A.2.2. Failing agreement, either party may apply to a nominator to make an appointment and refer the dispute within 7 days of the date of the notice to refer. The nominator is the President or Vice-President or Chairman or Vice-Chairman of the selection in the Appendix from the following bodies:

- the RIBA
- the RICS
- the Construction Confederation
- the National Specialist Contractors Council.

If no selection is made, the default nominating body is the RIBA.[17]

It may be that the parties wish to name the Adjudicator in the Contract. This has the advantage of savings in the time necessary to make an appointment. Furthermore, there are opportunities for the Adjudicator to develop better knowledge of what is happening on the project. However, it has the disadvantage that the named person may not be suitable for all types of dispute. The JCT has recognized that some parties may wish to name the Adjudicator by providing guidelines[18] on necessary amendments to the Contract.

17.2.2 JCT Adjudication Agreement

Nobody is to be nominated or appointed unless he will enter an agreement in the terms of the *JCT Adjudication Agreement*, a standard form produced by the JCT: Clause 41A.2.1. There is a similar agreement for use where the Adjudicator is named in the Contract. Its main terms are:

1. The Adjudicator shall perform the duties allocated to him under Clause 41A of the Contract.
2. The parties are jointly and severally liable for the Adjudicator's fee stated in the agreement and reasonable expenses.

---

16 See Para. 9(1) of the Scheme.
17 See Section 1.5.2 dealing with Appendix entry against Clause 41A.2.
18 See pp. 55–56 of the Guidance Notes to Amendment 18 of the JCT 80.

3. In the event of the Adjudicator becoming unavailable to act on account illness or other cause, he shall notify the Contracting Parties to that effect immediately.
4. The Adjudicator's appointment may be terminated jointly by the Contracting Parties at any time by written notice. In such an event, except where he failed to give his decision within the relevant timescale, the Adjudicator shall be entitled to any part of his fee still outstanding plus expenses reasonably incurred before the termination. The Adjudicator shall not be entitled to recover his fee and expenses where he failed to deliver his decision on time.

17.2.3 Referral of the dispute

The Referring Party must refer the dispute to the Adjudicator within 7 days of the date of the notice to refer or on the date of execution of the *JCT Adjudication Agreement* by the Adjudicator, whichever is later. The referral documentation is to include:

- particulars of the dispute;
- a summary of his contentions on the matter;
- the remedy and relief the Adjudicator should consider.

The referral and the accompanying documentation are to be served simultaneously on the other party. Within 7 days of the date of the referral, the other party (the Responding Party) must respond with a statement of his own contentions and any other material that he wishes the Adjudicator to consider. It is a major criticism of the adjudication process that whilst the Referring Party may have spent months preparing his case, the Responding Party has only 7 days to respond no matter how complex the issues raised are.

Clause 41A.4.2 specifies the means of communicating with the Adjudicator for the purpose of both referrals and responses from the responding party. It should be by actual delivery, FAX, or registered post or recorded delivery. If sent off by registered post or recorded delivery, the referral or its accompany documentation is deemed to have been received 48 hours after the date of posting. Sundays and Public Holidays are to be ignored in the reckoning of the 48 hours. Anything sent by FAX must also be sent off forthwith by first class post or actual delivery. All this is subject to the EDI supplement if applicable.

17.2.4 Role of the Adjudicator

Clause 41A.5.3 avoids uncertainty about the position in law of the process to be embarked upon by providing that the Adjudicator's role is as specified in the Construction Act and that he is neither an arbitrator or an expert, thus avoiding problems of the type that arose in *Cameron Ltd v. John Mowlem and Co.*[19] In that case it was unsuccessfully argued that the decision of an adjudicator should be treated as an arbitration award for the purposes of registration and enforcement as a court judgment.

Subject to very limited rights to extend the time, the Adjudicator must reach his decision within 28 days of receipt of the referral and its accompanying documentation. The reference to Clause 41A.4.1 in Clause 41A.5.3 means that the 28 days begin to run from

---

19 (1989) 52 BLR 25; see also *Cape Durasteel v. Rosser & Russell* (1995) 46 Con. LR 75 in which a procedure given the label 'adjudication' was found to meet the requirements of a written arbitration agreement.

receipt of the documentation from the Referring Party and not those of the Responding Party. This explains the need for Clause 41A.5.1, which requires the Adjudicator to confirm this date to both parties. The Referring Party and the Adjudicator may extend time for up to 14 days. However, with the agreement of the Contracting Parties, time may be extended by any amount. Considering that the agreement of the Parties must be reached after the referral and not as part of the contract, such agreement is highly unlikely unless, possibly, where the Adjudicator requests further information from both parties and they both need the time to provide it.

17.2.5 Powers of the Adjudicator

Subject to a requirement that he must act impartially, the Adjudicator has almost unbridled powers. He may set his own procedure and take any steps to ascertain the facts and law as they pertain to the dispute. The only possible limitation is that he is entitled to only reasonable expenses. This means that if he is too extravagant regarding the level of, and remuneration for hired skills and expertise, he may not be able to recover some of the expenses. Specific powers include:

- to use his own knowledge and/or expertise;
- to review decisions, certificates and the like under the Contract;
- to require the parties to provide further information;
- to require the parties to carry out tests or open up work;
- to visit the site and workshops in which work is being or has been carried out for the Contract;
- to obtain information from employees of the parties provided he gives notice;
- to obtain information or professional advice from third parties provided he gives notice and furnishes estimates of costs involved;
- where the Contract provides for payment of interest, to decide the circumstances in which it is to be paid.

17.2.6 Immunity of the Adjudicator

Clause 41A.8 provides that the Adjudicator, his employees and agents shall not be liable to the Contracting Parties for any act or omission in furtherance of the adjudication provided the act or omission was not in bad faith. This implies that neither the Contractor nor the Employer is entitled to claim against these people for mistakes made in the adjudication process. However, this immunity does not bind third parties to the Contract who may be able to establish liability under the 'reliance' or 'assumption of responsibility' doctrine in *Hedley Byrne*.[20]

17.2.7 Costs of the Adjudication

The parties are to meet their own costs: Clause 41A.5.7. The Adjudicator's decision must specify how payment of his fee and reasonable expenses is to be apportioned between the

---

20 See n. 10.

parties. If the decision does not do this, each party is responsible for one half of the costs. However, they are jointly and severally liable for them. This means that the Adjudicator may recover all these costs from either party and leave either to pursue the other for contribution.

17.2.8 Enforcing the decision of an adjudicator

The effect of the adjudication legislation is that the parties are under a contractual obligation to give effect to an adjudicator's decision pending final settlement of their dispute in litigation, arbitration or by agreement. Clause 41A.7 of the JCT 98 provides to this intended effect in these terms:

> 41A.7.1   The decision of the Adjudicator shall be binding on the Parties until the dispute or difference is finally determined by arbitration or by legal proceedings or by an agreement in writing between the Parties after the decision of the Adjudicator has been given.

> 41A.7.2   The parties shall, without prejudice to their other rights under the Contract, comply with the decisions of the Adjudicator, and the Employer and the Contractor shall ensure that the decisions of the Adjudicator are given effect.

> 41A.7.3   If either Party does not comply with the decision of the Adjudicator the other Party shall be entitled to take legal proceedings to secure such compliance pending any final determination of the referred dispute or difference pursuant to clause 41A.7.1.

The award may therefore be enforced as any other contractual obligation. Two methods of enforcing adjudicators' decisions were considered in *Drake and Scull Engineering Ltd v. McLaughlin and Harvey plc.*,[21] which concerned adjudication before the Construction Act. These are through a Summary Judgment order[22] or mandatory injunction issued by a court.

17.2.8.1 Summary Judgment

The party wishing to enforce may commence an action in court for breach of Clause 41A.7.2 and apply for a Summary Judgment order. The basis of such an application is that: (i) the defendant has no real prospect of successfully defending his failure to comply with the decision; and (ii) there is no other reason why the case should go for trial. The court is therefore asked to make an order in the applicant's favour without a full trial. Such an order may then be enforced as any other court judgment. However, where there is an arbitration agreement, the other party may contest the application by applying to the court to stay the proceedings to arbitration. In the past, particularly in relation to applications to enforce payment obligations, the court could refuse to grant stay of proceedings on the grounds that there was no real dispute capable of being stayed to arbitration. In effect, the courts were distinguishing between genuine disputes and refusal to pay contractual debts.

---

21 (1992) 60 BLR 102 (hereafter *Drake and Scull*).
22 This was formerly Order 14 of the Rules of the Supreme Court. It is now referred to as 'Rule 24' of the Civil Procedure Rules.

However, recent developments have made the availability of Summary Judgment orders in these circumstances very doubtful. As explained in Section 17.1, the courts now adopt such a wide definition of 'dispute' that even a simple refusal to pay a contractual debt qualifies as a dispute. In any case, s. 9(5) of the Arbitration Act 1996 provides that upon an application by a party to legal proceedings, who is also a party to an arbitration agreement governing the issue in the proceedings, to stay them to arbitration, the court must stay them unless satisfied that the arbitration agreement is 'null and void, inoperative, or incapable of being performed'.[23] To entertain the application, the court must be satisfied that the applicant:

- is party to a valid arbitration agreement covering the matter before the court;
- is a party to the litigation;
- has acknowledged the legal proceedings against him;
- has not taken any step in the proceedings to answer the substantive claim;
- served notice of the application upon the other parties to the proceedings.

An application for stay of legal proceedings to arbitration under the Arbitration Act 1996 was considered by the Court of Appeal in *Halki*.[24] The court stated that there is a dispute once money is claimed unless it is admitted that the sum claimed is due and payable. The court then concluded that the effect of s. 9 of the Act is that, once there is a dispute in this wide sense and it is within the arbitration agreement, stay of the proceedings must be granted. In *Macob Civil Engineering Ltd v. Morrison Construction Ltd*,[25] the first case concerning statutory adjudication, it was contended that the effect of *Halki* is that a Summary Judgment order should not be available to enforce a payment award of an adjudicator unless the arbitration agreement in the contract was carefully drafted so as effectively to exclude from its ambit disputes about enforcement of decisions of adjudicators. That litigation was to enforce the decision of an adjudicator requiring a main contractor to make immediate payment to a sub-contractor. It is submitted that such exclusion has been achieved in JCT 98. Article 7A excludes from the ambit of the arbitration agreement any dispute or difference 'in connection with the enforcement of any decision of an Adjudicator appointed to determine a dispute or difference arising thereunder'.

The court did not grant Summary Judgment in *Makob v. Morrison* because the claimant did not apply for it. However, the court was clearly inclined to distinguish *Halki* as applicable to only applications other than those for enforcement of decisions of adjudicators. Summary Judgment was granted for the first time to enforce the decision of an adjudicator under the statutory regime in *A&D Maintenance and Construction v. Pagehurst Construction Services Ltd*.[26] Here again, the adjudicator's decision was that a main contractor should make immediate payment to a sub-contractor. The readiness of the courts to enforce adjudicators' decisions was highlighted by the fact that there was no formal written contract in this case, a ground upon which the main contractor contested the jurisdiction of the adjudicator in the enforcement proceedings. The main contractor's case was that, as there was no formal written contract, the Construction Act did not apply and

---

23 The meaning of this phrase is considered in Section 17.4.2.
24 See n. 3.
25 [1999] BLR 93 (hereafter *Macob v. Morrison*).
26 (1999) CILL 1518 (*A & D v. Pagenhurst* hereafter).

that, therefore, there was no right to adjudication. Judge Wilcox decided that the adjudicator had jurisdiction by applying s. 107(5) of the Act. That section provides:

> An exchange of written submissions in adjudication proceedings or in arbitral or legal proceedings in which the existence of an agreement otherwise than in writing is alleged by one party against another party and not denied by the other party in his response constitutes as between those parties an agreement in writing to the effect alleged.

The defendant not having denied the contract at the first opportunity, the parties were treated as having entered into an agreement in writing.

17.2.8.2 Mandatory injunction

The other method of enforcement involves exercise of the court's jurisdiction to issue a mandatory injunction requiring compliance with the award of an adjudicator. Such an injunction was issued in *Drake and Scull* itself. However, in *Macob v. Morrison* Dyson J declined to issue it to enforce a payment award on the grounds that it is not appropriate to payment of contractual debts between contractual parties. He had in contemplation the fact that failure to comply with such an injunction constitutes contempt of court, a criminal offence carrying possible sanctions of a fine or imprisonment. *Drake and Scull* was distinguished in *Macob v. Morrison* on the grounds that the earlier case involved a decision requiring payment of the amount in dispute to a third party for safe-keeping as a trustee stakeholder pending final determination of the dispute by arbitration.[27] He cited, as examples of situations where an injunction could be appropriate, decisions of an adjudicator requiring a party to return to the site to recommence work, to provide access or inspection facilities, to open up work, or to carry out specified work.

Under paragraph 23(1) of the SCC, the Adjudicator may, if he thinks fit, order any of the parties to comply peremptorily with his decision or any part of it. Although the Adjudicator cannot impose sanctions for failure to comply with a peremptory order as a court can for not complying with the equivalent court order, paragraph 24 of the SCC is designed to provide quick and easy access to court for enforcement of such an order by a mandatory injunction. That paragraph imports into the adjudication legislation s. 42 of the Arbitration Act 1996, which gives the court jurisdiction to make an order requiring a party to comply with a peremptory order of an arbitration tribunal. The effect of the importation is that, under the SCC, the following now applies to an adjudicator's decision:

> (1) Unless otherwise agreed by the parties, the court may make an order requiring a party to comply with a peremptory order made by the adjudicator.

> (2) An application for an order under this section may be made–
>   (a) by the adjudicator (upon notice to the parties),
>   (b) by a party to the adjudication with the permission of the adjudicator (and upon notice to the other parties).

> (4) No order shall be made under this section unless the court is satisfied that the person to whom the adjudicator's order was directed has failed to comply with it within the time prescribed in the order or, if no time was prescribed, within a reasonable time.

---

27 The particular contract expressly gave the adjudicator the power to make such a decision. An adjudicator would not now have such power unless the contract has a similar provision.

(5) The leave of the court is required for any appeal from a decision of the court under this section.

In *Macob v. Morrison* it was argued that reference to arbitration of a dispute arising out of the decision of an adjudicator under the Scheme amounts to agreement otherwise by the parties, the only exception to the court's jurisdiction to enforce the order. Rejecting this argument, the court pointed out that it has jurisdiction to grant a mandatory injunction even without a peremptory order from the adjudicator and, therefore, questioned the rationale for importation of s. 42 of the Arbitration Act 1996. It was suggested that if the intention was that an adjudicator's peremptory order would make the court more inclined to grant a mandatory injunction, it is mistaken because it is for the court alone to decide whether a mandatory injunction is appropriate. In *Outwing Construction Ltd v. H. Randell and Son Ltd*,[28] the second case to arise in England and Wales about statutory adjudication, His Honour Judge Humphrey Lloyds, QC questioned the need for the adjudicator in that case to have issued a peremptory order to comply with his decision. The omission in the JCT 98 to provide expressly for peremptory orders does not therefore make decisions of adjudicators under that contract any less enforceable than decisions under the Scheme.

17.2.8.3 Effect of challenging the Adjudicator's decision

The main issue settled by *Macob v. Morrison* concerned whether, for the purpose of enforcement, a 'decision' of an adjudicator was one not challenged by the party against whom it was made. It was common ground that the principle in *Halki* does not require stay of proceedings to arbitration if they were challenged on the merits of the adjudicator's decision, i.e. an argument that his answer, although a valid decision, was wrong. This is because s. 108(3) contemplates temporary binding effect until final review of the 'decision' in arbitration. However, the enforcement proceedings were challenged on the grounds that the 'decision' was invalid and a nullity and, therefore, incapable of having any binding effect. Admitting the ingenuity in this argument, Mr Justice Dyson nevertheless rejected it in these terms:

> If it had been intended to qualify the word 'decision' in some way, then this could have been done. Why not give the word its plain and ordinary meaning? I confess that I can think of no good reason for not doing so, and none was suggested to me in argument. If his decision on the issue referred to him is wrong, whether because he erred on the facts or the law, or because in reaching his decision he made a procedural error which invalidates the decision, it is still a decision on the issue. Different considerations may well apply if he purports to decide a dispute which was not referred to him at all.

He concluded that a decision the validity of which is challenged is nevertheless a decision within the meaning to the Act. However, this was entirely *obiter* because the defendants, by referring the adjudicator's decision to arbitration, had already elected to treat it as a valid but wrong decision. Dyson J held that they could not be allowed, for the purposes of the reference to arbitration, to argue that it was a valid decision but, for the enforcement proceedings, to contend that it was invalid. He left no doubt that he had adopted a purposive approach, i.e. one that would advance the cause for which Parliament had

---

28 [1999] BLR 156 (hereafter *Outwing v. Randell*)

passed the relevant part of the Construction Act. All the cases so far involving proceedings to enforce decisions of adjudicators under the statutory regime echoed the appropriateness of the purposive approach. In *Outwing v. Randell*, the court stated that it was justified to shorten normal court procedures so as to give effect to Parliament's intention that adjudicators' decisions, if not complied with, were to be enforced without delay. In *A&D v. Pagehurst* the decision was made within a month of the date of commencement of the proceedings. In *Rentokil Ailsa Environmental Ltd v. Eastend Civil Engineering Ltd*,[29] the first Scottish case on statutory adjudication, an adjudicator decided that the defendant should make payment to the pursuers (Scottish equivalent of claimant). After making the payment to the pursuers' solicitors the defendants sought a court order to freeze the funds in the hands of the solicitors. It was decided that though the court has general jurisdiction to issue such an order, it was not appropriate because, otherwise, it would subvert the whole point of the adjudication legislation.

17.2.9 Advantages of statutory adjudication

1. Its most valued feature is the speediness of the decision. It is often argued that such a decision, albeit temporarily binding, enables the parties to devote their energy to execution of the project rather than to the demanding procedures of more permanent resolution methods.
2. It avoids the conflict of interest inherent in the traditional role of the project architect/engineer as the first-tier tribunal for resolution of disputes between employers and contractors.
3. It is intended as a relatively cheap process, with minimum lawyer or expert witness involvement. Whether this objective will be achieved in practice remains to be seen.
4. It is expected that, once a dispute has been adjudicated, the parties would, in most cases, accept the decision as a final determination of their dispute, thus clearing the air for the parties to concentrate on completing the project. Anecdotal evidence and two reported cases[30] suggest that the step of simply invoking adjudication often compels the parties to agree a compromise before any further steps are taken.
5. If either party cannot accept the decision of the adjudicator as a final, it can always be reviewed in arbitration or litigation.

17.2.10 Disadvantages

1. Its most serious criticism is that, with the complexity of construction disputes, the restrictive timetable for the process can result in serious injustice, which can only worsen adversarialism.
2. It tips the balance in favour of large and well-resourced parties who can more easily accommodate the demanding timetable requirements.
3. It affords the opportunities for one party to ambush the other, i.e. one party prepares a full and sophisticated submission of his case over a long period which then has be responded to by the other party in 1–2 weeks.

---

29 (1999) CILL 1506 (hereafter *Rentokil v. Eastend*).
30 *Outwing v. Randell; Rentokil v. Eastend*.

4. The legislation fails to recognize that not all disputes can be suitably resolved, even on a temporary basis, by adjudication.
5. Implementation of certain decisions that are subsequently reversed in litigation or arbitration can lead to bizarre results in certain situations, e.g. a decision that work be taken down for alleged poor quality and it is decided otherwise after the contractor has implemented the decision.
6. Where the party in whose favour a payment decision is given subsequently becomes insolvent, the availability of the final review in arbitration or litigation is only academic.
7. It is not particularly suitable for multi-party disputes.
8. It undermines the well-understood role of the project architect/engineer.

17.3 Litigation

Article 7B and Clause 41C.1 provide that where 'Clause 41B applies' has been deleted in the Appendix, the parties are to resolve their disputes by legal proceedings, i.e. the law courts are approached to resolve them and not an arbitrator.

17.4 Arbitration

Arbitration may be defined as a private procedure for settling disputes whereby a dispute between parties is decided judicially by an impartial individual or a panel of individuals appointed for that purpose. An individual so appointed is referred to as an 'arbitrator' whilst his decision is referred to as his 'award'. An arbitrator's award is legally binding on all the parties to the arbitration proceedings to whom it is addressed.

An arbitrator does not have to possess any particular skills or qualifications unless they are specified in the agreement to resolve disputes by arbitration. In most cases, a person is appointed to act as an arbitrator on the strength of his expertise and experience in the subject matter of the dispute. In addition, as he is expected to act in a judicial capacity, he must possess some knowledge of the law, particularly the law of contract, tort and evidence. Most practising arbitrators are fellows of the Chartered Institute of Arbitrators, which undertakes the training of arbitrators in all fields. Most construction arbitrators are professionally qualified in a construction field and/or the law. Research[31] suggests that, with construction disputes, arbitrators qualified in the law and a construction discipline are preferred to those qualified in only one field.

Most of the law on arbitration is enshrined in the Arbitration Act 1996, which applies to England, Wales and Northern Ireland. All references to sections in the rest of this chapter are therefore to sections of that Act unless the contrary is expressly indicated.

17.4.1 JCT 98 Arbitration Agreement

Article 7A contains an arbitration agreement covering any dispute or difference 'as to any matter or thing of whatsoever nature arising under this Contract or in connection therewith' except:

---

31 Ndekugri I., and Jenkins, H., 'Construction Arbitration: A Survey' (1994) ICLR 388, pp. 366–83.

- disputes in connection with the enforcement of any decision of an adjudicator appointed to determine a dispute arising under the Contract;[32]
- disputes arising under Clause 31 for which legislation provides a method for their resolution;
- disputes arising under Clause 3 of the VAT Agreement.

In addition, as explained in Section 17.1, the Contract provides that, in stated circumstances, some matters are not subject to review.

17.4.2 Enforcement of the Arbitration Agreement

An arbitration agreement is a contract. A party to the agreement who commences litigation proceedings would therefore be acting in breach of contract. However, the common law remedies are either inappropriate (damages) or rarely granted (specific performance). S. 9 now provides an appropriate remedy. As explained in Section 17.2.8.1, upon an application by a party to the proceedings to stay them to arbitration, the court must grant a stay unless it is satisfied that 'the arbitration agreement is null and void, inoperative, or is incapable of being performed'. Until the meanings of the terms in that phrase are clarified in future cases, there will be some uncertainty. However, there is some guidance from existing case law. *Harbour Insurance v. Kansa*[33] and other cases cited in it establish the principle of separability of the arbitration agreement. S. 7 has now put this principle on a statutory footing. An arbitration agreement is therefore not rendered 'null and void' simply by the fact of the underlying contract being null and void. There is dearth of authority on when an arbitration agreement becomes 'inoperative'. What is certain is that failure of the parties to comply with time limits, determination of the underlying contract, or elapse of time within which the claim should have been brought does not have this effect.[34] However, it has been held that refusal of the courts of another country to enforce it may make it inoperative.[35] There are similar difficulties with the meaning of 'incapable of being performed'. There is authority for the proposition that neither the incapacity of a party to settle the award[36] nor lack of funds to finance the proceedings[37] has this effect.

The Arbitration Act 1996 originally contained ss. 85–7, which bestows upon courts a discretionary authority regarding the granting of a stay in relation to a domestic arbitration agreement. Because of concerns that these sections may offend against the Treaty of Rome by discriminating between English and other EU parties, they have not been brought into force.

17.4.3 Arbitration procedure

Under s. 33, the Arbitrator has a general duty to act fairly and impartially as between the parties, giving each party reasonable opportunity to put forward his case and to respond to that of his opponent. In particular, he must adopt appropriate procedures for achieving this

---

32 For explanation of the rationale behind this exclusion see para. 17.2.8.
33 [1993] 1 Lloyd's Rep. 455.
34 *Radio Publicity v. Cie Luxembougeoise* [1936] 2 All ER 721 (failure to keep time limits and determination of the underlying contract); *The Merak* [1964] and [1965] 2 Lloyds' Rep. 527.
35 *Philip Alexander Securities and Futures Ltd v. Bamberger and Others*, 8 May 1996, unreported (Waller J)
36 *The Rena K* [1979] QB 377.
37 *Paczy v. Haendler* [1981] 1 Lloyd's Rep. 302.

objective. On the question of procedure, Article 7A of the JCT 98 provides that it is to be in accordance with Clause 41B and the JCT 98 edition of the Construction Industry Model Arbitration Rules (CIMAR).[38] Rule 6 of CIMAR requires the Arbitrator to consult the parties before deciding on the procedure. Subject to the above considerations, the procedure to follow is entirely at the Arbitrator's discretion. CIMAR comprises 14 Rules governing various aspects of arbitration procedure. The matters needing special attention include:

- choice from three recognized types of procedure
- commencement of the proceedings
- appointment of the Arbitrator
- multi-party disputes.

17.4.3.1 Choice of three types of procedure

CIMAR describes three types of procedures that the Arbitrator may decide to adopt: (i) Short Hearing Procedure; (ii) Documents Only Procedure; and (iii) Full Procedure.

The Short Hearing Procedure (Rule 7) is appropriate where the Arbitrator can determine the dispute by inspecting work, materials or the operation of machinery. It requires the parties to submit to the Arbitrator written statements of their case. The hearing should not take more than a day unless the parties agree to extend it. An example of a dispute for which this procedure would be appropriate is whether work is in accordance with the Contract.

The Documents Only Procedure (Rule 8) involves the Arbitrator making his award only on the basis of written statements of claim. This is appropriate where the issues in dispute are such that there is no need for oral evidence or the amount involved does not warrant it. As an example, consider a dispute whether an instruction issued by the Architect is authorized under the Conditions. There is no reason why the Arbitrator cannot determine such a dispute by examining the instruction, the Conditions and the parties' written submissions.

The Full Procedure is appropriate where neither of the two procedures described above is appropriate. With this procedure, the parties must exchange statements of claim and defence before the hearing. Rule 9 provides guidelines with which the statements must comply. An example of dispute for which this type of procedure would be suitable is a complex claim for delays and disruption from a variety of different causes.

17.4.3.2 Commencement of arbitration proceedings

The limitation acts also apply to arbitration, i.e. if the proceedings are not commenced within the appropriate limitation period, the matter can no longer be pursued.[39] Although the courts have power under s. 12(1) to extend time, it is exercised only very sparingly. This makes it of paramount importance that parties know what type of act amounts to commencement of the proceedings. Under s. 14(1), the parties may agree when proceedings are to be considered commenced. Sub-rule 2.1 of CIMAR states that proceedings are commenced in respect of a dispute when one party serves on the other a

---

38 These were produced at the instigation of the Society of Construction Arbitrators who viewed as unsatisfactory the use of different arbitration rules within the construction industry.
39 See Section 17.4.5.2.

written notice of arbitration. The notice should specify the dispute or disputes and require the other party to agree to the appointment of an Arbitrator. The notice may further specify names of persons proposed for appointment as the Arbitrator but this may be left to subsequent correspondence.

17.4.3.3 Appointment of the Arbitrator

The parties have 14 days from the date of the notice of arbitration, or of a previously appointed Arbitrator ceasing to act, to agree the person to act as the Arbitrator. Obviously, this period may be extended by agreement between the parties. On receipt of the notice, the other party will often respond by proffering his own list of names of possible arbitrators. Fourteen days is a very short time to check CVs and availability of arbitrators. CIMAR recognizes the possibility of the parties failing to agree with the Arbitrator by providing that, in such event, either party may apply to a person empowered under the Contract to make the appointment on the parties' behalf. With the JCT 98, this is the person named as the Appointor in the Appendix.[40] The options in the Appendix are the President or Vice-President of: (i) the Royal Institute of British Architects (RIBA) or; (ii) the Royal Institution of Chartered Surveyors (RICS) or; (iii) the Chartered Institute of Arbitrators (CIArb). If no selection is made the Appointor is the RIBA by default. Research[41] indicates most arbitrators are appointed by the designated institutions rather than the parties themselves. This reflects the considerable distrust created by disputes. If the designated Appointor fails to make the appointment, either party may apply to the court to exercise its jurisdiction under s. 18(3) to make an appointment.

17.4.3.4 Multi-party disputes

The reality of construction disputes is that there are often several parties with some interest in the subject matter of a dispute. As an illustration, consider the cause of delay. It could be the Contractor, the Architect, or a sub-contractor. Furthermore, because of sub-contracting, a party who is found liable under one contract may wish to pass that liability down the chain of contracts to the party ultimately responsible. However, there is the problem that, with an arbitration from one contract, third parties to that contract have neither the right nor the obligation to be a party to the proceedings. An Architect therefore has no right or obligation to join in an arbitration between the Employer and a Contractor. This thus raises the problem of separate proceedings on the same or related disputes and, consequently, the risk of inconsistent outcomes, e.g. the finding in the Employer–Contractor proceedings is that a defect is attributable to a design error whilst the Employer–Architect proceedings conclude that it was caused by poor workmanship of the Contractor. This used to be grounds for the courts to refuse to stay proceedings to arbitration, the rationale being that, as the court has powers to join all the relevant parties in the same action, it was better for the court to decide the matter.[42]

S. 9, which governs the court's consideration of applications for stay of litigation proceedings, does not allow the court any discretion to refuse a stay for this reason. S. 35(1) provides that the parties are free to consolidate related proceedings or conduct concurrent proceedings. Under s. 35(2), in the absence of such agreement, the arbitrator

---

40 See Section 1.5.2.
41 See n. 31.
42 See *Tauton-Collins v. Cromie* [1964] 1 WLR 637 (CA).

has no power to consolidate proceedings. As a result of incorporation of CIMAR, the parties to a JCT 98 contract give their consent to consolidation of proceedings. Under Rule 3, the Arbitrator is empowered to order consolidation of certain types of related proceedings with or without conditions. Sub-rule 2.6 allows an appointing body, when approached to make an appointment, to consider referring the dispute to an arbitrator already appointed for a related dispute and to do so where appropriate. Parties may also add disputes not included in the original notice of arbitration. To complete the circle, consolidation of proceedings and running concurrent hearings are allowed in defined circumstances. All disputes between the Contractor and the Employer notified before the appointment of the Arbitrator must be consolidated. The Arbitrator may consolidate additional disputes from the same arbitration agreement notified after his appointment with the proceedings if appropriate. However, where the separate proceedings involve different parties, the Arbitrator *may* order concurrent hearings only where the disputes arise from the same project and raise a common issue. In this type of situation, he must not consolidate the proceedings without the agreement of all concerned. With consolidated proceedings, the Arbitrator, unless the parties otherwise agree, must deliver a single award. Separate awards are required where there have only been concurrent hearings. It is obvious from the description of the scheme for consolidation that it will be workable only if CIMAR is incorporated into the related contracts and the appointing bodies cooperate with each in the interest of consolidation. Other obstacles to consolidation are discussed in Section 17.5.14.

17.4.4 Enforcement of an arbitrator's award

The successful party has two ways of enforcing the award: (i) he can sue on it; (ii) he can register it as a judgment or order of a court under s. 66 and enforce it as such. It is often an express term of the arbitration agreement that the award will be binding on the parties. For example, under Clause 41B.3, the award of the Arbitrator is final and binding on the parties subject to any court determination of preliminary points of law under s. 45 or determination of points of law arising out of an award under s. 69. Even in the absence of such express provisions, a term to the same effect would normally be implied. Failure to abide by the award would therefore constitute an actionable breach of contract. Under s. 66, an arbitration award may, with the leave of the High Court, be entered as a judgment of the court and enforced as such. This alternative, which has the advantage of convenience, is available where the award is clear and without doubt as to its validity. The unsuccessful party may seek to prevent this method of enforcement by asking the court to set it aside for lack of substantive jurisdiction, serious irregularity or a point of law arising from the award.[43]

17.4.5 Court intervention into arbitration

The aim of an arbitration agreement is to provide for the ability to choose a private tribunal, in preference to the courts, to settle any disputes between the parties to the agreement. However, for two reasons, arbitration agreements do not completely remove the possibility of the courts getting involved in their dispute. First, any agreement that parties to a contract should never go to court in connection with disputes under the contract would be null and void on public policy grounds. The only exception is a form of agreement commonly referred to as a *Scott v. Avery*[44] clause, after the name of the case in which it first appeared. This form, which does not actually oust the jurisdiction of the courts, makes the award of an arbitrator a condition precedent to litigation. A defendant who is party to this form of agreement is entitled, as a right, to stay of proceedings, unless he has prejudiced this right either by a waiver or improper conduct. However, s. 9(5) provides that the court may ignore a *Scott v. Avery* clause and hear the case where the arbitration agreement is null and void, inoperative or incapable of being performed. Second, in addition to the court's power to enforce arbitration awards as explained in Section 17.4.4, various parts of the Arbitration Act invest the courts with other supervisory powers over arbitrations. However, these powers have to be looked at in the light of the general principle stated in s. 1(c) of Part I of the Act that 'in matters governed by this Part the court should not intervene except as provided by the Part'. Generally, the policy in the legislation is to keep interference by the court to a minimum and to support the arbitration where it is allowed to intervene. The more important of these powers are next explained.

17.4.5.1 Appointment of arbitrators
Most arbitration agreements expressly state the manner in which the arbitrator is to be appointed. If the contractual method proves unsuccessful, the court can appoint the arbitrator upon an application from either party to do so.[45] In the event of an appointed arbitrator being unable or refusing to act, dying, or otherwise failing to act after he has been validly appointed, a replacement has to be appointed. Subject to the express provisions of the arbitration agreement on this eventuality, this vacancy can be filled along the same lines as used in the initial appointment, i.e. agreement, third party appointment, or appointment by the High Court.

17.4.5.2 Extension of time for commencement of arbitration
The statutes of limitation apply to both litigation and arbitration.[46] This means that arbitration would be statute-barred if a notice to refer to arbitration were not given within 6 years (for simple agreements) or 12 years (for agreements executed as deeds) from when the right to arbitrate accrued. The notice to refer to arbitration is the equivalent of the writ in litigation. However, an arbitration agreement may stipulate different limitation periods. However, under s. 12, the High Court has discretion to extend time if a claimant is out of time. The section expressly states that the discretion is to be exercised only if the court is satisfied that: (i) the circumstances were outside the reasonable contemplation of the parties when they agreed the provision in question and it would be just to extend time; or (ii) the conduct of a party makes it unjust to hold the other party to the strict term of the provision limiting the period for commencement of arbitration. In practice, the courts are generally reluctant to interfere with arbitration agreements in this manner.

---

44 (1856) 5 H.L.C. 811.
45 S. 18.
46 S. 13(1).

17.4.5.3 Removal of an arbitrator

S. 24 empowers the courts to remove an arbitrator in four situations:

1. There are justifiable doubts as to his impartiality.
2. He does not posses the qualifications required by the arbitration agreement.
3. He is physically or mentally incapable.
4. He has refused or failed properly to conduct the proceedings or to use all reasonable dispatch in conducting the proceedings or making an award, thus causing substantial injustice.

17.4.5.4 Resignation of an arbitrator

Under s. 25, the parties may agree with the arbitrator his entitlement to his fees and expenses and his liability if he resigns. If they do not so agree, the arbitrator may apply to the court for relief from any liability and an order in respect of his fees.

17.4.5.5 Interlocutory orders

These are orders issued by the arbitrator on procedural matters. For example, under s. 39, if both parties agree, the arbitrator can order on a provisional basis, payment of a sum of money or the disposition of property between the parties or an interim payment in respect of costs. Any such order can be subject to the arbitrator's final assessment in his award. If the party to whom an order is addressed fails to comply, it can be enforced by the court unless the parties excluded such jurisdiction of the court. The JCT 98 does not contain such an exclusion.

17.4.5.6 Determination of recoverable costs

This is the process whereby the make-up of the amount demanded by a successful party as costs is assessed as to their validity. Where they are excessive, they will be determined on the basis of criteria set out in s. 63(5). The main guideline is that recoverable costs shall be determined on the basis of a reasonable amount in respect of all costs reasonably incurred and that any doubt as reasonableness is to be resolved in favour of the paying party.

17.4.5.7 Determination of preliminary points of law

Under s. 45, unless the parties otherwise agreed, the court has jurisdiction to determine any question of law arising in the course of the proceedings if a party applies to it and the court is satisfied that the question of law substantially affects the rights of one or more of the parties. Even where the court is so satisfied, it must not consider the application unless either of two conditions is satisfied:

1. the other parties to the proceedings consent that the court may consider it; or
2. the Arbitrator consents and the court is convinced that the application was brought without delay and that the determination is likely to produce a substantial saving of costs.

Clause 41B.4 of the JCT 98 provides that the parties have given their consent to this type of application to the court. The validity of this type of term, i.e. terms giving consent to appeals and applications in the arbitration agreement itself, was upheld in *Vascroft*

(Contractors) Ltd v. Seeboard plc.[47] There is therefore an automatic right under the JCT 98 to apply to the court to determine preliminary points of law.

Under s. 87, the parties may enter an agreement to exclude the jurisdiction of the court to determine preliminary questions of law. However, for reasons already explained in Section 17.4.2, that section has not been brought into force. However, the last part of s. 45(1) provides that where the parties agree that the arbitrator shall not give reasons for his award, that is equivalent to excluding the court's jurisdiction. This is the case irrespective of whether or not a domestic arbitration agreement is involved. S. 45(2) requires the applicant to state in the application the question to be determined and, unless all the parties to the proceedings consent to the application, why the court must determine it. It follows from the above discussion that the application must contain the information required by the court to decide whether the conditions have been met.

Leave of the court is required to appeal against its decision as to whether or not any of the conditions for the exercise of its jurisdiction to determine the question has been met. There is a similar condition on appeal against the court's determination of the question of law, but with the added requirement that such leave is not to be given unless the court considers that the question is one of general importance or that there are other special reasons why the Court of Appeal should consider it. It is to be noted that the Court of Appeal itself has no discretion to grant leave if the lower court declines to do so.

17.4.5.8 *Appeals against an arbitration award*

Under ss. 67 (jurisdiction), 68 (serious irregularity affecting the arbitrator, the proceedings or the award) and 69 (point of law arising out of the award), a party may appeal against an arbitrator's award. Ss. 67 and 68 are mandatory but s. 69 applies subject to the agreement of the parties otherwise. By Clause 41B.4 of the JCT 98, the parties have given their consent in advance to appeals on points of law arising from an award.[48]

Upon appeal, the court may confirm, vary, or set aside an award, or remit it to the Arbitrator for reconsideration together with the court's opinions on the question of law that was the subject of the appeal. However, with challenges for serious irregularity, under s. 68, the court will normally avoid setting aside an award if it can do so. This is because setting aside the award would entail re-commencement of the arbitration with a new arbitrator, with all the implications for costs and delays. Where an award has been remitted, the Arbitrator, unless the order otherwise directs, must make his award within 3 months of the date of the remission order.

All appeals are subject to s. 70, which sets out that the appeal is first subject to the requirement that:

- any available process of appeal to, or review by the Arbitrator has been exhausted;
- any necessary application for correction of the award under s. 57 has been made;
- the appeal is brought within 28 days of the date of the award.

There are additional conditions depending on the grounds for the appeal.

---

47 (1996) 78 BLR 132.
48 As explained in Section 17.4.5.7 of this book, the parties therefore have automatic rights of appeal on points of law arising out of an award.

Substantive jurisdiction or serious irregularity

The substantive jurisdiction of the Arbitrator is defined under s. 30(1) as:

- whether there is a valid arbitration agreement;
- whether the Arbitrator has been properly appointed;
- whether the dispute is within the arbitration agreement; and
- the agreed procedure was used in referring it to arbitration.

Serious irregularity is defined in s. 68(2) as any of the following that the court considers to have caused or will cause substantial injustice to the applicant:

1. failure of the Arbitrator to comply with his s. 33 obligations, e.g. failing to be impartial and following an unfair procedure;
2. the Arbitrator exceeding his powers;
3. failure of the Arbitrator to follow agreed procedure;
4. failure of the Arbitrator to deal with the issues submitted to him;
5. the Appointor exceeding his powers;
6. uncertainty or ambiguity as to the effect of the award;
7. the award was obtained by fraud or other means unacceptable on public policy grounds;
8. failure of the award to comply with agreed form;
9. any irregularity in the proceedings that is admitted by the Arbitrator or the Appointor.

Where the appeal is on any of these grounds, s. 73 imposes the additional condition that the objection as to jurisdiction or serious irregularity must have been made as soon as it became known. If that was not done, a party may not raise the objection later unless he can show that, at the time he took part or continued to take part in the proceedings, he did not know and could not, with reasonable diligence, have discovered the grounds for this objection.

Questions of law

For appeals on a point of law, leave of the court is required first, unless all parties to the proceedings consent that the appeal must go ahead.[49] Where leave of the court has to be obtained, s. 69(3) lists the following factors that must apply before the court can grant it:

1. The determination of the question will substantially affect the rights of one or more of the parties.
2. The Arbitrator was asked to determine the question of law.
3. From the facts in the award, either the Arbitrator's determination was obviously wrong or the question is one of general importance and the determination of the Arbitrator is at least open to serious doubt.
4. It is just and fair for the court to interfere by determining the question.

These requirements have been derived in part from guidelines formulated by the House of Lords in *Pioneer Shipping Ltd v. BTP Tioxide Ltd: The Nema*[50] regarding the exercise of this discretion under superseded legislation. There is a restricted right of appeal to the

---

49 S. 69(2).
50 [1982] AC 724; the guidelines, referred to the 'Nema' Guidelines, were reaffirmed in *Antaios Compania Naviera SA v. Salen Rederiena AB: The Antaios* [1985] AC 191.

Court of Appeal from the High Court. The restriction arises from the requirement that the High Court must grant leave to appeal.

17.5 Arbitration or litigation

Pre-1998 versions of the JCT 80 provided for arbitration as a mandatory dispute resolution technique. Arbitration is now optional. The change was made partly to avoid the effect of the decision in *Northern Regional Health Authority v. Crouch*[51] that the court has no power to review certificates, opinions, decisions and the like of an architect where an arbitration agreement invests arbitrators with such powers. To make the decision whether to opt for arbitration, there is need to understand its advantages and disadvantages when compared with litigation. They are therefore next compared on a number of issues.

17.5.1 Expertise

Proponents of arbitration, as against litigation, advance the argument that an arbitrator is normally selected for his expert knowledge of the subject matter of the dispute, whereas a judge rarely has any expert knowledge of the technical complexities of the construction industry. There are three counter-arguments commonly put forward. The first is that the Technology and Construction Court (formerly the Official Referces Court), which is the division of the High Court that hears the vast majority of construction contract disputes, has accumulated expertise in this area over the years. Second, it is claimed that the issues usually involved are more of a legal nature rather than of construction contract management and construction practices. Third, technical experts can always be called as witnesses to assist the judges in coping with the technical issues arising, although it has to be pointed out that that course of action can involve extra costs. Finally, anecdotal evidence suggests that, because of the inevitable disparity in familiarity with the relevant procedures, some non-lawyer arbitrators may have trouble in controlling proceedings where very experienced professional advocates represent the parties.

17.5.2 Advocacy

Only practising Barristers and Solicitors obtaining a rights of audience certificate have the right to appear before a High Court as advocates unless the litigant wishes to appear and represent himself in person. Anybody can appear before an arbitrator on behalf of the parties unless they agree otherwise.[52] There is the counter-argument that arbitration has spawned a new type of professional, the lay advocate, who offers specialist skills in the presentation of cases on behalf of parties to arbitration proceedings. The lay advocate is not necessarily cheaper than the traditional legal representatives and will normally not be as proficient in the law.

---

51 See n. 54 below; this decision was overruled in *Beaufort*, see n. 54.
52 S. 36.

17.5.3 Costs

It is claimed that the cost of arbitration is usually much less than the cost of litigation. Indeed, ss. 1 and 33 make it an obligation of an arbitrator to avoid unnecessary delay and expense. The other school of thought is that this belief is based purely on fiction rather than fact. The points of the counter-argument as to costs are:

- parties to arbitration usually have exactly the same type and level of legal representation as in litigation;
- the arbitrator and the hearing venue have to be paid for by the parties whilst the court and judges come 'free';
- secretarial and other support to record the proceedings of the arbitration are frequently a source of cost.

17.5.4 Simplicity

Procedure in the Court is now governed by the Civil Procedure Rules (CPR). Although they are intended to be simpler than the Rules of the Supreme Court that they replaced, they are still quite complex. By contrast, arbitration is intended to be simplicity itself. By s. 34, unless the parties agree their own procedure, the procedure to be followed in arbitration is at the discretion of the arbitrator subject to his mandatory obligations under s. 33 to adopt fair, impartial and appropriate procedures. It is often argued that the arbitrator can adopt a procedure that best fits the case he is called upon to settle and thus avoid the complexities of litigation procedures.

There are two counter-arguments against this point. First, it is said that an arbitrator's discretion on procedure is not as unfettered as is often imagined and that where an arbitrator falls to follow proper procedure, his award would be appealable to the courts. For example, in *Modern Engineering (Bristol) Ltd v. C. Miskin & Son Ltd*[53] the arbitrator was faced with a point of law as to whether an architect's certificate could be reviewed. After listening to one side's argument on the point, the arbitrator made up his mind without giving the other side the opportunity to put forward their argument. Though his opinion was the correct position, the court held that he should have listened to the counter-argument and that his failure to do so constituted misconduct (now referred to as 'serious irregularity' under the 1996 Act). He was therefore removed for misconduct and his award set aside under the 1950 Arbitration Act. Second, many arbitration agreements incorporate model procedures published by various bodies, e.g. CIMAR and the ICE Arbitration Procedure, some of which are quite complex.

17.5.5 Expedition

The law's delays are notorious but, following the reforms instigated by Lord Woolff, the position should improve under the new Civil Procedure Rules. Arbitration can be quicker. The reasons usually given for this relate to the simplicity issue already discussed. Much depends on the attitude of the parties and the robustness and experience of the arbitrator.

53 [1981] 1 Lloyds Rep. 135.

17.5.6 Convenience

It is common practice for an arbitrator to arrange for the arbitration proceedings to be held at times and places to suit the convenience of both parties. An action in court is listed subject to the availability of a judge and a court primarily and then, possibly, the availability of the legal representatives of the parties.

17.5.7 Courtesy

Rarely will an arbitrator forget the fact that, in effect, he has been appointed by the parties and that, therefore, he owes them a moral duty to extend to them the normal courtesies required in commercial and professional communication. By contrast, some believe that judges can be abrupt. However, judges are professionally trained and well experienced in controlling proceedings whereas arbitrators are, by contrast, amateurs who, if inexperienced, can be dominated by the parties' professional advocates.

17.5.8 Privacy

A lawsuit must normally be heard in open court with the Press and public free to attend. Litigation may therefore involve unpleasant publicity or result in the disclosure of trade secrets. With arbitration, the entire hearing takes place in private, thus avoiding the publicity and the disadvantages of the nervousness usually commonly induced by proceedings in court. However, if an arbitrator's award is appealed then the appeal is heard in open court and is liable to be reported in the Press.

17.5.9 Future business relations

Litigation is inherently adversarial, thus giving rise to risks of reduction in the chances of future contracts between the parties. Even where they enter into other contracts, the distrust resulting from the experience of litigation could easily poison such subsequent contractual relationships. Contractual relationships in general may even suffer because a reputation for litigious behaviour is very easily established within the industry as a whole. The counter-argument is that arbitration can be equally adversarial.

17.5.10 Powers of the arbitrator

In *Northern Regional Health Authority v. Derek Crouch Construction Ltd*[54] which arose from a JCT 63 contract, it was held by the Court of Appeal that, where an arbitration agreement expressly conferred powers on an arbitrator to open up, review or revise any decision, opinion, direction, certificate or valuation of the architect/engineer, the High Court did not possess similar powers. This meant that many cases could be resolved only by arbitration. However, the House of Lords overruled *Crouch* in *Beaufort Developments (NI) Ltd v Gilbert-Ash (NI) Ltd.*[55] Their Lordships stated that, unless a contractual

---

54 (1984) 26 BLR 1 (hereafter *Crouch*).
55 (1998) 88 BLR 1.

provision was expressed to be final and conclusive on an issue, the court always had jurisdiction to determine the contractual entitlements of the parties. They explained further that it was the arbitrator who had no such power unless the parties positively gave it to him, thus necessitating arbitration clauses of the type considered in *Crouch*.

17.5.11 Summary relief

In litigation, a party may apply for Summary Judgment and/or Interim Payment under the Civil Procedure Rules. The essence of a Summary Judgment application is that: (i) the defendant has no real prospect of successfully defending the claim or a particular issue; and (ii) there is no other reason why the case or issue should go for full trial. The court is therefore asked to decide for the applicant without a full trial. Usually both the evidence to support the application and that rebutting it are provided in the form of sworn statements. An Interim Payment order (under Rule 25), which is applicable to only monetary claims, is made where the court is satisfied from the sworn evidence that, although a full trial is necessary, the applicant would be awarded a substantial amount. In such cases, the court may order immediate payment of the part of the claim indubitably due pending a final trial. However, if the outcome at trial is that the amount paid was not owed in part or at all, the court can order an appropriate refund.

Contractors usually applied for these orders to enforce payment obligations under the contract. The customary response of employers was to apply for stay of proceedings to arbitration if the contract contained an arbitration agreement, suggesting that arbitrators would not be able to grant equivalent relief. With Interim Payment applications, the courts had no difficulty ordering Interim Payment and staying the proceedings to arbitration.[56] Similarly, on an application for Summary Judgment, the court could grant it for sums for which there is no reasonable prospect of being successfully defended and stay the proceedings to arbitration.[57]

As explained in Section 17.2.8.1, the *Halki* litigation highlighted the fact that the court's discretion to refuse stay of proceedings is now severely limited. It is therefore argued that parties may refuse to use arbitration because the court would be unable to grant these types of summary relief.[58] However, there is the counter-argument that the new arbitration legislation gives arbitrators powers to grant equivalent relief, e.g. a provisional award under s. 39, where the parties have agreed that the arbitrator is to have such power.

17.5.12 Finality

It is often argued that the scope for challenging the awards of arbitrators is much less than that for the decisions of the courts as the statutory framework places a number of restrictions on appeals from arbitrations.[59] Whilst this finality may be attractive to some

---

56 See *Imodco Ltd v. Wimpey Major Projects and Taylor Woodrow International Ltd* (1987) 40 BLR 1.
57 *Ellis Mechanical Services Ltd v. Wates Construction Ltd* (1976) 2 BLR 60; *Associate Bulk Carriers v. Koch Shipping* (1978) 7 BLR 18; *R. M. Douglas Construction Ltd v. Bass Leisure Ltd* (1990) 53 BLR 119
58 In *Macob v. Morrison*, see n. 25, Dyson J suggested that a court may still issue a Summary Judgment order despite *Halki* to enforce the decision of an adjudicator (see Section 17.2.8.1).
59 See Section 17.4.5.8.

parties, Wallace[60] cautions that any misuse of the extensive powers of arbitrators under the Act could lead to serious injustice against which the courts would be powerless to intervene.

17.5.13 National sovereignty

It is often argued that many foreign governments would rather use the private forum of arbitration to resolve their disputes than submit to the jurisdiction of the courts of another country.

17.5.14 Multiple parties

The duty to submit a difference to arbitration is imposed by the contract between the parties. It follows from the doctrine of privity of contract that no third party is obliged to, nor has a right to take part in the proceedings even if the dispute concerns him. The reality of construction disputes is that there are often several parties involved in the cause of the dispute. For example, who is responsible for delay in a construction contract? It could be the architect, the contractor, or a sub-contractor. In litigation, the court has jurisdiction under the Civil Liability (Contribution) Act 1978 to join any party concerned either for indemnity or for a contribution. In arbitration, an arbitrator has no such powers. This inability to join third parties was a proper ground on which a court could refuse stay of proceedings when the court had discretion whether or not to stay to arbitration.[61] The position now is therefore that there is an increased risk of parallel proceedings from the same facts where the dispute involves multiple parties.

Attempts have been made to allow consolidation of disputes in the same arbitration proceedings. Some model arbitration procedures expressly empower appointing institutions, when approached to make an appointment, to consider appointing an arbitrator to whom a related dispute has already been referred. In addition, the relevant contracts and arbitration procedures authorize the arbitrator to consolidate proceedings or run them concurrently where desirable. However, because of the fragmentation of the industry, multiplicity of appointing institutions leading to gaps in knowledge of ongoing proceedings and frequent amendments to contracts, it may not be always possible to achieve this end. Possible sources of difficulty include gaps in the contractual and organizational framework, conflicting claims, compromise settlements, and disagreements as to whether the disputes are related.

Fragmentation of the industry works against the necessary dovetailing of the relevant contracts to enforce consolidation. The late Dr John Parris[62] wrote that very few suppliers will accept arbitration in place of litigation, not to talk of consolidation of arbitration proceedings. If this obstacle is overcome at an industry-wide level, there is still the problem of frequent amendments to the standard forms. Furthermore, the multiplicity of appointing institutions and rivalry among them not only promotes gaps in knowledge of

---

60 Duncan Wallace, I.N., 'First Impressions of the 1996 Arbitration Act' (1997) ICLR 71, pp. 71–116.
61 See *Taunton-Collins v Cromie*, n. 42.
62 Parris, J., *Arbitration: Principles and Practice* (Granada, St Albans, 1983), at. p. 46.

ongoing proceedings but also stands in the way of the desired level of cooperation. For example, research[63] suggests that the professional institutions tend to appoint only their members regardless of the nature of dispute to be referred.

With careful drafting of the relevant contracts, arbitrators can be given powers to consolidate proceedings. However, whether this power will be exercised is debatable. For example, an arbitrator may decline a request to exercise such powers because, in his opinion, the consolidation could cause injustice to one or more of the parties. This possibility was recognized by the Court of Appeal in *Abu Dhabi Gas Liquefaction Co. Ltd v. Eastern Bechtel Corporation*.[64] In that case the Court ordered that related disputes involving an employer, a main contractor and sub-contractors on an engineering project could be heard by the same arbitrator but indicated that they would consider any application to appoint a second arbitrator for particular issues if consolidation of such issues would be prejudicial to any party.

It is not unusual for the common party in related disputes to make a claim in one dispute which conflicts with his claims in the other. For example, in *Monk (A) & Co. Ltd v. Devon County Council*[65] the main contractor (Monk) denied liability for extra costs to their sub-contractors. As the disputes with the sub-contractors had not yet been resolved, Monk had to recognize the possibility that eventually the disputed liability might attach to them. With this in mind, they claimed in respect of this liability in their dispute with the employer. A related difficulty is that sometimes there may be a compromise in one dispute which turns out to be completely unacceptable to the other party in the second dispute. For instance, the main contractor may reach a compromised settlement under the main contractor for reasons of future business relations with the employer. Such a settlement may be totally unsatisfactory to the sub-contractor.

The common condition for consolidation is that the disputes raise issues which are substantially the same or connected with each other. Whether or not the disputes raise the same or connected issues is likely to pose questions to which there will not always be clear answers. It would therefore appear that there would have to be a preliminary arbitration, involving possibly a different arbitrator, where objections to joint proceedings are raised on any of these grounds.

17.5.15 Compound interest

Under s. 49, an arbitrator has unfettered discretion to award simple or compound interest on money payable but not paid. In contrast, the court will only do so in very particular circumstances.[66]

---

63 See, n. 31.
64 (1982) 21 BLR 117.
65 (1978) 10 BLR 9.
66 *Westdeutsche Landesbank Girozentrale v. Islington* [1996] 2 WLR 802.

17.5.16 Legal aid

If a party qualifies for legal aid, he is likely to prefer litigation to arbitration because such aid is not available for arbitrations. A party not entitled to legal aid but who is confident of winning may prefer arbitration because he would be entitled to recover costs against the losing party. In litigation this may not be possible where the losing party was assisted by legal aid.

Table of Cases

Table of Clause References

Table of Statutes

Subject Index